AF712851

Cones and Duality

Cones and Duality

Charalambos D. Aliprantis
Rabee Tourky

Graduate Studies
in Mathematics

Volume 84

American Mathematical Society
Providence, Rhode Island

Editorial Board

David Cox (Chair)
Walter Craig
N. V. Ivanov
Steven G. Krantz

2000 *Mathematics Subject Classification.* Primary 46A40, 46B40, 47B60, 47B65; Secondary 06F30, 28A33, 91B28, 91B99.

For additional information and updates on this book, visit
www.ams.org/bookpages/gsm-84

Library of Congress Cataloging-in-Publication Data

Aliprantis, Charalambos D.
Cones and duality / Charalambos D. Aliprantis, Rabee Tourky.
p. cm. — (Graduate studies in mathematics, ISSN 1065-7339 ; v. 84)
Includes bibliographical references and index.
ISBN 978-0-8218-4146-4 (alk. paper)
1. Cones (Operator theory). 2. Linear topological spaces, Ordered. I. Tourky, Rabee, 1966– II. Title.

QA329.A45 2007
515′.724—dc22 2007060758

Copying and reprinting. Individual readers of this publication, and nonprofit libraries acting for them, are permitted to make fair use of the material, such as to copy a chapter for use in teaching or research. Permission is granted to quote brief passages from this publication in reviews, provided the customary acknowledgment of the source is given.

Republication, systematic copying, or multiple reproduction of any material in this publication is permitted only under license from the American Mathematical Society. Requests for such permission should be addressed to the Acquisitions Department, American Mathematical Society, 201 Charles Street, Providence, Rhode Island 02904-2294, USA. Requests can also be made by e-mail to reprint-permission@ams.org.

© 2007 by the American Mathematical Society. All rights reserved.
The American Mathematical Society retains all rights
except those granted to the United States Government.
Printed in the United States of America.

∞ The paper used in this book is acid-free and falls within the guidelines
established to ensure permanence and durability.
Visit the AMS home page at http://www.ams.org/

10 9 8 7 6 5 4 3 2 1 12 11 10 09 08 07

To the great Russian mathematician and economist

Leonid Vitaliyevich Kantorovich (1912–1986),

the 1975 Nobel Prize co-recipient in economics,

... whose brilliant ideas have shaped the field of ordered vector spaces and are present throughout this book.

Contents

Preface

Ordered vector spaces made their debut at the beginning of the twentieth century. They were developed in parallel (but from a different perspective) with functional analysis and operator theory. Before the 1950s ordered vector spaces appeared in the literature in a fragmented way. Their systematic study began in various schools around the world after the 1950s. We mention the Russian school (headed by L. V. Kantorovich and the Krein brothers), the Japanese school (headed by H. Nakano), the German school (headed by H. H. Schaefer), and the Dutch school (headed by A. C. Zaanen). At the same time several monographs dealing exclusively with ordered vector spaces appeared in the literature; see for instance [**55, 56, 71, 75, 89, 91**]. The special class of ordered vector spaces known as *Riesz spaces* or *vector lattices* has been studied more extensively; see the monographs [**14, 15, 66, 68, 86, 88, 93**].

The theory of ordered vector spaces plays a prominent role in functional analysis. It also contributes to a wide variety of applications and is an indispensable tool for studying a variety of problems in engineering and economics; see for instance [**29, 31, 35, 36, 38, 42, 47, 49, 54, 64, 65, 76**]. The introduction of Riesz spaces and more broadly ordered vector spaces to economic theory has proved tremendously successful and has allowed researchers to answer difficult questions in general price equilibrium theory, economies with differential information, the theory of perfect competition, and incomplete assets economies.

The goal of this monograph is to present the theory of ordered vector spaces from a contemporary perspective that has been influenced by the study of ordered vector spaces in economics as well as other recent applications. We try to imbue the narrative with geometric intuition, which is in keeping with a long tradition in mathematical economics. We also approach the subject with our own personal presentiment that the special class of Riesz spaces is somehow "perfect" and thus loosely conceive of general ordered vector spaces as "deviations" from this "perfection." The book also contains material that has not been published in a monograph form before. The study of this material was initially motivated by various problems in economics and econometrics.

The material is spread out in eight chapters. Chapter 8 is an Appendix and contains some basic notions of functional analysis. Special attention is paid to the properties of linear topologies and the separation of convex sets. The results in this chapter (some of which are presented with proofs) are used throughout the monograph without specific mention.

Chapter 1 presents the fundamental properties of wedges and cones. Here we discuss Archimedean cones, lattice cones, extremal vectors of cones, bases of cones, positive linear functionals and the important decomposability property of cones known as the *Riesz decomposition property*. Chapter 2 introduces cones in topological vector spaces. This chapter illustrates the variety of remarkable results that can be obtained when some link between the order and the topology is imposed. The most important interrelationship between a cone and a linear topology is known as *normality*. We discuss normal cones in detail and obtain several characterizations. In normed spaces, the normality of the cone amounts to the norm boundedness of the order intervals generated by the cone. In Chapter 2 we also introduce ideals and present some of their useful order and topological properties.

Chapter 3 studies in detail cones in finite dimensional vector spaces. The results here are much sharper. For instance, as we shall see, every closed cone of a finite dimensional vector space is normal. The reader will find in this chapter a study (together with a geometrical description) of the polyhedral cones as well as a discussion of the properties of linear inequalities—including a proof of "the Principle of Linear Programming." The chapter culminates with a study of pull-back cones and establishes the following "universality" property of $C[0,1]$: every closed cone of a finite dimensional vector space is the pull-back cone of the cone of $C[0,1]$ via a one-to-one operator from the space to $C[0,1]$.

Chapter 4 investigates the fixed points and eigenvalues of an important class of positive operators known as *Krein operators*. A *Krein space* is an ordered Banach space having order units and a closed cone. A positive operator T on a Krein space is a *Krein operator* if for any $x > 0$ the vector $T^n x$ is an order unit for some n. Many integral operators are Krein operators. These operators possess some useful fixed points that are investigated in this chapter.

Chapters 5, 6, and 7 contain new material that, as far as we know, has not appeared before in any monograph. Chapter 5 develops in detail the theory of $\mathcal{K}$-lattices. An ordered vector space L is called a *$\mathcal{K}$-lattice*, where $\mathcal{K}$ is a super cone of L, i.e., $\mathcal{K} \supseteq L_+$, if for every nonempty subset A of L the collection of all L_+-upper bounds of A is nonempty and has a $\mathcal{K}$-infimum. As can be seen immediately from this definition, the notion of a $\mathcal{K}$-lattice has applications to optimization theory. Chapter 5 also introduces the notion of the Riesz–Kantorovich transform for an m-tuple of order bounded operators that is used to investigate the fundamental duality properties of ordered bounded operators from an ordered vector space to a Dedekind complete Riesz space. Subsequently, using the theory of $\mathcal{K}$-lattices, we define an important order extension of $\mathcal{L}_b(L, N)$, the ordered vector space of all order bounded operators from L to a Dedekind complete Riesz space. This extension allows us to enrich the lattice structure of $\mathcal{L}_b(L, N)$ in a useful manner.

Chapter 6 specializes the theory of $\mathcal{K}$-lattices to the space of all ordered bounded linear functionals on an ordered vector space. Among other things, this chapter introduces an important order extension of L' called the "super topological dual of L'" and studies its fundamental properties. Moreover, in this chapter the reader will find several interesting optimization results. In essence, Chapter 6 brings, via the concept of a $\mathcal{K}$-lattice, the theory of ordered vector spaces to the theory of linear minimization. In other words, this chapter can be viewed as contributing new functional analytic tools to the study of linear minimization problems.

Finally, in Chapter 7 we present a comprehensive investigation of piecewise affine functions. It turns out that their structure is intimately related to order and lattice properties that are discussed in detail in this chapter. The main result here is that the collection of piecewise affine functions coincides with the Riesz subspace generated by the affine functions. Piecewise affine (or piecewise linear) functions are very important in approximation theory and have been studied extensively in one-dimensional settings. However, even until now, in dimensions more than one there seems to be no satisfactory theory of piecewise affine functions. They are defined on finite dimensional spaces, and no attempt has been made to generalize their theory to infinite dimensional settings. This provides the opportunity for several future research directions.

At the end of each section there is a list of exercises of varying degrees of difficulty designed to help the reader comprehend the material in the section. There are almost three hundred and fifty exercises in the book. Hints to selected exercises are also given. The inclusion of the exercises makes the book, on one hand, suitable for graduate courses and, on the other hand, a reference source on ordered vector spaces and cones. It is our hope (and belief) that this monograph will not only be beneficial to mathematicians but also to other scientists in many disciplines, theoretical and applied, as well.

We take this opportunity to thank our late colleague Yuri Abramovich for reading an early draft of the manuscript and making numerous suggestions and corrections. The help provided to us by Monique Florenzano during the writing of the book is greatly appreciated. Besides correcting several faulty proofs, she recommended many important structural changes that improved the exposition of the book. Special thanks are due to Grainne Begley, Daniela Puzzello and Francesco Ruscitti for reading the manuscript carefully and correcting numerous (mathematical and nonmathematical) mistakes. We express our appreciation to our graduate students Iryna Topolyan, for reading the entire manuscript and finding an infinite number of mistakes, and Qianru Qi, for her help with the drawing of certain figures in the monograph.

C. D. Aliprantis acknowledges with many thanks the financial support he received from the National Science Foundation under grants SES-0128039, DMI-0122214, and DMS-0437210, and the DOD Grant ACI-0325846. R. Tourky acknowledges with many thanks the financial support he received from the Australian Research Council under grant A00103450.

C. D. ALIPRANTIS, West Lafayette, Indiana, USA
R. TOURKY, Queensland, Brisbane, AUSTRALIA

January 2007

The "isomorphism" notion

A typical mathematical field is usually described by a class of sets that are endowed with a "structure" concept that characterizes the subject matter of the field. Schematically, a typical branch of study in mathematics consists of pairs $(X, \mathfrak{S})$, where X is a set and $\mathfrak{S}$ is the "structure" imposed on the set X that is characteristic to the area. The structure $\mathfrak{S}$ can be expressed in terms of algebraic or topological properties or a mixture of the two. Here are a few examples of mathematical areas.

(1) *Groups*: Here for the typical object $(X, \mathfrak{S})$, the structure $\mathfrak{S}$ represents the algebraic structure on X given by the operation of multiplication $(x, y) \mapsto xy$ and of the inverse function $x \mapsto x^{-1}$.

(2) *Topological spaces*: Here for the typical object $(X, \mathfrak{S})$, the structure $\mathfrak{S}$ represents the topology of the set X.

(3) *Vector spaces*: Here for the typical object $(X, \mathfrak{S})$, the structure $\mathfrak{S}$ represents the algebraic structure imposed on X by means of the addition $(x, y) \mapsto x + y$ and the scalar multiplication $(\lambda, x) \mapsto \lambda x$.

(4) *Topological vector spaces*: Here for the typical object $(X, \mathfrak{S})$, the structure $\mathfrak{S}$ represents the mixture of the vector space structure of X and the topological structure of X that makes the algebraic operations of X continuous.

(5) *Ordered vector spaces*: Here for the typical object $(X, \mathfrak{S})$, the structure $\mathfrak{S}$ represents the algebraic structure of X together with the vector ordering on X.

Once one deals with the objects of a mathematical field, one would like to have a way of identifying two objects of the field that look "alike."

This is done with the "isomorphism" concept. The idea is very simple: We say that two objects $(X_1, \mathfrak{S}_1)$ and $(X_2, \mathfrak{S}_2)$ from a mathematical field are **isomorphic** if there exists a one-to-one surjective (i.e., onto) function $f\colon X_1 \to X_2$ (called an **isomorphism**) such that $f(\mathfrak{S}_1) = \mathfrak{S}_2$. The last identity should be intrepreted in the sense that f "preserves" the structures $\mathfrak{S}_1$ and $\mathfrak{S}_2$ of the sets X_1 and X_2, respectively. For instance, if X_1 and X_2 are isomorphic vector spaces via the isomorphism f, then (besides f being one-to-one and onto) it also satisfies $f(\alpha x + \beta y) = \alpha f(x) + \beta f(y)$ for all $x, y \in X_1$ and all scalars α and β. Likewise, by saying that two ordered vector spaces X_1 and X_2 are isomorphic via f, we mean that $f\colon X_1 \to X_2$ is a one-to-one surjective (linear) operator such that $f(x) \geq f(y)$ holds in X_2 if and only if $x \geq y$ holds in X_1.

The word "isomorphism" is the English version of the Greek word "*ισομορφισμός*," which etymologically is the composition of the Greek words "*ίσος*" (which means equal, even, the same) and "*μορφή*" (which means form, figure, shape, appearance, structure). So, when we say that two mathematical objects are "isomorphic," we simply express the fact that they have the same (or similar or identical) shape (or form or appearance) and the concept of an "isomorphism" simply designates the state of being "isomorphic."

Chapter 1

Cones

A wedge is any nonempty subset of a vector space that is closed under vector addition and scalar multiplication by nonnegative real numbers. A cone is a wedge that does not contain any negative multiple of its nonzero vectors. In every ordered vector space the set of vectors that dominate zero is a cone called the positive cone. Conversely, every cone induces a vector ordering on the vector space in which it is the positive cone. Therefore, cones and vector orderings are related in a one-to-one manner. In other words, an ordered vector space is a vector space equipped with cone.

This chapter studies the basic properties of cones. Like the rest of this book, the chapter talks about cones in the language of ordered vector spaces. The emphasis is to articulate features related to the order structure of an ordered vector space in terms of the geometry of the positive cone and to translate geometric properties of cones back into the properties of the vector ordering. The material in the chapter is presented without explicitly invoking any topological concepts.

In the first section, we introduce wedges and cones. In the second section, we study the special class of Archimedean cones. The important class of lattice cones (cones of vector lattices) is discussed in Section 1.3 and the basic lattice identities of the lattice cones are stated and proved here. Sections 1.4, 1.5, 1.6, and 1.7 introduce and investigate order bounded operators, positive linear functionals, cone bases, faces, and extremal rays.

The chapter concludes with a thorough discussion of the decomposition and interpolation properties of cones in vector spaces. These important properties are central to duality theory and are shared by lattice cones as well as certain function spaces. These properties provide a rich duality theory that can be described without invoking topological notions. The decomposition and interpolation properties play an important role and will be featured frequently throughout the book.

1.1. Wedges and cones

In this section we shall present some basic properties of wedges and cones in vector spaces. We start by recalling that a subset C of a vector space is called **convex** if $x, y \in C$ and $0 \leq \lambda \leq 1$ imply $\lambda x + (1-\lambda)y \in C$.

Definition 1.1. *A nonempty subset $\mathcal{W}$ of a vector space is said to be a* ***wedge*** *if it satisfies the following two properties:*

(1) $\mathcal{W} + \mathcal{W} \subseteq \mathcal{W}$,[1]

(2) $\alpha\mathcal{W} \subseteq \mathcal{W}$ *for all* $\alpha \geq 0$.

Clearly, every wedge is a convex set. Moreover, if $\mathcal{W}$ is a wedge, then it is not difficult to see that the set $\mathcal{W} \cap (-\mathcal{W})$ is a vector subspace. When this vector subspace is trivial, the wedge is called a cone. Precisely, we have the following definition.

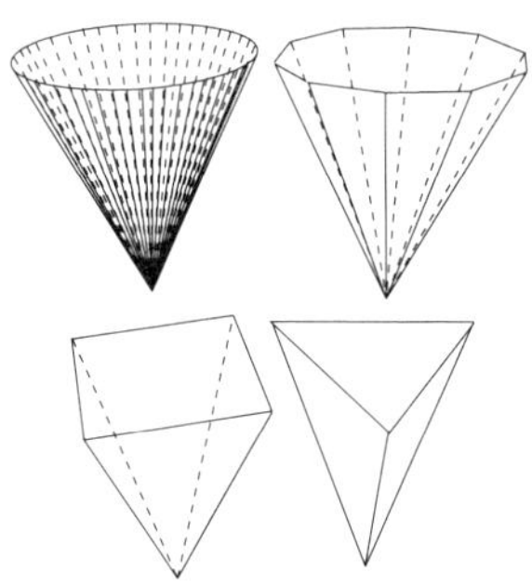

Definition 1.2. *A nonempty subset $\mathcal{K}$ of a vector space is said to be a* ***cone*** *if it satisfies the following three properties:*

(1) $\mathcal{K} + \mathcal{K} \subseteq \mathcal{K}$,

(2) $\alpha\mathcal{K} \subseteq \mathcal{K}$ *for all* $\alpha \geq 0$,

(3) $\mathcal{K} \cap (-\mathcal{K}) = \{0\}$.

A ***cone with vertex*** *x is any set of the form $x + \mathcal{K}$, where $\mathcal{K}$ is a cone.*[2]

Obviously, every cone is a wedge but a wedge need not be a cone. For instance, every vector subspace is a wedge but only the trivial vector subspace is a cone. Clearly, every cone (being a wedge) is automatically a convex set.

[1] Throughout this monograph we shall use the following standard notation. If A and B are subsets of a vector space X and $\alpha, \beta \in \mathbb{R}$, then we let

$$\alpha A + \beta B = \{x \in X : \exists\ a \in A \text{ and } b \in B \text{ such that } x = \alpha a + \beta b\}.$$

[2] Many authors define a cone to be simply a nonempty subset $\mathcal{K}$ of a vector space such that $\alpha x \in \mathcal{K}$ holds for all $x \in \mathcal{K}$ and all $\alpha \geq 0$. In the literature a wedge is quite often referred to as a **convex cone**. A cone as defined in our Definition 1.2 is also known as a **pointed convex cone** or as a **pointed convex cone with vertex at zero**or even as a **cone with vertex at zero**.

Recall that a binary relation $\geq$ on a set X is said to be an **order relation** on the set X (and X equipped with $\geq$ is called a **partially ordered set**) if it satisfies the following three properties:

(i) **reflexivity**: $x \geq x$ for all $x \in X$,

(ii) **antisymmetry**: $x \geq y$ and $y \geq x$ imply $x = y$,

(iii) **transitivity**: $x \geq y$ and $y \geq z$ imply $x \geq z$.

If a binary relation $\geq$ on a set X satisfies properties (i) and (iii), i.e., if $\geq$ is reflexive and transitive but not necessarily antisymmetric, then $\geq$ is called a **pre-order relation** on the set X (and X equipped with $\geq$ is called a **pre-partially ordered set**).

Definition 1.3. *In a pre-ordered set, the notation $x > y$ means $x \geq y$ and $x \neq y$. We shall also express $x \geq y$ by saying that x **dominates** y, and when $x > y$ is true, we shall say that x **strictly dominates** y.*

Alternatively, the symbol $x \geq y$ will be denoted by $y \leq x$, and $x > y$ will be denoted by $y < x$. A subset A of a pre-ordered set is said to **majorize** another set B if for each $b \in B$ there exists some $a \in A$ such that $a \geq b$.

Wedges are associated with vector pre-orderings and cones with vector orderings. An order relation (resp. a pre-order relation) $\geq$ on a vector space L is said to be a **vector ordering** (resp. a **vector pre-ordering**) if, in addition to being reflexive, antisymmetric and transitive (resp. reflexive and transitive), $\geq$ is also compatible with the algebraic structure of L in the sense that if $x \geq y$, then

(a) $x + z \geq y + z$ for each $z \in L$ and

(b) $\alpha x \geq \alpha y$ for all $\alpha \geq 0$.

A vector space L equipped with a vector ordering $\geq$ is called an **ordered vector space** or a **partially ordered vector space**, denoted $(L, \geq)$ or simply L if $\geq$ is well understood. In an ordered vector space $(L, \geq)$ any vector satisfying $x \geq 0$ is known as a **positive vector** and the collection of all positive vectors $L_+ = \{x \in L\colon\ x \geq 0\}$ is referred to as the **positive cone** or simply as the **cone** of L. Notice that L_+ is indeed a cone in L. Similar definitions can be given for a vector pre-ordering. The cone L_+ also will be denoted by L^+.

An arbitrary cone $\mathcal{K}$ of a vector space X defines a vector ordering $\geq$ on X by letting $x \geq y$ whenever $x - y \in \mathcal{K}$. Equivalently, $x \geq y$ whenever $x \in y + \mathcal{K}$ or $y \in x - \mathcal{K}$; see the companion figure. It is easy to check that $\geq$ is indeed a vector ordering whose positive cone coincides with $\mathcal{K}$, that is, $X_+ = \mathcal{K}$. Consequently,

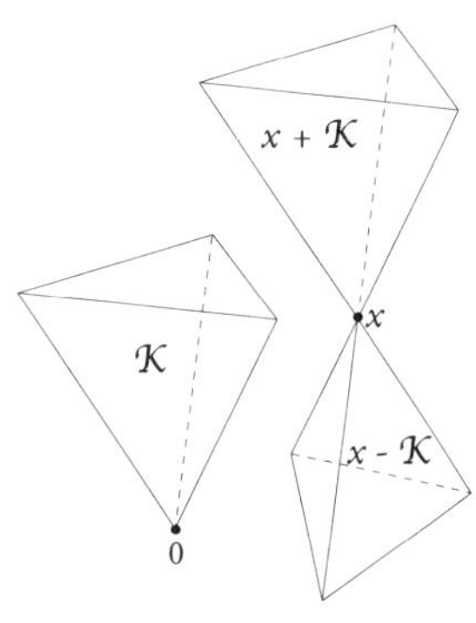

the vector orderings and cones are in one-to-one correspondence. That is, every cone $\mathcal{K}$ of a vector space X induces a vector ordering $\geq$ on X whose positive vectors are precisely the vectors in $\mathcal{K}$. If we wish to emphasize that the vector order on X is generated by the cone $\mathcal{K}$, then we shall denote it by $\geq_{\mathcal{K}}$.

Similarly, if $\geq$ is a vector pre-ordering on a vector space X, then the set of positive vectors $X_+ = \{x \in X\colon\ x \geq 0\}$ is a wedge. Conversely, every wedge $\mathcal{W}$ of X gives rise to a vector pre-ordering $\geq$ on X (by letting $x \geq y$ if $x - y \in \mathcal{W}$) that satisfies $X_+ = \mathcal{W}$. Again, if there is a need to emphasize that the vector pre-ordering on X is generated by the wedge $\mathcal{W}$, then we shall denote it by $\geq_{\mathcal{W}}$. In summary, cones give rise to vector orderings and wedges to vector pre-orderings.

It turns out that wedges are simply sums of vector subspaces with cones. The proof of the next result is straightforward and is omitted.

Lemma 1.4. *A nonempty convex subset $\mathcal{W}$ of a vector space is a wedge if and only if it is the sum of a vector subspace and a cone, that is, if and only if there exists a vector subspace V and a cone $\mathcal{K}$ such that $\mathcal{W} = V + \mathcal{K}$.*

Moreover, whenever $\mathcal{W}$ is a wedge, we have the following.

(1) *If we let $V = \mathcal{W} \cap (-\mathcal{W})$ and $\mathcal{K} = \{0\} \cup (\mathcal{W} \setminus V)$, then V is a vector subspace and $\mathcal{K}$ is a cone satisfying $\mathcal{W} = V + \mathcal{K}$.*

(2) *If $\mathcal{W} = V + \mathcal{K}$, where V is a vector subspace and $\mathcal{K}$ is a cone such that $V \cap \mathcal{K} = \{0\}$, then $V = \mathcal{W} \cap (-\mathcal{W})$.*

(3) *The wedge $\mathcal{W} = V + \mathcal{K}$ is a cone, i.e., $V = \{0\}$, if and only if $\mathcal{W}$ contains no straight line passing through the origin.*

Now let $\mathcal{W}$ be a wedge of a vector space X, and let x be any vector in the vector subspace generated by $\mathcal{W}$. Choose vectors $x_1, \ldots, x_n \in \mathcal{W}$ and real numbers $\alpha_1, \ldots, \alpha_n$ such that $x = \sum_{i=1}^n \alpha_i x_i$. If we let $I_1 = \{i\colon\ \alpha_i \geq 0\}$ and $I_2 = \{i\colon\ \alpha_i < 0\}$, then we have

$$x = \sum_{i=1}^{n} \alpha_i x_i = \sum_{i \in I_1} \alpha_i x_i - \sum_{i \in I_2} (-\alpha_i) x_i \in \mathcal{W} - \mathcal{W}.$$

From this, it easily follows that the vector subspace generated by $\mathcal{W}$ is precisely $\mathcal{W} - \mathcal{W}$.

Definition 1.5. *A wedge $\mathcal{W}$ of a vector space X is said to be **generating** if $X = \mathcal{W} - \mathcal{W}$, i.e., if the vector subspace generated by $\mathcal{W}$ coincides with X.*

It should be clear that a wedge is generating if and only if it majorizes the space. That is, we have the following simple result.

Lemma 1.6. *A wedge $\mathcal{W}$ of a vector space X is generating if and only if for each $x \in X$ there exists some $y \in \mathcal{W}$ satisfying $y \geq_{\mathcal{W}} x$.*

Let $\mathcal{W}$ be a wedge of a vector space X. We shall say that a vector $e \in \mathcal{W}$ is an **order unit** (or more precisely a $\mathcal{W}$**-order unit** in case the wedge $\mathcal{W}$ must be indicated) if for each $x \in X$ there exists some $\lambda > 0$ such that $x \leq \lambda e$. Clearly, if $\mathcal{W}$ has an order unit, then the wedge $\mathcal{W}$ is automatically generating. Also, if $e \in \mathcal{W}$ is an order unit, then so are αe for all $\alpha > 0$ and $e + x$ for each $x \in \mathcal{W}$.

The order units of a wedge coincide with its internal points.[3]

Lemma 1.7. *Let $\mathcal{W}$ be a wedge of some vector space X. A vector $e \in \mathcal{W}$ is an order unit if and only if it is an internal point of $\mathcal{W}$.*

Proof. Assume first that e is an internal point of $\mathcal{W}$ and let $x \in X$. Then there exists some $\alpha > 0$ such that $e + \alpha(-x) \geq 0$. This implies $x \leq \frac{1}{\alpha} e$, so that e is an order unit.

For the converse, suppose that e is an order unit and let $x \in X$. Fix some $\lambda_0 > 0$ such that $\lambda_0(-x) \leq e$. Now notice that for each $0 \leq \lambda \leq \lambda_0$ we have

$$\lambda(-x) = (\tfrac{\lambda}{\lambda_0})\lambda_0(-x) \leq \tfrac{\lambda}{\lambda_0} e \leq e\,.$$

Consequently, $e + \lambda x \geq 0$ for all $0 \leq \lambda \leq \lambda_0$, which shows that e is an internal point of $\mathcal{W}$. ■

There are some useful sets associated with an ordered vector space that will play an important role in our study in this book. They are introduced next. If x and y are two vectors in an ordered vector space L satisfying $x \leq y$, then the **order interval** $[x, y]$ is the set defined by $[x, y] = \{z \in L\colon\ x \leq z \leq y\}$. If $x \leq y$ is not true, then we let $[x, y] = \emptyset$, so that the order interval is defined for all vectors x and y.[4] Clearly, for all x and y we have

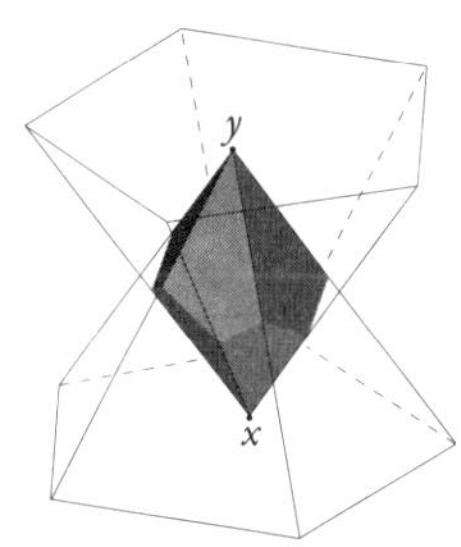

$$[x, y] = (x + L^+) \cap (y - L^+) = x + [0, y - x]\,.$$

In particular, each order interval is a convex set.

Definition 1.8. *A subset A of an ordered vector space L is said to be* ***full*** *(or* ***order-convex****) if for every $x, y \in A$ we have $[x, y] \subseteq A$.*

[3] Recall that a point $a \in A$, where A is a subset of a vector space X, is called an **internal point** of A if for each $x \in X$ there exists some $\lambda_0 > 0$ such that $a + \lambda x \in A$ for all $0 \leq \lambda \leq \lambda_0$. Notice that a point $a \in A$ is an internal point if and only if for each $x \in X$ there exists some $0 < \alpha_0 < 1$ such that $\alpha a + (1 - \alpha)x \in A$ for all $\alpha_0 < \alpha < 1$.

If A is a convex set, then a vector $a \in A$ is an internal point of A if and only if for each $x \in X$ there exists some $\alpha > 0$ such that $a + \alpha x \in A$.

[4] This "order interval" definition is applicable to any pre-ordered set as well.

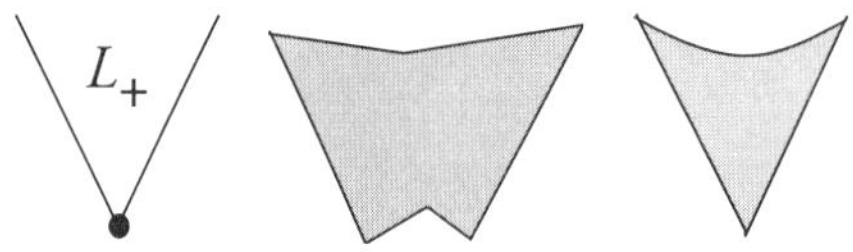

Full sets that are not convex

Clearly, the intersection of a family of full sets is a full set. From this, we easily see that every subset B of L is included in a smallest (with respect to inclusion) full subset—called the **full hull** of B and denoted $[B]$. A moment's thought reveals that

$$[B] = (B + L^+) \cap (B - L^+) = \bigcup_{x,y \in B} [x, y] .$$

From the first formula we conclude that

(1) the full hull of a convex set is convex, and

(2) the full hull of a circled set is circled.[5]

A subset A of an ordered vector space L is said to be **bounded above** if there exists some vector x (called an **upper bound** of A) such that $a \leq x$ holds true for all $a \in A$. Similarly, A is **bounded below** if there exists some vector x (called a **lower bound** of A) satisfying $x \leq a$ for all $a \in A$. A subset A of L is **order bounded** if A is bounded from above and below or, equivalently, if it is included in an order interval.

A vector u in an ordered vector space L is called the **least upper bound** (or the **supremum**) of a nonempty subset A of L, written $u = \sup A$, if

(a) u is an upper bound of A, i.e., $a \leq u$ holds for all $a \in A$, and

(b) u is the least upper bound of A in the sense that for any upper bound v of A we have $u \leq v$.

It should be clear from the above definition that a nonempty subset of an ordered vector space can have at most one supremum. The definition of the **greatest lower bound** (or the **infimum**) of a nonempty set A, denoted $\inf A$, is introduced analogously.

The classical lattice notations for the supremum and infimum of a finite set $\{x_1, \ldots, x_n\}$ are

$$\sup\{x_1, \ldots, x_n\} = \bigvee_{i=1}^{n} x_i \quad \text{and} \quad \inf\{x_1, \ldots, x_n\} = \bigwedge_{i=1}^{n} x_i .$$

We also write $\sup\{x, y\} = x \vee y$ and $\inf\{x, y\} = x \wedge y$.

[5] Recall that a subset S of a vector space is called **circled** (or **balanced**) whenever $\lambda x \in S$ holds for all $x \in S$ and all $\lambda \in \mathbb{R}$ with $|\lambda| \leq 1$.

Of course, in any ordered vector space $x \wedge x = x \vee x = x$. In very special ordered vector spaces any finite set of points has a supremum and an infimum. Other ordered vector spaces that are equally rare have the property that some pairs of vectors do not have a supremum or an infimum while other pairs do have least upper and greatest lower bounds. In most ordered vector spaces for each pair of points x, y that are not comparable (i.e., $x \not\geq y \not\geq x$) the supremum $x \vee y$ and infimum $x \wedge y$ do not exist.

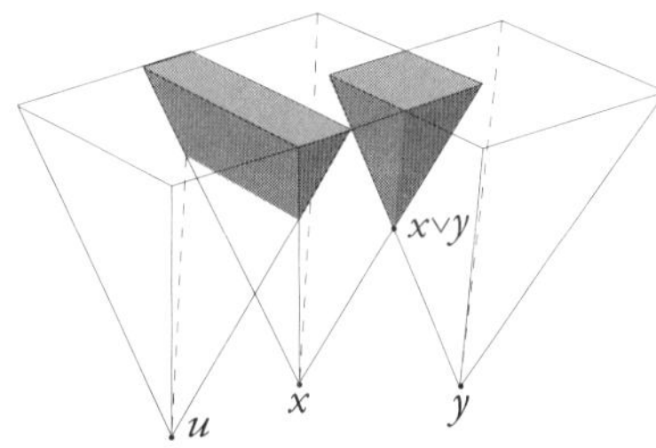

There is a useful geometric interpretation of the supremum and the infimum of a set of two points. Suppose that $\mathcal{K}$ is a cone of a vector space L. Then for any two vectors $x, y \in L$ there exists some $z \in L$ satisfying $(x+\mathcal{K}) \cap (y+\mathcal{K}) = z+\mathcal{K}$ if and only if $z = x \vee y$. Similarly, there exists some $w \in L$ such that $(y-\mathcal{K}) \cap (x-\mathcal{K}) = w-\mathcal{K}$ if and only if $w = x \wedge y$. Notice that in the companion figure $x \vee y$ exists but $x \vee u$ does not exist. We shall return to the lattice properties in Section 1.3.

We continue by recalling the notions of monotone nets. A net $\{x_\alpha\}$ in a partially ordered set is said to be **increasing**, written $x_\alpha \uparrow$ (resp. **decreasing**, written $x_\alpha \downarrow$) if $\alpha \geq \beta$ implies $x_\alpha \geq x_\beta$ (resp. $x_\alpha \leq x_\beta$). The symbol $x_\alpha \uparrow x$ means that $x_\alpha \uparrow$ and $x = \sup\{x_\alpha\}$. Likewise the notation $x_\alpha \uparrow \leq x$ (resp. $x \leq x_\alpha \downarrow$) means that the net $\{x_\alpha\}$ is increasing (resp. decreasing) and $x_\alpha \leq x$ (resp. $x \leq x_\alpha$) holds for each index α. The meaning $x_\alpha \downarrow x$ is analogous. The decreasing and increasing nets are referred to as **monotone nets**.

A nonempty subset D of a partially ordered set is said to be **directed upward**, in symbols $D\uparrow$, if for every pair $x, y \in D$ there exists some $z \in D$ satisfying $x \leq z$ and $y \leq z$. The **directed downward** sets are defined similarly.

Next, we shall indicate how one can obtain new ordered vector spaces from old ones. First, notice that every subspace of an ordered vector space can be viewed as an ordered vector space in its own right. Indeed, if Y is a vector subspace of an ordered vector space X, then it should be clear that Y equipped with the induced order from X is automatically an ordered vector space. Its positive cone is simply the cone $Y_+ = Y \cap X_+$. Unless otherwise stated, the vector subspaces of ordered vector spaces will be considered as ordered vector spaces with the induced vector ordering.

Another way of constructing ordered vector spaces is by means of Cartesian products. If $\{(L_i, L_i^+)\}_{i \in I}$ is a family of ordered vector spaces, then it should be easy to see that the product $L^+ = \prod_{i \in I} L_+^i$ is a cone of the

Cartesian product $L = \prod_{i \in I} L_i$, called the **product cone** of the family of cones $\{L_i^+\}_{i \in I}$, and so (L, L^+) is an ordered vector space. Notice that $(x_i)_{i \in I} \geq (y_i)_{i \in I}$ holds in L if and only if $x_i \geq y_i$ holds in L_i for each i. An important particular case of a Cartesian product ordered vector space is the ordered vector space $\mathbb{R}^\Omega$, where Ω is a nonempty set—and, as above, $f \geq g$ holds in $\mathbb{R}^\Omega$ if and only if $f(\omega) \geq g(\omega)$ for each $\omega \in \Omega$. More generally, if L is an ordered vector space and I is any nonempty index set, then the Cartesian product L^I is likewise an ordered vector space, where $(x_i)_{i \in I} \geq (y_i)_{i \in I}$ in L^I is equivalent to $x_i \geq y_i$ in L for each $i \in I$.

Finally, we shall close the section with the notion of an order isomorphism. Recall that a function $T \colon X \to Y$ is called an **operator** (or a **linear transformation**) if $T(\alpha x + \beta y) = \alpha T(x) + \beta T(y)$ holds for all $x, y \in X$ and all scalars $\alpha, \beta \in \mathbb{R}$.

Definition 1.9. *Let L and M be two ordered vector spaces. An operator $T \colon L \to M$ is called an **order isomorphism** (or an **order-embedding**) if*

(a) *T is one-to-one and*

(b) *$x \geq 0$ holds in L if and only if $Tx \geq 0$ holds in M.*

*If there exists an onto (i.e., surjective) order isomorphism from L to M, then L and M are called **order isomorphic** ordered vector spaces.*[6]

That is, two ordered vector spaces are order isomorphic if and only if there exists a one-to-one correspondence between the elements of the ordered vector spaces that preserves both the algebraic and the order structures. From the point of view of the theory of ordered vector spaces two order isomorphic ordered vector spaces are viewed as identical objects.

We shall also say that an ordered vector space L is **order-embeddable** into another ordered vector space M if there exists an order isomorphism from L to M. In this case, the subspace $T(L)$ ordered with the induced vector ordering from M can be identified with L and for this reason $T(L)$ is referred to as a **copy** of L in M.

Exercises

1. Show that a nonempty convex subset of a vector space is a wedge if and only if it is closed under nonnegative scalar multiplication.
2. If $\mathcal{W}$ is a wedge, then prove that $\mathcal{W} \cap (-\mathcal{W})$ is a vector subspace.
3. If two vectors x and y of a cone satisfy $x+y = 0$, then show that $x = y = 0$.

[6] Since the structures under consideration here are the vector orderings, according to the general notion of an isomorphism (as explained following the Preface of the book) the "order isomorphism" can be very well referred to simply as an "isomorphism."

4. If $\mathcal{W}_1$ and $\mathcal{W}_2$ are two wedges of a vector space, then show that $\lambda\mathcal{W}_1$ for $\lambda \in \mathbb{R}$, $\mathcal{W}_1 \cap \mathcal{W}_2$, and $\mathcal{W}_1 + \mathcal{W}_2$ are all wedges. Is the sum of two cones a cone?

5. If two cones $\mathcal{K}_1$ and $\mathcal{K}_2$ of a vector space satisfy $\mathcal{K}_1 \cap (-\mathcal{K}_2) = \{0\}$, then show that $\mathcal{K}_1 + \mathcal{K}_2$ is a cone.

6. Prove that if $\mathcal{K}$ is a cone and $\mathcal{W}$ is a wedge of a vector space, then $\mathcal{K} \cap \mathcal{W}$ is a cone—and so, in particular, the intersection of two cones is a cone.

7. Consider the wedge $\mathcal{W}$ of $\mathbb{R}^2$ defined by $\mathcal{W} = \{(x, y) \in \mathbb{R}^2\colon\ x + y \leq 0\}$. According to Lemma 1.4 we know that $\mathcal{W}$ has a decomposition of the form $\mathcal{W} = V + \mathcal{K}$, where V is a vector subspace and $\mathcal{K}$ is a cone. Find three different such decompositions of $\mathcal{W}$.

8. Establish that if e is an order unit of an ordered vector space L, then so are $e + x$ and λe for all $x \in L_+$ and $\lambda > 0$.

9. Following the terminology in this section, we shall say that a vector subspace X of an ordered vector space L is **majorizing** if for each $u \in L$ there exists some $x \in X$ such that $u \leq x$. (A majorizing vector subspace is also called a **cofinal** vector subspace.)

For an ordered vector space L establish the following.

(a) A vector subspace X of L majorizes L if and only if $L = X - L_+$.

(b) The cone L_+ is generating if and only if the vector space $L_+ - L_+$ majorizes L.

(c) If a vector subspace contains an order unit, then it majorizes L.

(d) A vector subspace that majorizes L need not have an order unit.

10. For an operator $T\colon X \to Y$ between vector spaces show the following.

(a) If $\mathcal{W}$ is a wedge of X, then $T(\mathcal{W})$ is a wedge of Y.

(b) If $\mathcal{W}$ is a wedge of Y, then $T^{-1}(\mathcal{W})$ is a wedge of X.

(c) If T is one-to-one and $\mathcal{W}$ is a cone of X, then $T(\mathcal{W})$ is a cone of Y.

(d) If T is one-to-one and $\mathcal{W}$ is a cone of Y, then $T^{-1}(\mathcal{W})$ is a cone of X.

11. Prove Lemma 1.4.

12. Can a cone contain a straight line?

13. Establish that a positive vector e of an ordered vector space is an order unit if and only if the vector zero is an internal point of the convex set $e - L_+ = \{x \in L\colon\ x \leq e\}$. [*Hint*: Notice that e is an internal point of L_+ if and only if 0 is an internal point of $e - L_+$ and then apply Lemma 1.7.]

14. If $[a, b]$ and $[c, d]$ are order intervals of an ordered vector space, then establish that $[a, b] + [c, d] = [a + c, b + d], \lambda[a, b] = [\lambda a, \lambda b]$ for all $\lambda \geq 0$, and $\lambda[a, b] = [\lambda b, \lambda a]$ for all $\lambda < 0$.

15. Prove that translations, arbitrary intersections, and scalar multiples of full sets are full sets.

16. Show that if B is a nonempty subset of an ordered vector space, then its full hull $[B]$ is given by

$$[B] = (B + L^+) \cap (B - L^+) = \bigcup_{x,y \in B} [x, y]\,.$$

Use the formula $[B] = (B + L^+) \cap (B - L^+)$ to conclude that
(i) the full hull of a nonempty convex set is convex and
(ii) the full hull of a circled set is circled.

17. Let B be a nonempty subset of an ordered vector space consisting of "incomparable" vectors, i.e, if $x, y \in B$ and $x \neq y$, then neither $x \geq y$ nor $y \geq x$ is true. Find the full hull of B. [*Hint*: Note that $[x, x] = \{x\}$.]

18. If a nonempty subset A of an ordered vector space L has a supremum (resp. an infimum), then show that for each $x \in L$ the set $x + A$ has a supremum (resp. an infimum) and the following lattice identities hold:

$$x + \sup A = \sup(x + A) \quad \text{and} \quad x + \inf A = \inf(x + A)\,.$$

Similarly, with an appropriate interpretation establish the following useful lattice identities:

$$x - \sup A = \inf(x - A) \quad \text{and} \quad x - \inf A = \sup(x - A).$$

$$\lambda \sup A = \sup(\lambda A) \quad \text{and} \quad \lambda \inf A = \inf(\lambda A) \ \text{ for all } \lambda \geq 0\,.$$

19. Assume that two non-empty subsets A and B of an ordered vector space have suprema. Then show that the set $A + B$ also has a supremum and that $\sup(A + B) = \sup A + \sup B$ holds true.

20. For each $n \in \mathbb{N}$ let $f_n\colon [0, 1] \to \mathbb{R}$ be the continuous function whose graph consists of the two line segments joining the points $(0, 0)$ with $(\frac{1}{n}, 1)$ and $(\frac{1}{n}, 1)$ with $(1, 1)$. Show that $f_n \uparrow \mathbf{1}$ holds in $C[0, 1]$, where $\mathbf{1}$ denotes the constant function one on $[0, 1]$.

Likewise, if $\{g_n\} \subseteq C[0, 1]$ is defined by $g_n(t) = \sqrt[n]{t}$, then show that $g_n \uparrow \mathbf{1}$ in $C[0, 1]$.

21. For each $n \in \mathbb{N}$ let $h_n\colon [0, 1] \to \mathbb{R}$ be the continuous function whose graph consists of the polygonal line joining the points $(0, 0)$, $(\frac{1}{2}, 0)$, $(\frac{1}{2} + \frac{1}{2n}, 1)$, and $(1, 1)$. Show that $h_n \uparrow$ holds true in $C[0, 1]$ and that the sequence $\{h_n\}$ does not have a supremum in $C[0, 1]$.

22. Show that every bounded from above nonempty subset A of $\mathbb{R}^\Omega$, where Ω is a fixed nonempty set, has a supremum in the ordered vector space $\mathbb{R}^\Omega$ and that $[\sup A](\omega) = \sup_{f \in A} f(\omega)$ holds for all $\omega \in \Omega$.[7]

Use this conclusion to find the suprema of the sequences appearing in Exercises 20 and 21 above in $\mathbb{R}^{[0,1]}$.

23. For two nets $\{x_\alpha\}$ and $\{y_\delta\}$ of an ordered vector space show the following.
(a) If $x_\alpha \uparrow x$ and $y_\delta \uparrow y$, then $x_\alpha + y_\delta \uparrow x + y$, $\lambda x_\alpha \uparrow \lambda x$ for all $\lambda \geq 0$, and $\lambda x_\alpha \downarrow \lambda x$ for all $\lambda < 0$.
(b) If $x_\alpha \uparrow x$, $y_\delta \uparrow$, and $x_\alpha + y_\delta \uparrow z$, then $y_\delta \uparrow z - x$.

24. Show that for two vectors x and y in an ordered vector space their infimum $x \wedge y$ exists if and only if their supremum $x \vee y$ exists—in which case we have the identity $x \wedge y + x \vee y = x + y$. [*Hint*: Use Exercise 18 above.]

---

[7] An ordered vector space for which every increasing bounded from above net has a supremum is called **Dedekind** (or **order**) **complete**. It is not difficult to see that, in this terminology, our conclusion can be stated as follows: *The ordered vector space* $\mathbb{R}^\Omega$ *is Dedekind complete.*

1.2. Archimedean cones

The objective here is to introduce and study the special class of Archimedean cones. We start with their definition.

Definition 1.10. *An ordered vector space L is said to be* ***Archimedean*** *(or we say that L has the* ***Archimedean property****) whenever it follows from $y \in L$, $x \in L_+$, and $ny \leq x$ for all $n = 1, 2, \ldots$ that $y \leq 0$.*

A cone $\mathcal{K}$ of a vector space L is said to be ***Archimedean*** *if the order induced by $\mathcal{K}$ on L makes L an Archimedean ordered vector space.*

The vector $x \in L_+$ in the preceding definition can be assumed to lie in L.

Lemma 1.11. *An ordered vector space L is Archimedean if and only if it follows from $x, y \in L$ and $ny \leq x$ for all $n = 1, 2, \ldots$ that $y \leq 0$.*

Proof. If the condition holds true, then L is clearly Archimedean. For the converse, assume that L is Archimedean and that for some $x, y \in L$ we have $ny \leq x$ for all $n \in \mathbb{N}$. This implies $(n+1)y \leq x$ or $ny \leq x - y$ for all $n \in \mathbb{N}$. Since $x - y \geq 0$, it easily follows that $y \leq 0$. ■

A classical example of a non-Archimedean ordered vector space is provided by the lexicographic plane. The **lexicographic plane** is the vector space $L = \mathbb{R}^2$ equipped with the following vector ordering (known as the **lexicographic ordering**):

$$(x_1, x_2) \geq (y_1, y_2) \text{ if either } x_1 > y_1 \text{ or else } x_1 = y_1 \text{ and } x_2 \geq y_2 .$$

The lexicographic ordering is induced by the cone

$$L_+ = \{(x, y) \in \mathbb{R}^2 \colon \text{ either } x > 0 \text{ or else } x = 0 \text{ and } y \geq 0\},$$

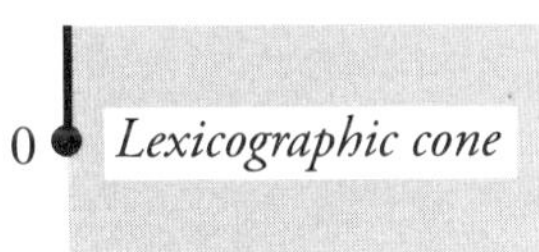

which is referred to as the **lexicographic cone**. To see that L equipped with the cone L_+ is a non-Archimedean ordered vector space, note, for instance, that $n(0,1) \leq (1,0)$ holds for all n while $(0,1) \notin -L_+$. Observe also that the lexicographic ordering is **total**. That is, any two arbitrary vectors $u, v \in L$ are comparable in the sense that either $u \geq v$ or $v \geq u$ is true.

The next two results contain some basic "order continuity" properties of Archimedean ordered vector spaces.

Theorem 1.12. *Let $x_1, \ldots, x_k$ be positive vectors in an Archimedean ordered vector space. Assume that the real sequences $\{\alpha_n^i\}$ $(i = 1, \ldots, k)$ satisfy $\alpha_n^i \xrightarrow[n\to\infty]{} \alpha_i$ in $\mathbb{R}$ for each $i = 1, \ldots, k$ and $\alpha_n^1 x_1 + \cdots + \alpha_n^k x_k \leq 0$ for each n. Then we have $\alpha_1 x_1 + \cdots + \alpha_k x_k \leq 0$.*

Proof. Fix any $m \in \mathbb{N}$ and then pick some n such that $\alpha_i - \alpha_n^i < \frac{1}{m}$ holds for all $i = 1, \dots, k$. Now from

$$\begin{aligned} \alpha_1 x_1 + \cdots + \alpha_k x_k &= (\alpha_1 - \alpha_n^1)x_1 + \cdots + (\alpha_k - \alpha_n^k)x_k + \alpha_n^1 x_1 + \cdots + \alpha_n^k x_k \\ &\leq (\alpha_1 - \alpha_n^1)x_1 + \cdots + (\alpha_k - \alpha_n^k)x_k \\ &\leq \tfrac{1}{m}x_1 + \cdots + \tfrac{1}{m}x_k\,, \end{aligned}$$

we see that $m(\alpha_1 x_1 + \cdots + \alpha_k x_k) \leq x_1 + \cdots + x_k$ holds for all $m \in \mathbb{N}$. Now invoke the Archimedean property to see that $\alpha_1 x_1 + \cdots + \alpha_k x_k \leq 0$. ■

An easy consequence of the preceding result is the following.

Corollary 1.13. *Let $x_1, \dots, x_k$ and $y_1, \dots, y_m$ be vectors in an Archimedean ordered vector space L. Assume that the sequences of real numbers $\{\alpha_n^i\}$ $(i = 1, \dots, k)$ and $\{\beta_n^j\}$ $(j = 1, \dots, m)$ satisfy*

(a) $\alpha_n^i \xrightarrow[n\to\infty]{} \alpha_i$ *for each* $i = 1, \dots, k$,

(b) $\beta_n^j \xrightarrow[n\to\infty]{} \beta_j$ *for each* $j = 1, \dots, m$, *and*

(c) $\alpha_n^1 x_1 + \cdots + \alpha_n^k x_k \leq \beta_n^1 y_1 + \cdots + \beta_n^m y_m$ *for each* n.

If the x_i and the y_j belong to $L_+ - L_+$, then

$$\alpha_1 x_1 + \cdots + \alpha_k x_k \leq \beta_1 y_1 + \cdots + \beta_m y_m\,.$$

Exercises

1. Show that an ordered vector space L is Archimedean if and only if for each $x \in L_+$ we have $\frac{1}{n}x \downarrow 0$.

2. Show that a totally ordered Archimedean ordered vector space is either trivial or order isomorphic to $\mathbb{R}$. [*Hint*: Assume L is nontrivial. Fix some $0 < u \in L$ and for each $x \in L$ let $T(x) = \sup\{\alpha \in \mathbb{R}\colon\ \alpha u \leq x\}$. Establish that $T(x)$ is well defined such that $T(x)u = x$, and infer from this that $T\colon L \to \mathbb{R}$ is an order isomorphism.]

3. Show that the Cartesian product of a family of Archimedean ordered vector spaces is an Archimedean ordered vector space.

4. Let $\mathcal{K}$ be a cone of a vector space X and let $Y = \mathcal{K} - \mathcal{K}$ be the vector subspace generated by the cone $\mathcal{K}$ in X. Show that the cone $\mathcal{K}$ is an Archimedean cone of X if and only if it is an Archimedean cone of Y.

5. Is a subcone of an Archimedean cone necessarily Archimedean?

6. A cone $\mathcal{K}$ of a vector space L is said to be **almost Archimedean** if it follows from $-\frac{1}{n}x \leq_{\mathcal{K}} y \leq_{\mathcal{K}} \frac{1}{n}x$ for all $n \in \mathbb{N}$ and some $x, y \in L$ that $y = 0$. An ordered vector space L is called **almost Archimedean** if its cone L_+ is almost Archimedean.

(a) Prove that every subcone of an arbitrary Archimedean cone is almost Archimedean—and conclude from this that every Archimedean cone is almost Archimedean.

(b) Give an example of an almost Archimedean ordered vector space that is not Archimedean.

1.3. Lattice cones

In this section we shall discuss briefly the class of Riesz spaces (or vector lattices). Since there is a vast literature dealing with Riesz spaces (see for instance [**14, 15, 66, 68, 86, 88, 93**]), we shall mainly exhibit some of their basic properties. However, we shall introduce more properties of vector lattices in subsequent sections as they are needed. We start with the definition of a Riesz space.

Definition 1.14. *An ordered vector space L is said to be a **Riesz space** (or a **vector lattice**) if every nonempty finite subset of L has a supremum (least upper bound) and an infimum (greatest lower bound).*

*The cone of a Riesz space is called a **lattice cone**. We shall also say that a cone $\mathcal{K}$ of a vector space X is a **lattice cone** if X partially ordered by the cone $\mathcal{K}$ is a Riesz space.*

In other words, the cone L_+ of an ordered vector space L is a lattice cone if and only if L is a Riesz space. Using an easy inductive argument, it is easy to see that

- *an ordered vector space L is a Riesz space if and only if every two vectors of L have a supremum and an infimum in L.*

Also, we shall say that a vector ordering $\geq$ on an arbitrary vector space X is a **lattice ordering** if its positive cone $X^+ = \{x \in X\colon\ x \geq 0\}$ is a lattice cone.

We also have the following useful characterization of Riesz spaces.

Lemma 1.15. *An ordered vector space L is a Riesz space if and only if for every pair of vectors $x, y \in L$ their infimum $x \wedge y$ exists in L. Moreover, if x and y are elements in a Riesz space, then*

$$x \vee y = -\big[(-x) \wedge (-y)\big] \quad \textit{and} \quad x \wedge y = -\big[(-x) \vee (-y)\big].$$

Likewise, an ordered vector space L is a Riesz space if and only if for every pair of vectors $x, y \in L$ their supremum $x \vee y$ exists in L.

Proof. Assume first that every pair in an ordered vector space L has an infimum, and let $x, y \in L$. Put $z = (-x) \wedge (-y)$. We shall prove that $x \vee y = -z$. To see this, note first that we have $z \leq -x$ and $z \leq -y$, or $x \leq -z$ and $y \leq -z$. That is, $-z$ is an upper bound of the set $\{x, y\}$. To see that $-z$ is the least upper bound of the set $\{x, y\}$, suppose that a vector t satisfies $x \leq t$ and $y \leq t$. This implies $-t \leq -x$ and $-t \leq -y$, and therefore

$-t \le (-x) \wedge (-y) = z$, i.e., $t \ge -z$. Hence $x \vee y = -z$ holds. The converse is straightforward. ■

There is another important geometrical characterization of cones that are lattice cones. This characterization will be used extensively in this work.

Theorem 1.16. *A cone $\mathcal{K}$ in a vector space X is a* ***lattice cone*** *if and only if for any $x, y \in X$ there exists some $z \in X$ satisfying*

$$(x + \mathcal{K}) \cap (y + \mathcal{K}) = z + \mathcal{K},$$

in which case $z = x \vee y$ (or, equivalently, for any $x, y \in X$ there exists some $w \in X$ satisfying

$$(x - \mathcal{K}) \cap (y - \mathcal{K}) = w - \mathcal{K},$$

in which case we have $w = x \wedge y$).

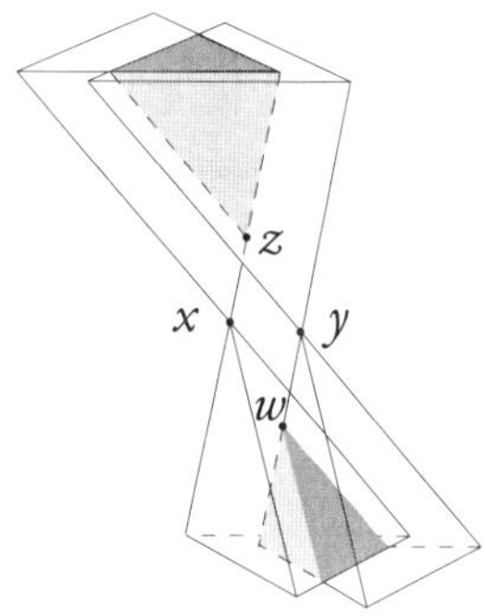

In every Riesz space there are three important vectors associated with each vector x. Its **positive part** x^+, its **negative part** x^-, and its **absolute value** $|x|$. These vectors are defined via the formulas

$$x^+ = x \vee 0, \quad x^- = (-x) \vee 0 \quad \text{and} \quad |x| = x \vee (-x).$$

Their geometrical meaning is shown in the figure below.

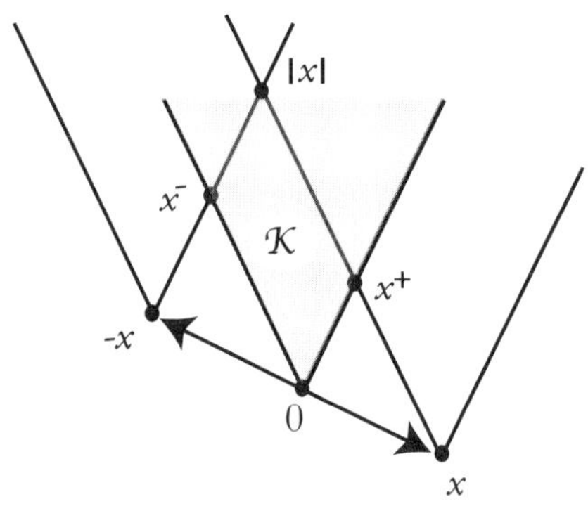

The following theorem contains a list of several important lattice identities that hold true for elements in Riesz spaces. These identities will be used freely throughout this monograph.

Theorem 1.17 (Lattice identities). *For any elements x, y, and z in a Riesz space we have the following:*

(1) $x + y \vee z = (x + y) \vee (x + z)$ *and* $x + y \wedge z = (x + y) \wedge (x + z)$,

(2) $x - y \wedge z = (x - y) \vee (x - z)$ *and* $x - y \vee z = (x - y) \wedge (x - z)$,

(3) $x \vee y = (x-y)^+ + y = (y-x)^+ + x$,

(4) $\lambda(x \vee y) = (\lambda x) \vee (\lambda y)$ *and* $\lambda(x \wedge y) = (\lambda x) \wedge (\lambda y)$ *for* $\lambda \geq 0$,

(5) $|\lambda x| = |\lambda|\,|x|$ *for all* $\lambda \in \mathbb{R}$,

(6) $x \vee y = \frac{1}{2}\big(x + y + |x-y|\big)$ *and* $x \wedge y = \frac{1}{2}\big(x + y - |x-y|\big)$,

(7) $x + y = x \vee y + x \wedge y$,

(8) $x = x^+ - x^-$ *and* $x^+ \wedge x^- = 0$,

(9) $|x| = x^+ + x^-$ *(and so* $|x| = 0$ *if and only if* $x = 0$*)*,

(10) $|x-y| = x \vee y - x \wedge y$,

(11) $|x+y| \vee |x-y| = |x| + |y|$,

(12) $|x| \vee |y| = \frac{1}{2}\big(|x+y| + |x-y|\big)$ *and* $|x| \wedge |y| = \frac{1}{2}\big||x+y| - |x-y|\big|$.

Proof. (1) The first identity should be interpreted as follows:

- *if two elements y and z in an ordered vector space have a supremum $y \vee z$, then for every element x the pair of vectors $x+y$ and $x+z$ also has a supremum and $(x+y) \vee (x+z) = x + (y \vee z)$.*

To see this, let $t = y \vee z$ and fix x. Clearly, $x+y \leq x+t$ and $x+z \leq x+t$. Now assume that $x+y \leq s$ and $x+z \leq s$. It follows that $y \leq s-x$ and $z \leq s-x$, and so $t = y \vee z \leq s-x$ or $x+t \leq s$. This shows that $x+t$ is the supremum of the set $\{x+y, x+z\}$, and this proves the validity of the first identity. The validity of the second identity can be proven in a similar manner.

(2) It follows from part (1) and Lemma 1.15.

(3) To establish the first identity in this part note that

$$(x-y)^+ + y = (x-y) \vee 0 + y = \big[(x-y)+y\big] \vee (0+y) = x \vee y\,.$$

The second identity can be proven in a similar manner.

(4) Assume that $\lambda > 0$. From $x \leq x \vee y$ and $y \leq x \vee y$, we get $\lambda x \leq \lambda(x \vee y)$ and $\lambda y \leq \lambda(x \vee y)$. Now assume that $\lambda x \leq z$ and $\lambda y \leq z$. It follows that $x \leq \frac{1}{\lambda} z$ and $y \leq \frac{1}{\lambda} z$, and so $x \vee y \leq \frac{1}{\lambda} z$. Consequently $\lambda(x \vee y) \leq z$, and from this we see that $(\lambda x) \vee (\lambda y) = \lambda(x \vee y)$. The other identity can be established with a similar argument.

(5) If $\lambda \geq 0$, then $|\lambda x| = (\lambda x) \vee (-\lambda x) = \lambda\big[x \vee (-x)\big] = |\lambda||x|$ and if $\lambda < 0$, then $|\lambda x| = (\lambda x) \vee (-\lambda x) = \big[(-\lambda)(-x)\big] \vee (-\lambda x) = (-\lambda)\big[(-x) \vee x\big] = |\lambda||x|$.

(6) For the first identity in this part note that

$$\begin{aligned} \tfrac{1}{2}\big(x+y+|x-y|\big) &= \tfrac{1}{2}\big[x+y+(x-y) \vee (y-x)\big] \\ &= \tfrac{1}{2}\big[(2x) \vee (2y)\big] = x \vee y\,. \end{aligned}$$

(7) Adding the identities in (6) yields the desired identity.

(8) Letting $y = 0$ in (7) yields

$$x = x \vee 0 + x \wedge 0 = x \vee 0 - (-x) \vee 0 = x^+ - x^- .$$

For the second identity in this part, note that

$$\begin{aligned} x^+ \wedge x^- &= (x^+ - x^-) \wedge 0 + x^- = x \wedge 0 + x^- \\ &= -\big[(-x) \vee 0\big] + x^- = -x^- + x^- = 0 . \end{aligned}$$

(9) Observe that

$$|x| = x \vee (-x) = (2x) \vee 0 - x = 2x^+ - (x^+ - x^-) = x^+ + x^- .$$

(10) Using the definition of the absolute value, (1) and (7), we get

$$\begin{aligned} |x - y| &= (x - y) \vee (y - x) = (2x) \vee (2y) - (x + y) \\ &= 2(x \vee y) - (x \vee y + x \wedge y) \\ &= x \vee y - x \wedge y . \end{aligned}$$

(11) Notice that

$$\begin{aligned} |x + y| \vee |x - y| &= \big[(x + y) \vee (-x - y)\big] \vee \big[(x - y) \vee (y - x)\big] \\ &= \big[(x + y) \vee (x - y)\big] \vee \big[(-x - y) \vee (y - x)\big] \\ &= \big[x + y \vee (-y)\big] \vee \big[-x + (-y) \vee y\big] \\ &= \big[x + |y|\big] \vee \big[-x + |y|\big] \\ &= \big[x \vee (-x)\big] + |y| = |x| + |y| . \end{aligned}$$

(12) For the first identity in this part note that

$$\begin{aligned} \tfrac{1}{2}\big[|x + y| + |x - y|\big] &= \tfrac{1}{2}\big[(x + y) \vee (-x - y) + |x - y|\big] \\ &= \tfrac{1}{2}\big[(x + y + |x - y|) \vee (-x - y + |x - y|)\big] \\ &= \tfrac{1}{2}\Big(\big[2(x \vee y)\big] \vee \big[2\{(-x) \vee (-y)\}\big]\Big) \\ &= x \vee y \vee (-x) \vee (-y) \\ &= \big[x \vee (-x)\big] \vee \big[y \vee (-y)\big] \\ &= |x| \vee |y| . \end{aligned}$$

Using (10), (7), and (11) and the preceding identity, we also get

$$\begin{aligned} &\big||x + y| - |x - y|\big| \\ &\quad = |x + y| \vee |x - y| - |x + y| \wedge |x - y| \\ &\quad = |x + y| \vee |x - y| - \big(|x + y| + |x - y| - |x + y| \vee |x - y|\big) \\ &\quad = 2\big(|x + y| \vee |x - y|\big) - \big(|x + y| + |x - y|\big) \\ &\quad = 2\big(|x| + |y|\big) - 2\big(|x| \vee |y|\big) = 2\big(|x| \wedge |y|\big) , \end{aligned}$$

and the proof is finished. ■

The interpretation given in the proof of property (1) of Theorem 1.17 also yields the following simple (but very useful) result.

Corollary 1.18. *A cone $\mathcal{K}$ of a vector space X is a lattice cone for the vector subspace $\mathcal{K}-\mathcal{K}$ generated by $\mathcal{K}$ in X if and only if for any two vectors $x, y \in \mathcal{K}$ their supremum $x \vee y$ exists in the partially ordered set $\mathcal{K}$.*[8]

Moreover, if $x, y, z, u \in \mathcal{K}$, then

$$(x-y) \vee (z-u) = (x+u) \vee (z+y) - (y+u)$$

holds in $\mathcal{K}-\mathcal{K}$.

We can generalize Theorem 1.16 by presenting another useful geometrical characterization of the Archimedean lattice cones.

Theorem 1.19. *In an Archimedean ordered vector space L, if $L_+ \cap (x+L_+)$ is a cone with vertex y, then $y = x^+$ [and so $L_+ \cap (x+L_+) = x^+ + L_+$].*

In particular, an Archimedean ordered vector space L is a Riesz space if and only if for each $x \in L_+$ the set $L_+ \cap (x+L_+)$ is a translation of a cone.

Proof. Assume that $L_+ \cap (x+L_+) = y+\mathcal{K}$, where $\mathcal{K}$ is a cone of L. Since $y \in y+\mathcal{K}$, we see that $y \in L_+$ (i.e., $y \geq 0$) and $y \in x+L_+$ (i.e., $y \geq x$). That is, y is an upper bound of the set $\{0, x\}$. To see that y is the least upper bound of the set $\{0, x\}$, suppose that another vector $z \in L$ satisfies $z \geq x$ and $z \geq 0$.

From $z = x+(z-x) \in x+L_+$ and $z \in L_+$, we get $z \in y+\mathcal{K}$. This implies $z-y \in \mathcal{K}$ and so $n(z-y) \in \mathcal{K}$ for each $n \in \mathbb{N}$. But then from $y+n(z-y) \in y+\mathcal{K}$ we get $y+n(z-y)-x \geq 0$ or $\frac{1}{n}(y-x)+z-y \geq 0$ for all n. Since $y-x \in L_+$ and $z-y \in L_+ - L_+$, it follows from the Archimedean property of L_+ (see Corollary 1.13) that $z-y \geq 0$. That is, $z \geq y$ and consequently $y = x^+$ holds true in L.

If this property is true for each vector x, then L must be a Riesz space; see Exercise 2 at the end of the section. ■

Recall that an ordered vector space for which every increasing bounded from above net has a supremum is called **Dedekind** (or **order**) **complete**. Likewise, an ordered vector space for which every increasing bounded from above sequence has a supremum is called **Dedekind σ-complete** or simply **σ-order complete**.

The Dedekind complete and Dedekind σ-complete Riesz spaces enjoy the following important properties.

[8] It is not difficult to see that the supremum of x and y exists in $\mathcal{K}$ if and only if it exists in X, in which case the suprema coincide.

Theorem 1.20. *In a Dedekind complete Riesz space*

(a) *every nonempty bounded from above set has a supremum and*

(b) *every nonempty bounded from below set has an infimum.*

Similarly, in a Dedekind σ-complete Riesz space every nonempty countable bounded from above (resp. from below) set has a supremum (resp. an infimum).

Proof. We consider the Dedekind complete case only. Let D be a nonempty bounded from above subset of a Dedekind complete Riesz space. Then the set of all finite suprema of elements from D can be viewed as an increasing net that is bounded from above. It is not difficult to see that the supremum of this net is also the supremum of D. ■

A vector subspace M of a vector lattice L is said to be

(1) a **Riesz subspace** or a **vector sublattice** if for each $x, y \in M$ we have $x \vee y$ and $x \wedge y$ in M (where the supremum and infimum are taken in L) and

(2) an **ideal** if $|y| \leq |x|$ and $x \in M$ implies $y \in M$.

An ideal is always a vector sublattice but a vector sublattice need not be an ideal.

For every nonempty subset A of a vector lattice, the symbol $A^\wedge$ will denote the collection of all vectors that can be written as infima of finite subsets of A. That is, a vector a belongs to $A^\wedge$ if there exist vectors $a_1, a_2, \ldots, a_k \in A$ such that $a = \bigwedge_{i=1}^k a_i$. Similarly, $A^\vee$ is the set consisting of all suprema of the finite subsets of A. We write $A^{\vee\wedge}$ for $(A^\vee)^\wedge$ and $A^{\wedge\vee}$ for $(A^\wedge)^\vee$. It always turns out that $A^{\vee\wedge} = A^{\wedge\vee}$. This is a consequence of the distributive law (see Exercise 11 at the end of the section) of the lattice operations in vector lattices and is left as an exercise.

If A is a nonempty subset of a Riesz space E, then the collection of all Riesz subspaces of E that include A is nonempty. The intersection of this collection of Riesz subspaces is the smallest Riesz subspace that includes A. This Riesz subspace is called the **Riesz subspace** (or the **vector sublattice**) **generated by** A and will be denoted $\mathcal{R}(A)$. The next result presents a description of $\mathcal{R}(A)$. Its proof is straightforward and is left as an exercise.

Lemma 1.21. *The Riesz subspace $\mathcal{R}(Y)$ generated by a vector subspace Y of a Riesz space coincides with $Y^{\wedge\vee}$ and also with $Y^{\vee\wedge}$. That is,*

$$\mathcal{R}(Y) = Y^{\wedge\vee} = Y^{\vee\wedge}.$$

Consequently, the Riesz subspace generated by a nonempty subset A of a Riesz space is $\mathcal{R}(A) = [A]^{\wedge\vee}$, where $[A]$ denotes the linear span of A.

Another interesting class of vector subspaces of an ordered vector space is that of lattice-subspaces.

Definition 1.22. *A vector subspace M of an ordered vector space L is said to be a* ***lattice-subspace*** *if M under the induced ordering from L is a vector lattice in its own right.*

That is, M is a lattice-subspace if for every $x, y \in M$ the least upper bound and greatest lower bound of the set $\{x, y\}$ exist in M when ordered by the cone $M \cap L^+$.

Every vector sublattice of a vector lattice is automatically a lattice-subspace, but a lattice-subspace need not be a vector sublattice. For details about lattice-subspaces see [**6, 69, 78, 79**].

An operator $T\colon L \to M$ between two vector lattices is said to be a **lattice homomorphism** (or a **Riesz homomorphism**) whenever we have $T(x \vee y) = T(x) \vee T(y)$ for all $x, y \in L$. A one-to-one lattice homomorphism $T\colon L \to M$ between two vector lattices is called a **lattice isomorphism** (or a **Riesz isomorphism**) and the Riesz subspace $T(L)$ of M is called a **lattice copy** of L in M. We also say that a Riesz space L is **lattice embeddable** into another Riesz space M if there exists a lattice isomorphism $T\colon L \to M$ (in which case $T(L)$ is a lattice copy of L in M). Two Riesz spaces are said to be **lattice isomorphic** (or **Riesz isomorphic**) if there exists a surjective lattice isomorphism from one to another.

The next result is simple but very useful.

Lemma 1.23. *Let $T\colon L \to M$ be a surjective one-to-one linear operator. If M is a vector lattice and we define the vector ordering $\geq$ on L by letting $u \geq v$ whenever $Tu \geq Tv$, then L is a vector lattice and $T\colon L \to M$ is a lattice isomorphism.*

Corollary 1.24. *Every copy of a vector lattice L in a partially ordered vector space M is a lattice-subspace of M.*

Exercises

1. Show that the lexicographic cone is a lattice cone.
2. Show that an ordered vector space L is a Riesz space if and only if for each $x \in L$ the supremum $x^+ = x \vee 0$ exists in L. [*Hint*: Use property (1) of Theorem 1.17 to establish that $x \vee y = (x - y)^+ + y$.]
3. Prove Corollary 1.18.
4. Let $L = C^1[0, 1]$ denote the vector space of all continuously differentiable functions on $[0, 1]$ and endow L with the pointwise vector ordering.
   (a) Show that L is an Archimedean ordered vector space with order units—and so the cone L_+ of L is generating.

(b) Prove that L is not a Riesz space.

(c) Give an example of two noncomparable functions f and g of L such that their supremum $f \vee g$ exists in L.

[*Hint*: For (b) show that the functions $\phi, \psi \in C^1[0,1]$, defined by $\phi(x) = x$ and $\psi(x) = 1 - x$, do not have a supremum in $C^1[0,1]$. For part (c) consider the functions $f(t) = (t - \frac{1}{2})^3$ and $g(t) = -(t - \frac{1}{2})^3$.]

5. For a Riesz space L show that the following statements are equivalent.

(a) The Riesz space L is Archimedean.

(b) The Riesz space L is almost Archimedean.

(c) For each $x \in L_+$ we have $\frac{1}{n}x \downarrow 0$.

(d) If $x, y \in L_+$ satisfy $ny \le x$ for all n, then $y = 0$.

(For the definition of an almost Archimedean ordered vector space see Exercise 6 of Section 1.2.)

6. Show that every Dedekind σ-complete (and hence every Dedekind complete) Riesz space is Archimedean. [*Hint*: Use the preceding exercise.]

7. If $x + y = u + v$ holds true in a Riesz space, then show that

$$x \vee u + y \vee v = x \wedge u + y \wedge v = x + y\,.$$

8. Show that a vector subspace X of a Riesz space is a Riesz subspace if and only if for each $x, y \in X$ we have $x \wedge y \in X$ (and also if and only if $x, y \in X$ implies $x \vee y \in X$). [*Hint*: Use the lattice identities $x \vee y = -[(-x) \wedge (-y)]$ and $x \wedge y = -[(-x) \vee (-y)]$.]

9. Let $\mathcal{W}$ be a wedge of a Riesz space. Show that the vector subspace generated by $\mathcal{W}$ (i.e., $\mathcal{W} - \mathcal{W}$) is a Riesz subspace if and only if for each $w_1, w_2 \in \mathcal{W}$ there exists some $w_3 \in \mathcal{W}$ such that $w_3 + w_1 \wedge w_2 \in \mathcal{W}$.

In particular, infer from this that if $w_1, w_2 \in \mathcal{W}$ implies $w_1 \wedge w_2 \in \mathcal{W}$, then $\mathcal{W} - \mathcal{W}$ is a Riesz subspace of L.

[*Hint*: Note that $(x_1 - x_2) \wedge (y_1 - y_2) = (x_1 + y_2) \wedge (y_1 + x_2) - (x_2 + y_2)$.]

10. Let X be a vector subspace of a Riesz space L and let $X_+ = X \cap L_+$. Establish that X is a Riesz subspace of L if and only if the cone X_+

(a) generates X, i.e., $X = X_+ - X_+$ and

(b) is closed under finite infima, i.e, $x, y \in X_+$ implies $x \wedge y \in X_+$.

11. Show that a Riesz space (as a lattice) satisfies the "**infinite distributive laws**." That is, show that if $\sup A$ exists for a nonempty subset A of a Riesz space L, then $\sup\{x \wedge a\colon\ a \in A\}$ exists for each $x \in L$ and

$$x \wedge \sup A = \sup\big\{x \wedge a\colon\ a \in A\big\}\,.$$

Likewise, if $\inf A$ exists for a nonempty subset A of some Riesz space L, then show that $\inf\{x \vee a\colon\ a \in A\}$ exists for each $x \in L$ and

$$x \vee \inf A = \inf\big\{x \vee a\colon\ a \in A\big\}\,.$$

In particular, if $x, x_1, x_2, \ldots, x_n$ are vectors in a Riesz space, then show that

$$x \wedge \Big[\bigvee_{i=1}^{n} x_i\Big] = \bigvee_{i=1}^{n} (x \wedge x_i) \quad \text{and} \quad x \vee \Big[\bigwedge_{i=1}^{n} x_i\Big] = \bigwedge_{i=1}^{n} (x \vee x_i)\,.$$

12. If x, y, and z are arbitrary vectors in a Riesz space, then show that

(a) (**Birkhoff's identity**)

$$|x \vee z - y \vee z| + |x \wedge z - y \wedge z| = |x - y|,$$

(b) (**Birkhoff's inequalities** [**30**, p. 296])

$$|x \vee z - y \vee z| \le |x - y| \quad \text{and} \quad |x \wedge z - y \wedge z| \le |x - y|.$$

13. Establish the validity of the following **lattice inequalities** in Riesz spaces.

(a) If two vectors x and y in a Riesz space satisfy $x \le y$, then

$$x^+ \le y^+ \quad \text{and} \quad y^- \le x^-.$$

(b) (**The triangle inequality**) If x and y are any two vectors in a Riesz space, then

$$\big||x| - |y|\big| \le |x + y| \le |x| + |y|.$$

(c) If $x, x_1, x_2, \ldots, x_n$ are positive vectors in a Riesz space, then

$$x \wedge (x_1 + x_2 + \cdots + x_n) \le x \wedge x_1 + x \wedge x_2 + \cdots + x \wedge x_n.$$

If $x \wedge x_i \wedge x_j = 0$ holds for all $i \ne j$, then equality holds.

(d) If $x_1, x_2, \ldots, x_n$ are vectors in a Riesz space, then

$$n\big(x_1^+ \wedge \cdots \wedge x_n^+\big) = n(x_1 \wedge \cdots \wedge x_n)^+ \le (x_1 + \cdots + x_n)^+.$$

14. Two vectors x and y in a Riesz space are called **disjoint** or **orthogonal**, in symbols $x \perp y$, if $|x| \wedge |y| = 0$. Establish the following disjointness properties in Riesz spaces.

(a) If $x \perp y$ and $x \perp z$ hold in a Riesz space, then $x \perp (\lambda y + \mu z)$ for all $\lambda, \mu \in \mathbb{R}$.

(b) Two vectors x and y in an arbitrary Riesz space are disjoint if and only if $|x+y| = |x-y|$. [*Hint*: Use property (12) of Theorem 1.17.]

(c) If $x \perp y$ holds in a Riesz space, then

$$|x + y| = |x - y| = |x| + |y| = \big||x| - |y|\big| = |x| \vee |y|.$$

[*Hint*: Use property (11) of Theorem 1.17.]

(d) Every subset of a Riesz space consisting of pairwise disjoint nonzero vectors is linearly independent.

15. Prove Lemma 1.23 and Corollary 1.24.

16. Show that two vector lattices are lattice isomorphic if and only if they are order isomorphic.

17. Let Σ be a σ-algebra of subsets of a nonempty set Ω and let $\mathrm{ca}(\Sigma)$ be the vector space of all countably additive signed measures on Σ having finite total variation. Recall that a set function $\mu\colon \Sigma \to [-\infty, \infty]$ that takes only at most one of the values $-\infty$ and ∞ is called a **signed measure** if $\mu(\emptyset) = 0$ and is σ-**additive** (or **countably additive**) in the sense that

$$\mu\Big(\bigcup_{n=1}^{\infty} A_n\Big) = \sum_{n=1}^{\infty} \mu(A_n)$$

holds for each pairwise disjoint sequence $\{A_n\}$ of Σ. A signed measure that takes only nonnegative extended values is called a **measure**. The **total variation** of a signed measure μ is the nonnegative extended real number $V(\mu)$ defined by

$$V(\mu)=\sup\Big\{\sum_{i=1}^{n}|\mu(A_n)|:\ A_i\in\Sigma \text{ for all } i \text{ and } A_i\cap A_j=\emptyset \text{ for } i\neq j\Big\}.$$

The signed measure μ is said to have **finite total variation** if $V(\mu)<\infty$.

On $\mathrm{ca}(\Sigma)$ define the binary relation $\geq$ by letting $\mu\geq\nu$ if $\mu(A)\geq\nu(A)$ for all $A\in\Sigma$. Show that $\geq$ is a vector ordering on $\mathrm{ca}(\Sigma)$ and that $\mathrm{ca}(\Sigma)$ under $\geq$ is a vector lattice whose lattice operations for each $\mu,\nu\in\mathrm{ca}(\Sigma)$ and each $A\in\Sigma$ are given by

$$[\mu\vee\nu](A)=\sup\{\mu(B)+\nu(A\setminus B):\ B\in\Sigma \text{ and } B\subseteq A\}$$

and

$$[\mu\wedge\nu](A)=\inf\{\mu(B)+\nu(A\setminus B):\ B\in\Sigma \text{ and } B\subseteq A\}.$$

18. Consider the vectors subspace V of $C[0,1]$ consisting of all linear functions, i.e., a function $x\in C[0,1]$ belongs to V if and only if there exist real numbers b and m such that for each $t\in[0,1]$ we have $x(t)=mt+b$. Show that V is a lattice-subspace of $C[0,1]$ but not a Riesz space.

19. Prove Lemma 1.21.

20. Let L denote the vector subspace of $\mathbb{R}^{[0,1]}$ consisting of all continuous functions of bounded total variation and let

$$\mathcal{K}=\{f\in L:\ f(x)\geq 0 \text{ for all } x\in[0,1]\} \text{ and}$$

and

$$L_+=\{f\in\mathcal{K}:\ f \text{ is nondecreasing}\}.$$

Establish the following.

(a) $\mathcal{K}$ is a lattice cone. Describe also the lattice operations of the Riesz space $(L,\mathcal{K})$.
(b) The Riesz space $(L,\mathcal{K})$ is not Dedekind σ-complete.
(c) L_+ is a lattice subcone of $\mathcal{C}$.
(d) The Riesz space (L,L_+) is Dedekind complete. Also describe the lattice operations of this Riesz space.

[*Hint*: See for details and more [**16**, Section 39, p. 366] and [**12**, Section 9.8, p. 365].]

1.4. Positive and order bounded operators

Recall that a function $T\colon X \to Y$ between two vector spaces is said to be

(1) **additive** if for all $x, y \in X$ we have $T(x+y) = T(x) + T(y)$,

(2) **homogeneous** whenever for all $x \in X$ and all $\lambda \in \mathbb{R}$ we have $T(\lambda x) = \lambda T(x)$,

(3) **linear** if it is both additive and homogeneous.

A linear function $T\colon X \to Y$ between two vector spaces is referred to as an **operator**.[9] It is very common in operator theory to denote the vector $T(x)$ by Tx and occasionally we shall be doing so. The vector space of all (linear) operators from a vector space X to a vector space Y will be denoted $\mathcal{L}(X, Y)$.

We now introduce some classes of operators between ordered spaces that are related to the order structures of the spaces.

An operator $T\colon L \to M$ between two ordered vector spaces is called

(a) **positive**, in symbols $T \geq 0$, if $T(x) \geq 0$ holds for all $x \in L^+$,

(b) **strictly positive**, in symbols $T \gg 0$, if $T(x) > 0$ for all $x > 0$,

(c) **regular** if T can be written as a difference of two positive operators,

(d) **order bounded** if T carries order bounded subsets of L to order bounded subsets of M.

It should be clear that every strictly positive operator is a positive operator, every positive operator is regular and that every regular operator is order bounded. So, if for any two ordered vector spaces L and M we denote the vector space of all order bounded operators from L to M by $\mathcal{L}_{\rm b}(L, M)$ and the vector space of all regular operators from L to M by $\mathcal{L}_{\rm r}(L, M)$, then we have the vector subspace inclusions

$$\mathcal{L}_{\rm r}(L, M) \subseteq \mathcal{L}_{\rm b}(L, M) \subseteq \mathcal{L}(L, M)\,.$$

The inclusion $\mathcal{L}_{\rm r}(L, M) \subseteq \mathcal{L}_{\rm b}(L, M)$ can be proper; see Exercise 3 at the end of this section. Also notice that the collection of all positive operators from L to M is a wedge of $\mathcal{L}(L, M)$. In case L_+ is a generating cone, this wedge is of course a cone.

As usual, the general additive functions are defined as follows.

Definition 1.25. *A mapping $T\colon \mathcal{W}_1 \to \mathcal{W}_2$ between two wedges is said to be **additive** whenever $T(x+y) = T(x) + T(y)$ holds for all $x, y \in \mathcal{W}_1$.*

---

[9] More precisely, it should be called a *linear operator.* In this monograph the term "operator" is synonymous with "linear operator."

The following is a basic extension result dealing with additive functions between cones. The result informs us that quite often the additive mappings extend to positive operators.

Lemma 1.26. *Suppose that L and M are two ordered vector spaces with M Archimedean. For an additive mapping $T\colon L^+ \to M^+$ we have the following.*

(1) *The mapping T extends to a positive operator from L to M.*

(2) *Any extension of T to all of L is uniquely determined on $L_+ - L_+$, the vector subspace generated by L_+.*

(3) *If L_+ is generating, then T extends uniquely to a positive operator from L to M.*

Proof. Assume that $T\colon L^+ \to M^+$ is an additive mapping. The extension of T to $X = L_+ - L_+$ is achieved as follows. For each $x \in X$ pick $x_1, x_2 \in L_+$ with $x = x_1 - x_2$ and put

$$S(x) = T(x_1) - T(x_2)\,. \tag{$\star$}$$

Notice that if $x = y_1 - y_2$ with $y_1, y_2 \in L_+$, then $y_1 + x_2 = x_1 + y_2$, and so by the additivity of $T\colon L_+ \to M_+$ we get

$$T(y_1) + T(x_2) = T(y_1 + x_2) = T(x_1 + y_2) = T(x_1) + T(y_2)\,,$$

and so $T(y_1) - T(y_2) = T(x_1) - T(x_2)$. This shows that the mapping $S\colon X \to M$ as defined by $(\star)$ is well defined—and also that it is the only possible extension of $T\colon L_+ \to M_+$ to a positive operator on all of X. We shall show that S is linear. That is, we shall prove that S is additive and homogeneous on X.

For the additivity of S note that if $x = x_1 - x_2$ and $y = y_1 - y_2$ holds with the $x_1, x_2, y_1, y_2 \in L_+$, then we have

$$\begin{aligned} S(x+y) &= S\big((x_1 + y_1) - (x_2 + y_2)\big) \\ &= T(x_1 + y_1) - T(x_2 + y_2) \\ &= T(x_1) + T(y_1) - \big[T(x_2) + T(y_2)\big] \\ &= \big[T(x_1) - T(x_2)\big] + \big[T(y_1) - T(y_2)\big] \\ &= S(x) + S(y)\,. \end{aligned}$$

In particular, the additivity of S implies that S is $\mathbb{Q}$-homogeneous, i.e., $S(rx) = rS(x)$ holds for all $x \in X$ and all rational numbers r.

Next, we shall show that S is homogeneous. For this, we need to prove first that S is monotone. That is, $x \geq y$ in X implies $S(x) \geq S(y)$ in M. Indeed, if $x \geq y$ holds in X, then $x - y \in L^+$ and the additivity of S yields

$$S(x) = S\big((x-y) + y\big) = S(x-y) + S(y) = T(x-y) + S(y) \geq S(y)\,.$$

Now fix $x \in L^+$ and let $\lambda \in \mathbb{R}$. Pick two sequences of rational numbers $\{r_n\}$ and $\{t_n\}$ such that $r_n \uparrow \lambda$ and $t_n \downarrow \lambda$. The inequalities $r_n x \leq \lambda x \leq t_n x$ and the monotonicity of S coupled with the $\mathbb{Q}$-homogeneity of S imply

$$r_n S(x) = S(r_n x) \leq S(\lambda x) \leq S(t_n x) = t_n S(x)$$

for each n. Since $S(\lambda x) \in M_+ - M_+$, $S(x) = T(x) \geq 0$ and M is Archimedean, it follows from Corollary 1.13 that $\lambda S(x) = S(\lambda x)$. Finally, if $\lambda \in \mathbb{R}$ and $x \in X$, then write $x = x_1 - x_2$ with $x_1, x_2 \in L_+$ and note that

$$\begin{aligned} S(\lambda x) &= S(\lambda x_1 + (-\lambda)x_2) = S\big(\lambda x_1\big) + S\big((-\lambda)x_2\big) \\ &= \lambda S(x_1) - \lambda S(x_2) = \lambda\big[T(x_1) - T(x_2)\big] = \lambda S(x)\,. \end{aligned}$$

Consequently, S is also homogeneous, and so $S\colon X \to M$ is the unique extension of the additive mapping $T\colon L_+ \to M_+$ to a positive operator.

Next, let Y be any algebraic complement of X, i.e., let Y be any vector subspace of L such that $L = X \oplus Y$, and let $S_1\colon Y \to M$ be any operator. For instance, we can let $S_1(y) = 0$ for all $y \in Y$. Then, the operator $S_2\colon L \to M$ defined by $S_2(x \oplus y) = S(x) + S_1(y)$ is a positive linear extension of $T\colon L_+ \to M_+$ to all of L. Now the desired conclusions follow immediately from the above discussion. ■

The rest of the section is devoted to presenting several important results concerning extensions of positive operators. Our first extension result is a generalization of the classical Hahn–Banach theorem (due to H. Hahn [**50**] and S. Banach [**27**]) to the setting of ordered vector spaces. This theorem plays a fundamental role in modern analysis and without any doubt its generalization will be of great importance to us here.

In order to state and prove this result, we need to recall a few definitions. First, recall that a function $p\colon X \to M$, where X is a (real) vector space and M is an ordered vector space, is called **sublinear** whenever it is *subadditive* and *positively homogeneous*, i.e., if

(a) $p(x+y) \leq p(x) + p(y)$ for all $x, y \in X$ and

(b) $p(\lambda x) = \lambda p(x)$ for all $x \in X$ and all $\lambda \geq 0$.

Also, a function $f\colon C \to M$, where C is a convex subset of a vector space and M is an ordered vector space, is said to be **convex** if for all $x, y \in C$ and all $0 \leq \alpha \leq 1$ we have

$$f(\alpha x + (1-\alpha)y) \leq \alpha f(x) + (1-\alpha) f(y)\,.$$

Clearly, every sublinear function is a convex function.

We are now ready to state and prove the most general version of the classical Hahn–Banach extension theorem.

Theorem 1.27 (Hahn–Banach). *Let X be a (real) vector space, M a Dedekind complete Riesz space, and let $p\colon X \to M$ be a convex function—in particular, let p be sublinear. If Y is a vector subspace of X and $S\colon Y \to M$ is an operator satisfying $S(y) \le p(y)$ for all $y \in Y$, then there exists some operator $T\colon X \to M$ such that*

(1) *$T = S$ on Y, i.e., T is a linear extension of S to all of X and*

(2) *$T(x) \le p(x)$ holds for all $x \in X$.*

Proof. The critical step is to show that S has a linear extension satisfying (2) on an arbitrary vector subspace generated by Y and one extra vector. If this is done, then an application of Zorn's lemma guarantees the existence of an extension of S to all of X with the desired properties.

To see this, let $u \notin Y$, and let $V = \{y + \lambda u\colon\ y \in Y \text{ and } \lambda \in \mathbb{R}\}$. If $T\colon V \to M$ is a linear extension of S, then

$$T(y + \lambda u) = S(y) + \lambda T(u)$$

must hold true for all $y \in Y$ and all $\lambda \in \mathbb{R}$. Put $w = T(u)$. To complete the proof, we must establish the existence of some $w \in M$ such that

$$S(y) + \lambda w \le p(y + \lambda u) \tag{1}$$

holds for all $y \in Y$ and $\lambda \in \mathbb{R}$. It is a routine matter to verify that (1) is true if and only if there exists some vector $w \in M$ satisfying

$$\tfrac{1}{\lambda}\big[S(x) - p(x - \lambda u)\big] \le w \le \tfrac{1}{\mu}\big[p(y + \mu u) - S(y)\big] \tag{2}$$

for all $x, y \in Y$ and all $\lambda, \mu > 0$. Now notice that (2) is true for some $w \in M$ if and only if

$$\sup_{x\in Y,\, \lambda>0} \tfrac{1}{\lambda}\big[S(x) - p(x - \lambda u)\big] \le \sup_{y\in Y,\, \mu>0} \tfrac{1}{\mu}\big[p(y + \mu u) - S(y)\big],$$

which in turn is equivalent to

$$\tfrac{1}{\lambda}\big[S(x) - p(x - \lambda u)\big] \le \tfrac{1}{\mu}\big[p(y + \mu u) - S(y)\big] \tag{3}$$

for all $x, y \in Y$ and all $\lambda, \mu > 0$. Rearranging terms, we see that (3) is equivalent to

$$S(\mu x + \lambda y) \le \mu p(x - \lambda u) + \lambda p(y + \mu u) \tag{4}$$

for all $x, y \in Y$ and all $\lambda, \mu > 0$. Thus, a linear extension of S to all of X exists if and only if (4) is true. For the validity of (4) note that if $x, y \in Y$

and $\lambda, \mu > 0$, then using the convexity of p, we obtain

$$\begin{aligned} S(\mu x + \lambda y) &= (\lambda+\mu) S\big(\tfrac{\mu}{\lambda+\mu}x + \tfrac{\lambda}{\lambda+\mu}y\big) \\ &\le (\lambda+\mu) p\big(\tfrac{\mu}{\lambda+\mu}x + \tfrac{\lambda}{\lambda+\mu}y\big) \\ &= (\lambda+\mu) p\big(\tfrac{\mu}{\lambda+\mu}[x-\lambda u] + \tfrac{\lambda}{\lambda+\mu}[y+\mu u]\big) \\ &\le (\lambda+\mu)\big[\tfrac{\mu}{\lambda+\mu}p(x-\lambda u) + \tfrac{\lambda}{\lambda+\mu}p(y+\mu u)\big] \\ &= \mu p(x-\lambda u) + \lambda p(y+\mu u)\,, \end{aligned}$$

as desired. This complete the proof of the theorem. ■

Recall that a vector subspace G of a Riesz space E is said to be a **Riesz subspace** (or a **vector sublattice**) whenever G is closed under the lattice operations of E, i.e., whenever for each pair $x, y \in G$ the vector $x \vee y$ (taken in E) belongs to G.

As a first application of the Hahn–Banach extension theorem we present the following useful extension property of positive operators.

Theorem 1.28. *Let $T\colon L \to M$ be a positive operator between two Riesz spaces with M Dedekind complete. Assume also that G is a Riesz subspace of L and that $S\colon G \to M$ is an operator satisfying $0 \le S(x) \le T(x)$ for all $x \in G^+$. Then S can be extended to a positive operator from L to M such that $0 \le S(x) \le T(x)$ holds for all $x \in L^+$.*

Proof. Define $p\colon L \to M$ by $p(x) = T(x^+)$, and note that p is sublinear and satisfies $S(x) \le p(x)$ for all $x \in G$. By Theorem 1.27 there exists a linear extension of S to all of L (which we denote by S again) satisfying $S(x) \le p(x)$ for all $x \in L$. Now if $x \in L^+$, then

$$-S(x) = S(-x) \le p(-x) = T\big((-x)^+\big) = T(0) = 0\,,$$

and so $0 \le S(x) \le p(x) = T(x)$ holds, as desired. ■

A subset A of a Riesz space is called **solid** whenever $|x| \le |y|$ and $y \in A$ imply $x \in A$. As defined earlier, a solid vector subspace of a Riesz space is referred to as an **ideal**. From the lattice identity $x \vee y = \frac{1}{2}(x + y + |x-y|)$, it follows immediately that every ideal is a Riesz subspace.

The next result deals with restrictions of positive operators to ideals.

Theorem 1.29. *If $T\colon L \to M$ is a positive operator between two Riesz spaces with M Dedekind complete, then for every ideal A of L the formula*

$$T_A(x) = \sup\big\{T(y)\colon\; y \in A \;\text{ and }\; 0 \le y \le x\big\}\,,\quad x \in L^+\,,$$

defines a positive operator from L to M. Moreover, we have

(a) $0 \le T_A \le T$,

(b) $T_A = T$ *on* A,

(c) *if* B *is another ideal with* $A \subseteq B$, *then* $T_A \le T_B$ *holds.*

Proof. Note first that

$$T_A(x) = \sup\{T(x \wedge y)\colon\ y \in A^+\}$$

holds for all $x \in L^+$. According to Lemma 1.26 it suffices to show that T_A is additive on L^+.

To this end, let $x, y \in L^+$. If $z \in A^+$, then the inequality

$$(x + y) \wedge z \le x \wedge z + y \wedge z$$

implies that $T\big((x+y) \wedge z\big) \le T(x \wedge z) + T(y \wedge z) \le T_A(x) + T_A(y)$, and hence

$$T_A(x + y) \le T_A(x) + T_A(y)\,.$$

On the other hand, the inequality $x \wedge u + y \wedge v \le (x + y) \wedge (u + v)$ implies

$$T_A(x) + T_A(y) \le T_A(x + y)\,.$$

Therefore, $T_A(x + y) = T_A(x) + T_A(y)$ holds, so that T_A is additive on L^+.

Properties (a), (b) and (c) are now easy consequences of the formula defining the operator T_A. ■

Recall that a vector subspace G of an ordered vector space L **majorizes** L whenever for each $x \in L$ there exists some $y \in G$ with $x \le y$ (or, equivalently, if for each $x \in L$ there exists some $y \in G$ with $y \le x$).

It is important to know that every positive operator whose domain is a majorizing vector subspace and whose values are in a Dedekind complete Riesz space always has a positive extension. This is a classical result due to L. V. Kantorovich [**57**].

Theorem 1.30 (Kantorovich). *Let* L *and* M *be two ordered vector spaces with* M *a Dedekind complete Riesz space. If* G *is a majorizing vector subspace of* L *and* $T\colon G \to M$ *is a positive operator, then* T *has a positive linear extension to all of* L.

Proof. Fix $x \in L$ and let $y \in G$ satisfy $x \le y$. Since G is majorizing, there exists a vector $u \in G$ with $u \le x$. Hence, $u \le y$ and the positivity of T implies $T(u) \le T(y)$ for all $y \in G$ with $x \le y$. In particular, it follows that $\inf\{T(y)\colon\ y \in G \text{ and } x \le y\}$ exists in M for each $x \in L$. Thus, a mapping $p\colon L \to M$ can be defined via the formula

$$p(x) = \inf\{T(y)\colon\ y \in G \text{ and } x \le y\}\,.$$

Clearly, $T(x) = p(x)$ holds for each $x \in G$ and an easy argument shows that p is also sublinear.

Now, by the Hahn–Banach Extension Theorem 1.27, the operator T has a linear extension S to all of L satisfying $S(z) \le p(z)$ for each $z \in L$. If $z \in L^+$, then $-z \le 0$, and so from

$$-S(z) = S(-z) \le p(-z) \le T(0) = 0\,,$$

we see that $S(z) \ge 0$. This shows that S is a positive linear extension of T to all of L. ∎

We close the section with an interesting local approximation property of positive operators—which also shows that there is an abundance of positive operators between Riesz spaces.

Theorem 1.31. *Let $T\colon L \to M$ be a positive operator between two Riesz spaces with M Dedekind σ-complete. Then for each $x \in L^+$ there exists a positive operator $S\colon L \to M$ such that*

(1) $0 \le S \le T$,

(2) $S(x) = T(x)$,

(3) $S(y) = 0$ *for all* $y \in L$ *with* $|y| \wedge x = 0$.

Proof. Let $x \in L^+$ be fixed and define $S\colon L^+ \to M^+$ by

$$S(y) = \sup\bigl\{T(y \wedge nx)\colon\ n = 1, 2, \dots\bigr\}\,.$$

(The supremum exists since M is Dedekind σ-complete and the sequence $\{T(y \wedge nx)\}$ is bounded above in M by Ty.) We claim that S is additive.

To see this, let $y, z \in L^+$. From $(y+z) \wedge nx \le y \wedge nx + z \wedge nx$ we get

$$T\bigl((y+z) \wedge nx\bigr) \le T(y \wedge nx) + T(z \wedge nx) \le S(y) + S(z)\,,$$

and so $S(y+z) \le S(y) + S(z)$. On the other hand, for each m and n we have $y \wedge nx + z \wedge mx \le (y+z) \wedge (n+m)x$, and thus

$$T(y \wedge nx) + T(z \wedge mx) \le T\bigl((y+z) \wedge (n+m)x\bigr) \le S(y+z)$$

holds for all n and m. This implies $S(y) + S(z) \le S(y+z)$, and hence $S(y+z) = S(y) + S(z)$, so that S is additive.

By Lemma 1.26 the mapping S extends uniquely to all of L as a positive operator. Now it is a routine matter to verify that the operator S satisfies the desired properties. ∎

Exercises

1. Let $T\colon L \to M$ be a surjective (onto) positive operator between two ordered vector spaces. Show that if e is an order unit of L, then $T(e)$ is an order unit of M.

2. Show that Lemma 1.26 is not true if M is not Archimedean. [*Hint*: Let $\phi\colon \mathbb{R} \to \mathbb{R}$ be an additive function that is not linear, i.e., not of the form $\phi(x) = cx$, and let M be the lexicographic plane. Consider the mapping $T\colon \mathbb{R}^+ \to M^+$ defined by $T(x) = \big(x, \phi(x)\big)$ for each x in $\mathbb{R}^+$. Note that T is additive and that if T could be extended to an operator from $\mathbb{R}$ to M, then ϕ should be linear.]

3. This exercise presents an example (due to H. P. Lotz) of an order bounded operator that is not a regular operator—and consequently the inclusion $\mathcal{L}_{\mathrm{r}}(L, M) \subseteq \mathcal{L}_{\mathrm{b}}(L, M)$ can be proper.

   Let $L = M = C[-1, 1]$ and define the operator $T\colon L \to M$ for all $f \in L$ and all $-1 \le t \le 1$ by

$$[Tf](t) = \begin{cases} f(\sin \frac{1}{t}) - f(\sin(t + \frac{1}{t})) & \text{if } \ 0 < |t| < 1, \\ 0 & \text{if } \ t = 0 \,. \end{cases}$$

   Show that T defines an order bounded operator that fails to be regular. (For details about this problem see also [**15**, Example 1.16, p. 13].)

4. Let $L = L_p[0, 1]$ with $0 < p < 1$ and let $G = L_1[0, 1]$. Clearly, $G \subseteq L$ and G is an ideal of L. (Here $f \ge g$ means that $f(x) \ge g(x)$ holds for almost all x with respect to the Lebesgue measure.) Now consider the operator $T\colon G \to \mathbb{R}$ defined by

$$T(f) = \int_0^1 f(x)\, dx \,.$$

   Show that T does not have a positive linear extension to all of L. [*Hint*: To see this, assume by way of contradiction that T is extendable to a positive operator from L to $\mathbb{R}$. In particular, this implies that if $f \in L$ is defined by $f(x) = \frac{1}{x}$, then the set of real numbers

$$D = \big\{T(g)\colon \ g \in G \ \text{and} \ 0 \le g \le f\big\}$$

   is bounded. On the other hand, if $g_n = f\chi_{(\frac{1}{n},1)}$, then $T(g_n) = \ln n \in D$ holds for each n. Therefore, D must be unbounded, a contradiction.]

5. A positive operator $T\colon G \to M$, where G is a vector subspace of an ordered vector space L and M is an ordered vector space, is said to have a **smallest extension** whenever there exists a positive extension S of T to all of L satisfying $S \le R$ for all other positive extensions R of T to all of L, in which case S is called the **smallest extension** of T.

   Now let L and M be two Riesz spaces with M Dedekind complete, let A be an ideal of L, and let $T\colon A \to M$ be a positive operator. Show that if T has a positive extension to all of L, then T has a smallest extension. Moreover, if in this case S is the smallest extension, then

$$S(x) = \sup\big\{Ty\colon \ y \in A \ \text{and} \ 0 \le y \le x\big\}$$

   holds for all $x \in L^+$.

6. Let L and M be two Riesz spaces with M Dedekind complete, and let A be an ideal of L. For each $T \in \mathcal{L}_{\mathrm{b}}(L, M)$ let $\Pi(T)$ denote the restriction

of T to A. Show that the positive operator $\Pi\colon \mathcal{L}_b(L,M) \to \mathcal{L}_b(A,M)$ satisfies

$$\Pi(S \vee T) = \Pi(S) \vee \Pi(T) \quad \text{and} \quad \Pi(S \wedge T) = \Pi(S) \wedge \Pi(T)$$

for all $S, T \in \mathcal{L}_b(L,M)$.[10] (In other words, Π is a lattice homomorphism.)

1.5. Positive linear functionals

The **algebraic dual** of a vector space X (i.e., the vector space of all linear functionals on X) will be denoted by X^*. That is, a function $f\colon X \to \mathbb{R}$ belongs to X^* if and only if $f(\alpha x + \beta y) = \alpha f(x) + \beta f(y)$ holds for all $x, y \in X$ and all $\alpha, \beta \in \mathbb{R}$. A linear functional $f \in L^*$, where L is now an ordered vector space, is said to be

(1) **positive**, denoted $f \geq 0$, if $f(x) \geq 0$ holds for all $x \in L^+$,[11]

(2) **strictly positive**, denoted $f \gg 0$, if $f(x) > 0$ holds for all $x > 0$,[12]

(3) **regular** if f can be written as a difference of two positive linear functionals,

(4) **order bounded** if f carries order intervals of L to bounded subsets of $\mathbb{R}$.

It should be clear that

(a) every strictly positive linear functional is positive,

(b) every positive linear functional is regular,

(c) every positive functional (and hence every regular linear functional) is order bounded.

The geometrical meanings of positive, strictly positive and nonpositive linear functional are as follows. For any linear functional f define the half spaces $H = \{x\colon f(x) \geq 0\}$ and $G = \{x\colon f(x) > 0\}$. Clearly, f is positive if and only if $L^+ \subseteq H$. It is strictly positive if and only if $L^+ \subseteq G$. Notice that for any vector z we have $H_z = z + H = \{x\colon f(x) \geq f(z)\}$ and, of course, $G_z = z + G = \{x\colon f(x) > f(z)\}$. Therefore, f is positive if and only if there

[10] An operator between spaces of operators is referred to as a **transformer**. So, the operator Π is an example of a transformer.

[11] Recalling that a set A of a vector space is **supported** at a point $a \in A$ by a linear functional f (called a **supporting functional** of A at a) if f attains its minimum value over A at a, i.e., $f(x) \geq f(a)$ holds for all $x \in A$, we see that the positive linear functionals are precisely the linear functionals that support the cone at zero.

[12] Recalling that an extreme point a of a convex subset A of a vector space is called an **exposed point** of A if there exists a linear functional f (called a linear functional that **exposes** A at a) if $f(x) > f(a)$ holds for all $x \in A \setminus \{a\}$, we see that the strictly positive linear functionals are precisely the linear functionals that expose the cone at zero.

exists some z such that $z + L^+ \subseteq H_z$ and it is strictly positive if and only if there exists some z such that $z + L^+ \subseteq G_z$.

Positive but not strictly Strictly positive Non-positive

The reader should also notice immediately that we can use the same positivity notions for wedges as for cones. If $\mathcal{K}$ is a cone (or a wedge) in a vector space X, then a linear functional $f \in X^*$ is $\mathcal{K}$-**positive** whenever $x \in \mathcal{K}$ implies $f(x) \geq 0$. The notions of $\mathcal{K}$-strictly positive, $\mathcal{K}$-regular, and $\mathcal{K}$-order bounded linear functionals are defined analogously. The geometrical meanings of positive, strictly positive and nonpositive linear functional are shown in the figure above.

Our first result in this section is a characterization of the positive linear functionals. But first, we need a definition. A nonempty subset A of an ordered vector space L is said to be **comprehensive from above** (or simply **comprehensive**) if $x \leq y$ with $x \in A$ implies $y \in A$. Equivalently, a nonempty subset A is comprehensive if $A + L_+ = A$.

Lemma 1.32. *A linear functional on an ordered vector space is positive if and only if it is bounded below on a comprehensive set.*

In particular, every linear functional that supports a comprehensive subset of an ordered vector space is positive.

Proof. Let $f\colon L \to \mathbb{R}$ be a linear functional on an ordered vector space. If f is positive, then f is bounded below from 0 on the comprehensive set L_+. For the converse, assume that there exists a constant $c \in \mathbb{R}$ such that $f(a) \geq c$ holds true for all a in a comprehensive set A. Fix some $a \in A$ and let $x \in L_+$. Then $a + \lambda x \in A$ for all $\lambda > 0$ and so from our condition we get $f(x) \geq \frac{c-f(a)}{\lambda}$ for all $\lambda > 0$. Letting $\lambda \to \infty$ yields $f(x) \geq 0$ for all $x \in L_+$, i.e., f is a positive linear functional. ∎

We now introduce the order and regular duals of an ordered vector space.

Definition 1.33. *For a given ordered vector space L, the vector spaces*

$$L^\sim := \mathcal{L}_{\mathrm{b}}(L, \mathbb{R}) \quad \textit{and} \quad L^r := \mathcal{L}_{\mathrm{r}}(L, \mathbb{R})$$

are called the **order dual** *and the* **regular dual**, *respectively, of* L.

In other words, the order dual $L^\sim$ is the vector space consisting of all order bounded linear functionals on L and L^r is the vector space consisting of all regular linear functionals on L. Clearly, we have the following vector subspace inclusions: $L^r \subseteq L^\sim \subseteq L^*$.

The proof of the next result is straightforward and is omitted.

Lemma 1.34. *If L is an ordered vector space, then the collection of all positive linear functionals on L is a wedge in L^* (and also in L^r and $L^\sim$).*

When the cone L_+ is generating, the wedge of all positive linear functionals is a cone.

We shall close the section with two results regarding the extension of positive linear functionals. The first one (due to H. Bauer [**28**] and I. Namioka [**71**]) presents a necessary and sufficient condition for a positive linear functional defined on a vector subspace to have a positive linear extension to the whole space.

Theorem 1.35 (Bauer–Namioka). *Let X be a vector subspace of an ordered vector space L and let $f\colon X \to \mathbb{R}$ be a positive linear functional, that is, $0 \le x \in X$ implies $f(x) \ge 0$. Then f has a positive linear extension to all of L if and only if there is a nonempty convex and absorbing*[13] *subset A of L such that*

$$x \in X \cap (A - L_+) \implies f(x) \le 1\,.$$

Proof. If f has a positive linear extension, say g, to all of L, then the set $A = \{x \in L\colon\ g(x) \le 1\}$ is nonempty convex and absorbing and satisfies (as can be easily checked) the desired property.

For the converse, assume that there exists a nonempty convex and absorbing subset A of L such that if $x \in X \cap (A - L_+)$, then $f(x) \le 1$. From $A \subseteq A - L_+$, it follows that the nonempty convex set $A - L_+$ is also absorbing. So, if we let $p\colon L \to \mathbb{R}$ be the Minkowski functional of $A - L_+$, i.e., for each $x \in L$ we have

$$p(x) = \inf\{\lambda > 0\colon\ x \in \lambda(A - L_+)\}\,,$$

then p is a sublinear function. Now notice that if $x \in X$ and some $\lambda > 0$ satisfies $x \in \lambda(A - L_+)$, then $\frac{x}{\lambda} \in X \cap (A - L_+)$ and so $\frac{f(x)}{\lambda} \le 1$ or $f(x) \le \lambda$. This implies that $f(x) \le p(x)$ for all $x \in X$. By the Hahn–Banach extension theorem there exists a linear extension $h\colon L \to \mathbb{R}$ of f satisfying $h(x) \le p(x)$ for all $x \in L$.

[13]Recall that a subset S of a vector space V is said to be **absorbing** if for every $u \in V$, there exists some $\lambda_0 > 0$ such that $\lambda u \in S$ holds true for all λ with $|\lambda| \le \lambda_0$.

To complete the proof, we shall show that h is positive. To see this, fix $x \in L_+$. Then for each $\alpha > 0$ we have $-\alpha x \in -L_+ \subseteq A - L_+$ and so $\alpha p(-x) = p(-\alpha x) \leq 1$. It follows that $-\alpha h(x) = h(-\alpha x) \leq p(-\alpha x) \leq 1$ and thus $-h(x) \leq \frac{1}{\alpha}$ for all $\alpha > 0$. Letting $\alpha \to \infty$, we get $-h(x) \leq 0$ or $h(x) \geq 0$. This shows that h is a positive linear functional. ∎

Our second extension theorem for positive linear functionals is the special case of Theorem 1.30 (when $M = \mathbb{R}$). We state it as a theorem simply because of its wide range of applications.

Theorem 1.36 (Kantorovich). *If X is a majorizing vector subspace of an ordered vector space L and $f\colon X \to \mathbb{R}$ is a positive linear functional, then f has a positive linear extension to all of L.*

Exercises

1. Show that an arbitrary intersection of a family of comprehensive from above (resp. from below) subsets of an ordered vector space is a comprehensive from above (resp. from below) set.
2. If L is an ordered vector space with an order unit u, then prove that the order bounded linear functionals on L are precisely the linear functionals on L that carry the order interval $[0, u]$ to bounded subsets of $\mathbb{R}$, i.e.,
$$L^\sim = \big\{f \in L^*:\ f([0,u]) \text{ is a bounded subset of } \mathbb{R}\big\}.$$
3. Assume that a Riesz space L admits a strictly positive linear functional. Show that if a nonempty subset A of L has a supremum, then there exists an at most countable subset B of A such that $\sup A = \sup B$.[14] [*Hint*: Fix a strictly positive linear functional f on L and assume that $x_\alpha \downarrow 0$ holds in L. Let $f(x_\alpha) \downarrow s$ in $\mathbb{R}$ and then pick an increasing sequence $\{\alpha_n\}$ of indices such that $f(x_{\alpha_n}) \downarrow s$. Now verify that if $x \leq x_{\alpha_n}$ holds for each n, then $x \leq x_\alpha$ for each α and deduce that $x \leq 0$. This shows that $x_{\alpha_n} \downarrow 0$. The general case can be reduced to this by replacing A by the collection of all finite suprema of A.]
4. Let L be an ordered vector space. Show that the collections $L^\sim$ (of all order bounded linear functionals on L) and L^r (of all regular linear functionals on L) are both vector subspaces of L and that $L^r \subseteq L^\sim$.
5. Prove Lemma 1.34.
6. If L is an ordered vector space and $f \in L^\sim$, then for each $x \in L_+$ show that
$$\sup f([0,x]) + \inf f([0,x]) = f(x).$$

[14] An ordered vector space where for every nonempty subset A that has supremum there exists an at most countable subset B of A with $\sup B = \sup A$ is known as an ordered vector space with the **countable sup property**. So, with this terminology, the result in this exercise states that: *every Riesz space that admits a strictly positive linear functional has the countable sup property.*

7. Let $\mathcal{K}$ be a cone of a vector space X and let $Y = \mathcal{K} - \mathcal{K}$, the linear span of $\mathcal{K}$. Show that a linear functional $f \in X^*$ is $\mathcal{K}$-order bounded if and only if its restriction to Y is $\mathcal{K}$-order bounded.

8. Let $L = B[0,1]$, the ordered vector space of all (uniformly) bounded real-valued functions on $[0,1]$, and let $X = C[0,1]$, the ordered vector space of all real-valued continuous functions on $[0,1]$. Establish the following.
   (a) X majorizes L.
   (b) If $f\colon X \to \mathbb{R}$ is the positive linear functional defined via the formula $f(x) = \int_0^1 x(t)\,dt$ (the Riemann integral), then show that f has a positive linear extension to all of L.

9. Let X denote the vector subspace of $L = \ell_\infty$ consisting of all convergent (real) sequences and consider the positive linear functional $f\colon X \to \mathbb{R}$ defined by $f(x_1, x_2, \ldots) = \lim_{n\to\infty} x_n$. Show that f has a positive linear extension to all of L.

10. Let L be the collection of all (uniformly) bounded real functions on $\mathbb{R}$ that vanish outside closed intervals. That is, a function $x\colon \mathbb{R} \to \mathbb{R}$ belongs to L if and only if there exist some $M > 0$ and an interval $[a,b]$ (both depending on x) such that $|x(t)| \le M$ for all $t \in \mathbb{R}$ and $x(t) = 0$ for each $t \notin [a,b]$. Establish the following.
   (a) The collection L under the usual pointwise algebraic operations and the pointwise ordering is a Riesz space—and so its cone is generating.
   (b) The Riesz space L has no order units.
   (c) The vector subspace X of L consisting of all continuous real functions on $\mathbb{R}$ with compact support (i.e., vanishing off closed intervals) majorizes L.
   (d) The function $f\colon X \to \mathbb{R}$, defined by $f(x) = \int_{-\infty}^{\infty} x(t)\,dt$, is a strictly positive linear functional on X. (See also Exercise 4 of Section 2.5.)
   (e) The strictly positive linear functional f has a positive linear extension to all of L.

11. Show that every ordered vector space with an order unit admits a nonzero positive linear functional. [*Hint*: Let L be an ordered vector space having an order unit e, let $X = \{\lambda e\colon\ \lambda \in \mathbb{R}\}$ and consider the nonzero positive linear functional $f\colon X \to \mathbb{R}$ defined by $f(\lambda e) = \lambda$. Now note that X majorizes L. Alternatively, notice first that e is an internal point of the convex set $L_+ \setminus \{0\}$ and then use the Separation Theorem 8.15 to separate by a nonzero linear functional the disjoint convex sets $\{0\}$ and $L_+ \setminus \{0\}$.]

12. Assume that X is a majorizing vector subspace of an ordered vector space L and let $f\colon X \to \mathbb{R}$ be a positive linear functional. Give an alternative proof of Kantorovich's Extension Theorem 1.36 by using Theorem 1.35. [*Hint*: Let $f\colon X \to \mathbb{R}$ be a positive linear functional. Consider the nonempty convex set $B = \{x \in X\colon\ f(x) \le 1\}$ and let $A = B - L_+$. Clearly, A is convex and we claim that A is also absorbing. To see this, let $x \in L$. Since X majorizes L, there exists some $y \in X$ such that $y \ge x$. So, if we let $u = y - x \in L_+$, then $x = y - u$. Next, choose some $\alpha_0 > 0$ such that $\alpha f(y) = f(\alpha y) \le 1$ (or $\alpha y \in B$) holds for all $0 \le \alpha \le \alpha_0$ and notice that for any such α we have $\alpha x = \alpha y - \alpha u \in B - L_+$ proving that A

is also absorbing. Now notice that $A - L_+ = B - L_+$ and that for each $x \in X \cap (A - L_+)$ we have $f(x) \leq 1$, and then invoke Theorem 1.35.]

1.6. Faces and extremal vectors of cones

We start our discussion with the notion of a face of a convex set.

Definition 1.37. *A nonempty convex subset F of a convex set C is called a* ***face*** *if $\alpha x + (1-\alpha)y \in F$ with $x, y \in C$ and $0 < \alpha < 1$ imply $x, y \in F$.*

The reader should verify easily that a convex subset F of a convex set C is a face of C if and only if it follows from $\frac{1}{2}x + \frac{1}{2}y \in F$ with $x, y \in C$ that $x, y \in F$.

Some of the basic properties of faces are incorporated in the next result. Its straightforward proof is left for the reader.

Lemma 1.38. *For a nonempty convex subset C of a vector space X we have the following.*

(1) *A face of a face of C is itself a face of C.*
(2) *A nonempty intersection of a family of faces of C is a face of C.*
(3) *If $f\colon X \to \mathbb{R}$ is a nonzero linear functional that attains a minimum (or a maximum) value over the convex set C, say m, then the set $F = \{x \in C\colon\ f(x) = m\}$ is a face of C.*

Recall that a point e of a convex set C is called an **extreme point** of C if $e = \alpha x + (1-\alpha)y$ with $x, y \in C$ and $0 < \alpha < 1$ imply $x = y = e$.[15] Clearly, a point e of a convex set C is an extreme point of C if and only if the singleton $\{e\}$ is a face of C.

The following simple (but very useful result) should be obvious.

Lemma 1.39. *Every extreme point of any face of a convex set C is an extreme point of C.*

Every cone has only one extreme point—namely the zero vector.

Lemma 1.40. *If $\mathcal{K}$ is a cone of a vector space, then zero is the only extreme point of $\mathcal{K}$.*

Proof. If x is a nonzero vector of $\mathcal{K}$, then the identity $x = \frac{1}{2}0 + \frac{1}{2}(2x)$ shows that x cannot be an extreme point of $\mathcal{K}$. On the other hand, if $0 = \lambda x + (1-\lambda)y$ with $0 < \lambda < 1$ and $x, y \in \mathcal{K}$, then it is easy to see that $x = y = 0$. That is, 0 is the one and only extreme point of $\mathcal{K}$. ■

[15] It is not difficult to see that a point e of a convex set C is an extreme point of C if and only if the set $C \setminus \{e\}$ is likewise a convex set.

The faces of a cone are very special sets.

Lemma 1.41. *Every face of a cone $\mathcal{K}$ is a subcone*[16] *of $\mathcal{K}$.*

Proof. Let F be a face of a cone $\mathcal{K}$ and let $x \in F$. If $\lambda \geq 1$, then from the convex combination $x = \frac{1}{\lambda}(\lambda x) + (1 - \frac{1}{\lambda})0$ and $\lambda x, 0 \in \mathcal{K}$, it follows that $0 \in F$ and $\lambda x \in F$ for all $\lambda \geq 1$. On the other hand, for $0 < \lambda < 1$ the convexity of F implies $\lambda x = \lambda x + (1 - \lambda)0 \in F$. Thus, $\lambda x \in F$ holds for all $x \in F$ and all $\lambda \geq 0$.

The preceding conclusion in conjunction with the convexity of F shows that F is a wedge. Now use the fact that F is a subset of the cone $\mathcal{K}$ to conclude that F is indeed a subcone of $\mathcal{K}$. ∎

Among the special faces of a cone are its extremal rays and they will be discussed next. Recall that if x is a vector in a linear space, then the set $\mathcal{A}_x = \{\alpha x\colon\ \alpha \geq 0\}$ is called the **half-ray** generated by x.

Definition 1.42. *Let $\mathcal{K}$ be a cone of a vector space X. A nonzero vector $e \in \mathcal{K}$ is said to be an **extremal vector** (or a **discrete vector**) of the cone $\mathcal{K}$ if $0 \leq x \leq e$ implies that $x = \lambda e$ for some $\lambda \geq 0$. In this case, the half-ray $\mathcal{A}_e$ is called an **extremal ray** of the cone $\mathcal{K}$.*

The geometrical meaning of the extremal rays can be seen in the companion figure, where four extreme rays are shown for the cone in the left figure and then two rays that are not extreme in the right figure. It should be obvious that if e is an extremal vector of a cone, then so are the vectors λe for $\lambda > 0$. It turns out that the extremal vectors of a cone are precisely the ones whose half-rays are faces of the cone.

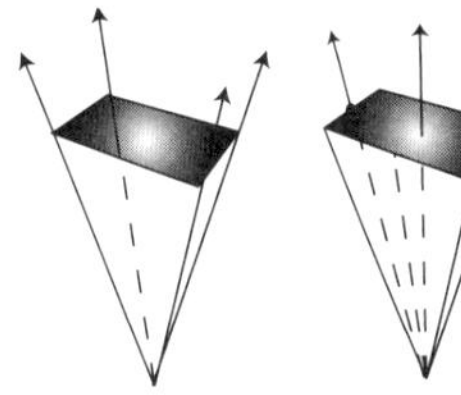

This and another important "algebraic" characterization of the extremal vectors are included in the next result.

Lemma 1.43. *For a nonzero vector e of a cone $\mathcal{K}$ the following statements are equivalent.*

(1) *The vector e is an extremal vector of the cone $\mathcal{K}$.*

(2) *If two vectors $x_1, x_2 \in \mathcal{K}$ satisfy $e = x_1 + x_2$, then the vectors x_1 and x_2 are linearly dependent.*

(3) *The half-ray $\mathcal{A}_e$ is a face of the cone $\mathcal{K}$.*

Proof. (1) $\Longrightarrow$ (2) Assume that two nonzero vectors $x_1, x_2 \in \mathcal{K}$ satisfy $e = x_1 + x_2$. From $0 < x_1 < e$ and $0 < x_2 < e$ (where $\leq$ denotes the vector

[16] As usual, a cone $\mathcal{K}_1$ is called a **subcone** of another cone $\mathcal{K}$ if $\mathcal{K}_1 \subseteq \mathcal{K}$.

ordering induced by the cone $\mathcal{K}$), it follows that there exist $\lambda > 0$ and $\mu > 0$ such that $x_1 = \lambda e$ and $x_2 = \mu e$. This implies $\mu x_1 - \lambda x_2 = 0$ so that x_1 and x_2 are linearly dependent.

(2) $\Longrightarrow$ (3) Assume that two vectors $x_1, x_2 \in \mathcal{K}$ and some $0 < \alpha < 1$ satisfy $\alpha x_1 + (1-\alpha)x_2 \in \mathcal{A}_e$, that is, there exists some $\lambda \geq 0$ such that $\alpha x_1 + (1-\alpha)x_2 = \lambda e$. If $\lambda = 0$, then $x_1 = x_2 = 0 \in \mathcal{A}_e$. Thus, we can suppose that $\lambda > 0$, and so $e = \frac{\alpha}{\lambda}x_1 + \frac{1-\alpha}{\lambda}x_2$.

From our hypothesis the vectors x_1 and x_2 are linearly dependent. We can assume that $x_2 \neq 0$ and that $x_1 = \mu x_2$ for some $\mu \geq 0$. Therefore, $x_2 = \frac{\lambda}{1-\alpha+\alpha\mu}e$ and $x_1 = \frac{\lambda\mu}{1-\alpha+\alpha\mu}e$ so that $\mathcal{A}_e$ is a face of $\mathcal{K}$.

(3) $\Longrightarrow$ (1) Assume that $0 < x < e$, i.e., $x > 0$ and $e - x > 0$. From $\frac{1}{2}x + \frac{1}{2}(e-x) = \frac{1}{2}e \in \mathcal{A}_e$ and our hypothesis, it follows that $x \in \mathcal{A}_e$. Consequently, for some $\lambda \geq 0$ we have $x = \lambda e$. ■

For the characterization of the extremal vectors of a polyhedral cone see Theorem 3.34. We shall close the section with a few more properties of extremal vectors.

Lemma 1.44. *Any three nonzero extremal vectors of a cone in a vector space that generate three distinct extremal rays are linearly independent.*

Proof. Let x_1, x_2, and x_3 be three nonzero extremal vectors of a cone $\mathcal{K}$ that generate three distinct extremal rays. Assume by way of contradiction that the vectors x_1, x_2, x_3 are linearly dependent.

It is easy to see (by relabeling the vectors if necessary) that there exist positive scalars λ and μ such that $x_1 = \lambda x_2 + \mu x_3$. But then a glance at Lemma 1.43 reveals that x_2 and x_3 must be linearly dependent from which it follows that $\mathcal{A}_{x_2} = \mathcal{A}_{x_3}$, a contradiction. Consequently, the vectors x_1, x_2, x_3 are linearly independent. ■

The preceding conclusion is not true for four or more extremal vectors. In other words, we have the following.

- *If $\mathcal{A}_{x_1}, \mathcal{A}_{x_2}, \ldots, \mathcal{A}_{x_k}$ are distinct extremal rays of a cone in a finite dimensional vector space X with $\dim X \geq k \geq 4$, then the vectors $x_1, x_2, \ldots, x_k$ need not be linearly independent.*

For an example, see Exercise 8 at the end of the section.

However, in spite of the above negative conclusion, any set of extremal vectors of a lattice cone that generates distinct extremal rays is linearly independent.

Theorem 1.45. *If $\mathcal{A}_{x_1}, \ldots, \mathcal{A}_{x_k}$ are nonzero distinct extremal rays of a lattice cone, then the vectors $x_1 \ldots, x_k$ are linearly independent.*

Proof. To see this, note that for $i \neq j$ we have $x_i \wedge x_j = 0$. Indeed, from $0 \leq x_i \wedge x_j \leq x_i$ and $0 \leq x_i \wedge x_j \leq x_j$ we get $x_i \wedge x_j = \lambda x_i$ and $x_i \wedge x_j = \mu x_j$, and so if $x_i \wedge x_j > 0$, then $\mathcal{A}_{x_i} = \mathcal{A}_{x_j}$, a contradiction. To complete the proof, note that in a vector lattice any collection of pairwise disjoint vectors is linearly independent; see part (d) of Exercise 14 of Section 1.3. ∎

For a complete version of this result see Theorem 3.21.

Exercises

1. Let S be a nonempty subset of a vector space and let $\operatorname{co} S$ be the convex hull of S. Show that every extreme point of $\operatorname{co} S$ belongs to S.
2. Prove Lemmas 1.38 and 1.39.
3. Let $\mathcal{K}$ be a cone of a vector space X. Assume that there exists a positive linear functional $f \in X^*$ such that $\mathcal{K} \cap \operatorname{Ker} f = \mathcal{A}_e$, where $e > 0$. Show that e is an extremal vector of the cone $\mathcal{K}$.
4. Show that a nonzero vector e of a cone $\mathcal{K}$ is an extremal vector of the cone $\mathcal{K}$ if and only if $\mathcal{K} \setminus \mathcal{A}_e$ is a convex set.
5. Show that the standard cone of $C[0,1]$ does not have any extremal rays.
6. Find the extremal rays of the standard cone of $\mathbb{R}^\Omega$.
7. Find all extremal vectors of the $L_p(\mu)$-spaces (ordered by $f \geq g$ whenever $f(\omega) \geq g(\omega)$ holds for μ-almost all ω) for any $0 \leq p \leq \infty$. [*Hint*: The extremal vectors are the positive multiples of the functions χ_A, where A is an atom. (Recall that an **atom** is any measurable set A of positive measure such that whenever B is a measurable set satisfying $B \subseteq A$, then either $\mu(B) = 0$ or $\mu(A \setminus B) = 0$.)]
8. Give an example of four distinct extremal vectors of a cone in $\mathbb{R}^4$ that are not linearly independent. [*Hint*: Consider the cone $\mathcal{K}$ of $\mathbb{R}^4$ given by $\mathcal{K} = \{(\lambda_1 + \lambda_4, \lambda_2 - \lambda_4, \lambda_3 + \lambda_4, 0)\colon\ \lambda_i \geq 0 \text{ for each } i\}$ and note that the four linearly dependent vectors $x_1 = (1,0,0,0)$, $x_2 = (0,1,0,0)$, $x_3 = (0,0,1,0)$, and $x_4 = (1,-1,1,0)$ are (up to positive scalar multiples) the only extremal vectors of $\mathcal{K}$.]
9. Does the infimum of two extremal vectors exist? [*Hint*: If two essentially distinct extremal vectors u and v have an infimum, then $u \wedge v = 0$.]

1.7. Cone bases

In this section, we shall discuss some special convex subsets of a cone known as *bases*. We start with their definition.

Definition 1.46. *Let $\mathcal{K}$ be a cone in a vector space. A nonempty convex subset $\mathcal{B}$ of $\mathcal{K} \setminus \{0\}$ is said to be a* **base** *for the cone $\mathcal{K}$ if for each $x \in \mathcal{K} \setminus \{0\}$ there exist $\lambda > 0$ and $b \in \mathcal{B}$ both uniquely determined such that $x = \lambda b$.*

The existence of a base is associated with the existence of strictly positive linear functionals.

Theorem 1.47. *A cone $\mathcal{K}$ of a vector space X has a base if and only if X admits a $\mathcal{K}$-strictly positive linear functional. Moreover, if $f\colon X \to \mathbb{R}$ is any $\mathcal{K}$-strictly positive linear functional, then for each fixed $\alpha > 0$ the convex set*

$$\mathcal{B} = \big\{x \in \mathcal{K}\colon \; f(x) = \alpha\big\}$$

is a base for the cone $\mathcal{K}$.

The intuition for this result is shown in the figure below.

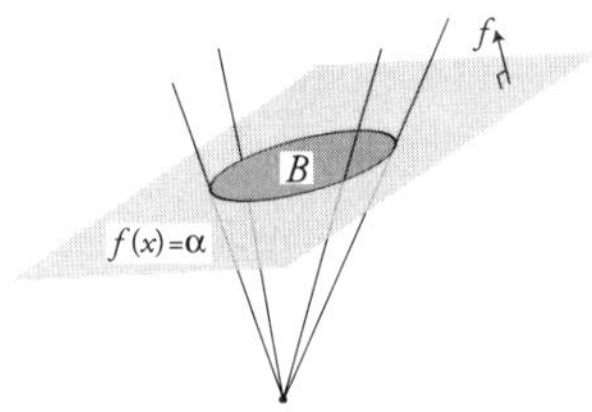

Proof. Let $\mathcal{B}$ be a base for a cone $\mathcal{K}$. Define the function $f\colon \mathcal{K} \to [0,\infty)$ by $f(0) = 0$ and for $x > 0$ let $f(x)$ be the unique positive real number for which there exists the unique vector $b \in \mathcal{B}$ with $x = f(x)b$. Clearly, f is strictly positive on $\mathcal{K}$.

To see that f is additive, let $x, y \in \mathcal{K} \setminus \{0\}$ and let $b, b_1 \in \mathcal{B}$ satisfy $x = f(x)b$ and $y = f(y)b_1$. Then the vector $b_2 = \frac{f(x)}{f(x)+f(y)}b + \frac{f(y)}{f(x)+f(y)}b_1 \in \mathcal{B}$ satisfies $x + y = [f(x) + f(y)]b_2$, and so by the definition of the base we get $f(x + y) = f(x) + f(y)$.

Therefore, f is also additive. By Lemma 1.26, the linear functional f extends to a $\mathcal{K}$-strictly positive linear functional on all of X.

For the converse, assume that $f\colon X \to \mathbb{R}$ is a $\mathcal{K}$-strictly positive linear functional. Fix $\alpha > 0$ and put $\mathcal{B} = \big\{x \in \mathcal{K}\colon \; f(x) = \alpha\big\}$. We claim that $\mathcal{B}$ is a base for $\mathcal{K}$. To see this, note first that if $x >_{\mathcal{K}} 0$, then $b = \frac{\alpha}{f(x)}x \in \mathcal{B}$ and $x = \frac{f(x)}{\alpha}b$ holds true. Now assume that $x = \lambda_1 b_1 = \lambda_2 b_2$, with $\lambda_i > 0$ and $b_i \in \mathcal{B}$ for each i. Then we have

$$\lambda_1 = \tfrac{\lambda_1}{\alpha} f(b_1) = \tfrac{1}{\alpha} f(\lambda_1 b_1) = \tfrac{1}{\alpha} f(\lambda_2 b_2) = \tfrac{\lambda_2}{\alpha} f(b_2) = \lambda_2\,,$$

and from this we get $b_1 = b_2$. This shows that $\mathcal{B}$ is a base of $\mathcal{K}$. ∎

We now turn our attention to extremal rays of cones. Recall that if x is a vector in a linear space, then the set $\mathcal{A}_x = \big\{\alpha x\colon \; \alpha \geq 0\big\}$ is called the **half-ray** generated by x. A half-ray $\mathcal{A}_x$ is an **extremal ray** of a cone $\mathcal{K}$ if the vector x is an extremal vector of the cone $\mathcal{K}$.

If a cone has a base, the extremal vectors of the cone are in essence the extreme points of the base.

Theorem 1.48. *Let $\mathcal{B}$ be a base of a cone $\mathcal{K}$. Then a vector $b \in \mathcal{B}$ is an extremal vector if and only if b is an extreme point of the convex set $\mathcal{B}$.*

Proof. Assume first that $b \in \mathcal{B}$ is an extremal vector of the cone $\mathcal{K}$ and let $b = \lambda b_1 + (1-\lambda)b_2$ with $b_1, b_2 \in \mathcal{B}$ and $0 < \lambda < 1$. Since $0 < \lambda b_1 \leq b$ and $0 < (1-\lambda)b_2 \leq b$ are both true, there exist two scalars $\alpha, \beta > 0$ such that $\lambda b_1 = \alpha b$ and $(1-\lambda)b_2 = \beta b$. This implies that $b = \frac{\lambda}{\alpha}b_1 = \frac{1-\lambda}{\beta}b_2$. Now a glance at the definition of a base guarantees that $b_1 = b_2 = b$. Therefore, b is an extreme point of $\mathcal{B}$.

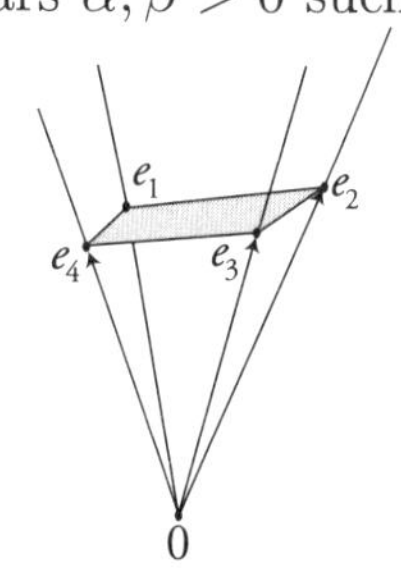

For the converse, assume that $b \in \mathcal{B}$ is an extreme point of the convex set $\mathcal{B}$. Let $0 < x < b$. We must show that x is a scalar multiple of b. To see this, start by observing that since $b - x > 0$ and $x > 0$ there exist $\lambda_1 > 0$, $\lambda_2 > 0$ and $b_1, b_2 \in \mathcal{B}$ such that $b + x = \lambda_1 b_1$ and $b - x = \lambda_2 b_2$. From

$$2b = (b+x) + (b-x) = \lambda_1 b_1 + \lambda_2 b_2 = (\lambda_1 + \lambda_2)\left[\tfrac{\lambda_1}{\lambda_1+\lambda_2}b_1 + \tfrac{\lambda_2}{\lambda_1+\lambda_2}b_2\right]$$

and the fact that $\mathcal{B}$ is a base of the cone $\mathcal{K}$, we get $b = \frac{\lambda_1}{\lambda_1+\lambda_2}b_1 + \frac{\lambda_2}{\lambda_1+\lambda_2}b_2$. Since b is an extreme point of $\mathcal{B}$, it follows that $b_1 = b_2 = b$. This implies $x = (b+x) - b = \lambda_1 b_1 - b = (\lambda_1 - 1)b$. (A cone having a base with four extreme points is shown in the above figure.) ∎

Corollary 1.49. *If $\mathcal{B}_1$ and $\mathcal{B}_2$ are two bases of a cone, then the sets of extreme points of $\mathcal{B}_1$ and $\mathcal{B}_2$ have the same cardinality.*

We shall close the section by presenting two examples of cones with bases. The first one is an "**ice cream cone**" in $\mathbb{R}^3$. One of them is defined by

$$\begin{aligned}\mathcal{K} &= \left\{(x,y,z) \in \mathbb{R}^3 \colon\ z \geq \sqrt{x^2+y^2}\right\} \\ &= \left\{\lambda(x,y,1) \in \mathbb{R}^3 \colon\ \lambda \geq 0 \text{ and } x^2+y^2 \leq 1\right\}.\end{aligned}$$

Clearly, the compact convex set $\mathcal{B} = \{(x,y,1) \colon\ x^2+y^2 \leq 1\}$ is a base for $\mathcal{K}$; see the shaded circle in the accompanying figure. In particular, notice that the linear functional $f \colon \mathbb{R}^3 \to \mathbb{R}$ defined by $f(x,y,z) = z$ is $\mathcal{K}$-strictly positive.

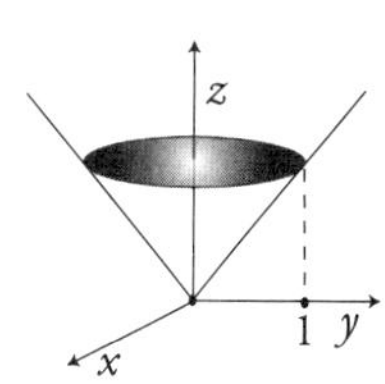

The corresponding infinite dimensional analogue ice cream cone in ℓ_2 of the preceding example is the cone given by

$$\begin{aligned}\mathcal{K}_1 &= \left\{(x_1,x_2,x_3,\ldots) \in \ell_2 \colon\ x_1 \geq \sqrt{\textstyle\sum_{n=2}^{\infty} x_n^2}\right\} \\ &= \left\{\lambda(1,x_2,x_3,\ldots) \in \ell_2 \colon\ \lambda \geq 0 \text{ and } \textstyle\sum_{n=2}^{\infty} x_n^2 \leq 1\right\}.\end{aligned}$$

The linear functional $g\colon \ell_2 \to \mathbb{R}$ defined via the formula $g(x_1, x_2, \ldots) = x_1$ is again $\mathcal{K}_1$-strictly positive—and so $\mathcal{B}_1 = \{(x_1, x_2, \ldots) \in \ell_2\colon\ x_1 = 1\}$ is a base for $\mathcal{K}_1$.

For an example of a Banach lattice whose cone does not have a base see Exercise 6 at the end of the chapter. For examples of general "ice cream cones" in normed spaces see Section 2.6. A general discussion of polyhedral cones will be presented in Section 3.3.

Exercises

1. Give an example of two bases of a cone in a topological vector space one of which is bounded and the other unbounded. [*Hint*: Consider an "open ice cream" cone in $\mathbb{R}^2$ or $\mathbb{R}^3$.]

2. Show that a subset $\mathcal{B}$ of a cone $\mathcal{K}$ of a vector space X is a base for the cone $\mathcal{K}$ if and only if there exists a $\mathcal{K}$-strictly positive linear functional $f\colon X \to \mathbb{R}$ such that $\mathcal{B} = \mathcal{K} \cap f^{-1}(\{1\})$.

3. Show that standard cones of the ordered vector spaces $C[0,1]$ and $L_p[0,1]$ $(1 \le p \le \infty)$ all have bases.

4. Let $\mathcal{K}_1$ and $\mathcal{K}_2$ be two cones of a vector space X such that the cone $\mathcal{K}_1 \cap \mathcal{K}_2$ is nontrivial. Assume also that a linear functional $f \in X^*$ is both $\mathcal{K}_1$- and $\mathcal{K}_2$-strictly positive. If we consider the bases

$$\mathcal{B}_1 = \{x \in \mathcal{K}_1\colon\ f(x) = 1\} \quad \text{and} \quad \mathcal{B}_2 = \{x \in \mathcal{K}_2\colon\ f(x) = 1\}$$

of $\mathcal{K}_1$ and $\mathcal{K}_2$, respectively, then show that

$$\mathcal{B}_1 \cap \mathcal{B}_2 = \{x \in \mathcal{K}_1 \cap \mathcal{K}_2\colon\ f(x) = 1\}.$$

(Conclude from this that $\mathcal{B}_1 \cap \mathcal{B}_2$ is a base for the cone $\mathcal{K}_1 \cap \mathcal{K}_2$.)

5. Let $\mathcal{B}$ be a base of a cone K. If $\alpha, \beta \in \mathbb{R}_+$ and $b_1, b_2 \in \mathcal{B}$ satisfy $\alpha b_1 < \beta b_2$, then show that $\alpha < \beta$. [*Hint*: Assume $\alpha > 0$ and that $\alpha b_1 < \beta b_2$. Pick $\lambda > 0$ and $b_3 \in \mathcal{B}$ (both uniquely determined) such that $\beta b_2 - \alpha b_1 = \lambda b_3$. This implies $\beta b_2 = (\lambda + \alpha)\big(\frac{\lambda}{\alpha+\lambda} b_3 + \frac{\alpha}{\alpha+\lambda} b_1\big)$, and from this it follows that $\beta = \lambda + \alpha > \alpha$.]

6. Let Ω be an uncountable set and let $L = B(\Omega)$ be the Banach lattice of all bounded real-valued functions on Ω equipped with the standard cone $L_+ = \{f \in B(\Omega)\colon\ f(\omega) \ge 0 \text{ for all } \omega \in \Omega\}$. Show that the closed cone L_+ does not have a base. [*Hint*: According to Theorem 1.47 we must show that L does not admit any strictly positive linear functional. If $\phi\colon L \to \mathbb{R}$ is a positive linear functional, then for each $\epsilon > 0$ the set $\{\omega \in \Omega\colon\ \phi(\chi_{\{\omega\}}) \ge \epsilon\}$ is a finite set. This implies that the set $\{\omega \in \Omega\colon\ \phi(\chi_{\{\omega\}}) \ge 0\}$ is at most countable. Since Ω is uncountable, it is easy to see that ϕ cannot be strictly positive.]

7. Let X be a finite dimensional vector space and let $\{x_1, \ldots, x_n\}$ be a (Hamel) basis of X. Consider the set

$$\mathcal{K} = \Big\{ \sum_{i=1}^{n} \lambda_i x_i \colon \ \lambda_i \geq 0 \text{ for each } i = 1, \ldots, n \Big\}.$$

Establish the following properties.

(a) The set $\mathcal{K}$ is a cone of X.

(b) If $x = \sum_{i=1}^{n} \lambda_i x_i$ and $y = \sum_{i=1}^{n} \mu_i x_i$, then $x \geq_{\mathcal{K}} y$ if and only if $\lambda_i \geq \mu_i$ holds in $\mathbb{R}$ for each $i = 1, \ldots, n$.

(c) The cone $\mathcal{K}$ is an Archimedean lattice cone whose lattice operations for each $x = \sum_{i=1}^{n} \lambda_i x_i$ and $y = \sum_{i=1}^{n} \mu_i x_i$ are given by

$$x \vee y = \sum_{i=1}^{n} \max\{\lambda_i, \mu_i\} x_i \quad \text{ and } \quad x \wedge y = \sum_{i=1}^{n} \min\{\lambda_i, \mu_i\} x_i \,.$$

(For this part see also Theorem 3.17.)

(d) The linear functional $f \colon X \to \mathbb{R}$, defined for each $x = \sum_{i=1}^{n} \lambda_i x_i$ by $f(x) = \sum_{i=1}^{n} \lambda_i$, is $\mathcal{K}$-strictly positive—and conclude from this that the cone $\mathcal{K}$ has a compact base.

(e) The extreme rays of $\mathcal{K}$ are precisely the half-rays $\mathcal{A}_{x_i}$, $i = 1, \ldots, n$.

(f) Consider $\mathbb{R}^n$ as a Riesz space with its natural order and define the mapping $T \colon \mathbb{R}^n \to X$ via the formula $T(\lambda_1, \ldots, \lambda_n) = \sum_{i=1}^{n} \lambda_i x_i$. Then T is a surjective lattice isomorphism—and so the Riesz space $(X, \mathcal{K})$ can be identified with $\mathbb{R}^n$.

8. Consider the $\mathbb{R}^3$ ice cream cone $\mathcal{K} = \big\{(x, y, z) \in \mathbb{R}^3 \colon \ z \geq \sqrt{x^2 + y^2}\big\}$. Find its extremal vectors.

1.8. Decomposability in ordered vector spaces

The purpose of this section is to discuss a very important property known as the decomposition property.[17]

Definition 1.50. *An ordered vector space L is said to have the* ***Riesz decomposition property*** *(or simply the* ***decomposition property****) if it follows from $x, u, v \in L_+$ and $x \leq u + v$ that there exist $x_1, x_2 \in L$ satisfying $0 \leq x_1 \leq u$, $0 \leq x_2 \leq v$ and $x = x_1 + x_2$.*

A cone $\mathcal{K}$ of a vector space is said to have the ***Riesz decomposition property*** *(or simply the* ***decomposition property****) if the ordered vector space $(L, \mathcal{K})$ has the Riesz decomposition property.*

The reader should notice at once the following simple result.

[17] In economics, the decomposition property is associated with the important economic notion of the *divisibility of commodities*.

Lemma 1.51. *An ordered vector space L has the Riesz decomposition property if and only if*

$$[0, x] + [0, y] = [0, x + y]$$

holds for all $x, y \in L_+$.

Now let us introduce the following notation. Let L be an ordered vector space. If A and B are subsets of L, let us write $A \leq B$ to mean that $a \leq b$ for all $a \in A$ and all $b \in B$. Also, the symbol $A \leq x$ means that $a \leq x$ for all $a \in A$. We shall say that a vector x in an ordered vector space is **intermediate** between two nonempty sets A and B if either $A \leq x \leq B$ or $B \leq x \leq A$ is true. The geometric interpretation of the intermediate vector x between the sets A and B is shown in the accompanying figure.

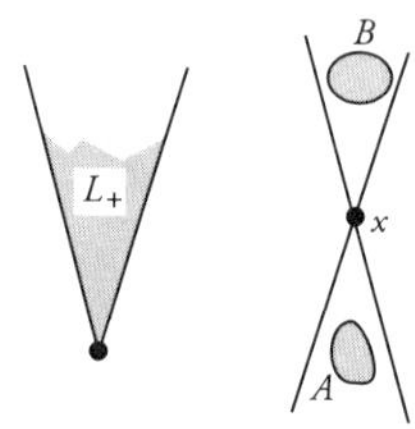

With this terminology at hand, the interpolation property in ordered vector spaces is introduced as follows.

Definition 1.52. *An ordered vector space L is said to have the* ***interpolation property*** *if for every pair of finite nonempty subsets A and B of L satisfying $A \leq B$ there exists an intermediate vector between A and B, i.e., there exists some $x \in L$ with $A \leq x \leq B$.*

A cone $\mathcal{K}$ of a vector space L is said to have the ***interpolation property*** *if the ordered vector space $(L, \mathcal{K})$ has the interpolation property.*

In the definition of the interpolation property we can assume that each of the sets A and B has two vectors.

Lemma 1.53. *An ordered vector space has the interpolation property if and only if every pair of subsets A and B consisting of two vectors each that satisfies $A \leq B$ has an intermediate point between them.*

Proof. Let L be an ordered vector space such that every pair of subsets A and B of L consisting of two vectors each that satisfies $A \leq B$ has an intermediate point between them. To establish that L has the interpolation property, it suffices to show that for any two nonempty finite subsets V and W with the same number of vectors satisfying $V \leq W$ there is an intermediate vector between them. We shall prove this using induction on the number n of elements of the sets V and W.

For $n = 1$ the conclusion is obvious and for $n = 2$ it is true by assumption. For the induction step, assume that any two subsets V and W with $n \geq 2$ vectors satisfying $V \leq W$ have an intermediate vector. It easily follows that any two nonempty finite subsets C and D each having at most n vectors and satisfying $C \leq D$ have an intermediate vector.

Let two subsets $V = \{v_1, \dots, v_n, v_{n+1}\}$ and $W = \{w_1, \dots, w_n, w_{n+1}\}$ of L satisfy $V \le W$. By the induction hypothesis, there exists a vectors x_1 such that $\{v_1, \dots, v_n\} \le x_1 \le \{w_1, \dots, w_n\}$. Since $\{v_1, \dots, v_n\} \le \{w_{n+1}, x_1\}$ and $n \ge 2$, the induction hypothesis guarantees the existence of a vector y_1 such that $\{v_1, \dots, v_n\} \le y_1 \le \{w_{n+1}, x_1\}$. In particular, note that

$$\{v_1, \dots, v_n\} \le y_1 \le W\,. \tag{$\star$}$$

Similarly, there exist vectors x_2 and y_2 satisfying

$$\{v_2, \dots, v_{n+1}\} \le x_2 \le \{w_1, \dots, w_n\} \text{ and } \{v_2, \dots, v_{n+1}\} \le y_2 \le \{w_{n+1}, x_2\}\,.$$

It follows that $\{y_1, y_2\} \le W$ and $\{v_2, \dots, v_{n+1}\} \le y_2 \le W$.

Next, it is not difficult to see that there exist vectors $y_3, \dots, y_{n+1}$ such that $\{y_1, y_2, y_3, \dots, y_n, y_{n+1}\} \le W$. By the conclusion shown in ($\star$), there exists a vector z such that $\{y_1, y_2, y_3, \dots, y_n\} \le z \le W$. Since $n \ge 2$, it follows that $\{y_1, y_2\} \le z$ and from this we see that $V \le z \le W$. This completes the induction and the proof. ■

The interpolation and the decomposition properties are equivalent. This was proven by F. Riesz [**81**].

Theorem 1.54 (F. Riesz). *For an ordered vector space L the following statements are equivalent.*

(a) *L has the interpolation property.*

(b) *L has the Riesz decomposition property.*

(c) *For any two collections $\{x_1, \dots, x_n\}$ and $\{y_1, \dots, y_m\}$ of positive vectors satisfying $\sum_{i=1}^n x_i = \sum_{j=1}^m y_j$ there exists another collection of positive vectors $\{z_{ij}\colon\ i=1, \dots, n;\ j=1, \dots, m\}$ such that*

$$x_i = \sum_{j=1}^m z_{ij} \text{ for } i=1, \dots, n \text{ and } y_j = \sum_{i=1}^n z_{ij} \text{ for } j=1, \dots, m\,.$$

(d) *If $x_1, x_2, y_1, y_2 \in L_+$ satisfy $x_1 + x_2 = y_1 + y_2$, then there exists a collection of positive vectors $\{z_{ij}\colon\ i, j = 1, 2\}$ such that*

$$x_i = z_{i1} + z_{i2} \text{ for } i=1,2 \text{ and } y_j = z_{1j} + z_{2j} \text{ for } j=1,2\,.$$

Proof. (a) $\Longrightarrow$ (b) Assume that $0 \le x \le u + v$ holds in L with $u, v \in L_+$. It should be clear that $\{0, x - u\} \le \{v, x\}$, and so by our hypothesis there exists some $x_1 \in L$ such that $\{0, x - u\} \le x_1 \le \{v, x\}$. It follows that $0 \le x_1 \le v$ and if we let $x_2 = x - x_1$, then from $x - u \le x_1 \le x$ we get $0 \le x_2 = x - x_1 \le u$. Clearly, $x_1 + x_2 = x$, and so L has the Riesz decomposition property.

(b) $\Longrightarrow$ (c) We shall use induction on m. For $m = 1$ the desired conclusion is obvious. Therefore, assume the result to be true for some m and all $n = 1, 2, \ldots$. Let

$$\sum_{i=1}^{n} x_i = \sum_{j=1}^{m+1} y_j \,,$$

where the vectors x_i and y_j are positive vectors. Since $\sum_{j=1}^{m} y_j \leq \sum_{i=1}^{n} x_i$ holds true, it follows from (b) that there exist vectors $u_1, \ldots, u_n$ satisfying $0 \leq u_i \leq x_i$ for each $i = 1, \ldots, n$ and $\sum_{i=1}^{n} u_i = \sum_{j=1}^{m} y_j$. Therefore, from our induction hypothesis, there exists a collection of positive vectors $\{z_{ij}\colon\ i = 1, \ldots, n;\ j = 1, \ldots, m\}$ such that

$$u_i = \sum_{j=1}^{m} z_{ij} \ \text{ for } i = 1, \ldots, n \quad \text{and} \quad y_j = \sum_{i=1}^{n} z_{ij} \ \text{ for } j = 1, \ldots, m\,.$$

For each $i = 1, \ldots, n$ put $z_{i,m+1} = x_i - u_i \geq 0$ and note that the collection of positive vectors $\big\{z_{ij}\colon\ i = 1, \ldots, n;\ j = 1, \ldots, m+1\big\}$ satisfies

$$x_i = \sum_{j=1}^{m+1} z_{ij} \ \text{ for } i = 1, \ldots, n \quad \text{and} \quad y_j = \sum_{i=1}^{n} z_{ij} \ \text{ for } j = 1, \ldots, m+1\,.$$

Thus, the conclusion holds true for $m + 1$ and all $n = 1, 2, \ldots$, and the validity of (c) has been established.

(c) $\Longrightarrow$ (d) Obvious.

(d) $\Longrightarrow$ (a) Assume that two subsets $A = \{a_1, a_2\}$ and $B = \{b_1, b_2\}$ of L satisfy $A \leq B$. From the identity

$$(b_1 - a_1) + (b_2 - a_2) = (b_1 - a_2) + (b_2 - a_1)$$

and the fact that each of the above four vectors in the parentheses is positive, it follows from our hypothesis that there exist positive vectors $z_{11}, z_{12}, z_{21}, z_{22}$ such that

$$b_1 - a_1 = z_{11} + z_{12}\,, \quad b_2 - a_2 = z_{21} + z_{22}\,, \quad b_1 - a_2 = z_{11} + z_{21}\,, \ \text{ and}$$

$$b_2 - a_1 = z_{12} + z_{22}\,.$$

To finish the proof, let $x = a_1 + z_{12}$ and note that $A \leq x \leq B$ holds. ■

Corollary 1.55. *Every Riesz space has the decomposition property. That is, every lattice cone in a vector space has the decomposition property.*

Proof. Assume that two nonempty finite sets A and B in a Riesz space L satisfy $A \leq B$. It follows that $\sup A \leq \inf B$, and from this we see immediately that L has the interpolation property, and hence the decomposition property. ■

An Archimedean ordered vector space with the decomposition property need not be a Riesz space. The next three examples will clarify the situation. The first one is due to F. Riesz [**81**].

Example 1.56. (F. Riesz) Let us say that a function $x\colon [0,1] \to \mathbb{R}$ is a **rational function** if there exist two polynomials p and q satisfying $q(t) \neq 0$ and $x(t) = \frac{p(t)}{q(t)}$ for all $t \in [0,1]$. Let L denote the collection of all rational functions. Under the usual pointwise algebraic operations L is a vector space—in fact, L is a vector subspace of $C[0,1]$. Thus, under the pointwise ordering, L is also an ordered vector space. Clearly, the constant function one is an order unit—and so the cone L_+ is generating. It is not difficult to see that L is not a Riesz space; see Exercise 9 at the end of the section. However, we claim that L satisfies the Riesz decomposition property.

To see this, assume that the three rational functions x, y_1 and y_2 satisfy $0 \leq x \leq y_1 + y_2 = z$ with $z > 0$. A moment's thought reveals that we can suppose without loss of generality that the rational functions x, y_1 and y_2 (and so z) are all polynomials. From this and the fact that z is a nonzero polynomial, we see that the set $\mathcal{Z} = \{t \in [0,1]\colon\ z(t) = 0\}$ is a finite set. Also, let $D = [0,1] \setminus \mathcal{Z} = \{t \in [0,1]\colon\ z(t) > 0\}$. Next for each $i = 1,2$ define $x_i\colon D \to \mathbb{R}$ via the formula

$$x_i(t) = \frac{x(t)y_i(t)}{z(t)}\,.$$

Clearly, the x_i are positive continuous functions on D and they satisfy

$$0 \leq x_i \leq y_i \ \text{ for } \ i = 1,2 \ \text{ and } \ x_1 + x_2 = x\,. \tag{$\star$}$$

We claim that the functions x_i can be extended to rational functions on all of $[0,1]$. If this is done, then a continuity argument shows that $(\star)$ holds true on $[0,1]$ and this will establish that L has the Riesz decomposition property.

If $\mathcal{Z} = \emptyset$, then there is nothing to prove. So, let $\mathcal{Z} = \{t_1, \ldots, t_k\}$. In this case, we can write $z(t) = (t-t_1)^{r_1} \cdots (t-t_k)^{r_k} p(t)$, where the r_i are natural numbers and p is a polynomial satisfying $p(t) \neq 0$ for all $t \in [0,1]$. Next, fix i and note that x and y_i vanish on $\mathcal{Z}$. This implies we can write $x(t)y_i(t) = (t-t_1)^{m_1} \cdots (t-t_k)^{m_k} q(t)$, where q is a polynomial satisfying $q(t_\ell) \neq 0$ for all $\ell = 1, \ldots, k$. In particular, for each $t \in D$ we have

$$x_i(t) = \frac{x(t)y_i(t)}{z(t)} = (t-t_1)^{m_1-r_1} \cdots (t-t_k)^{m_k-r_k} \frac{q(t)}{p(t)}\,. \tag{$\star\star$}$$

This implies that if for some ℓ we have $m_\ell - r_\ell < 0$, then $\lim_{t \to t_\ell} x_i(t) = \infty$, where the limit is one-sided if t_ℓ equals 0 or 1. However, since for each $t \in D$ (in view of $0 \leq \frac{y_i(t)}{z(t)} \leq 1$ and $0 \leq x(t) \leq z(t)$) we have $0 \leq x_i(t) \leq z(t)$, the latter conclusion is impossible. Hence, $m_\ell - r_\ell \geq 0$ holds true for each

$\ell = 1, \ldots, k$, and from $(\star\star)$ we see that each x_i is a rational function, as desired. (For another proof see Exercise 10 at the end of the section.) ■

The second example of an Archimedean ordered vector space with the decomposition property that is not a Riesz space is due to I. Namioka [**71**].

Example 1.57 (Namioka). We consider the Riesz space $C[0,2]$, where as usual the lattice operations $f \vee g$ and $f \wedge g$ are defined pointwise. The example is the vector subspace L of $C[0,2]$ defined by

$$L = \{f \in C[0,2]\colon\ f(0) + f(2) = f(1)\}\,.$$

The vector ordering on L is the one induced by $C[0,2]$, i.e., the pointwise ordering. Since the function u whose graph consists of the polygonal path joining the points $(0,1)$, $(1,2)$, and $(2,1)$ belongs to L and is an order unit, it follows that the cone L_+ is generating.

To see that L is not a Riesz space, consider the function $f \in L$ defined by $f(x) = x - 1$. We claim that the supremum of f and 0 does not exist in L. To establish this, assume by way of contradiction that the supremum $g = \sup\{f, 0\}$ exists in L. Then for each $0 \le \alpha < 1 < \beta \le 2$ the function $h_{\alpha,\beta}$ whose graph is the polygonal line joining the points $(0,0)$, $(\alpha,0)$, $(1,1)$, $(\beta, \beta-1)$ and $(2,1)$ belongs to L and satisfies $h_{\alpha,\beta} \ge 0$ and $h_{\alpha,\beta} \ge f$. This implies $g(x) = 0$ for all $0 \le x \le \alpha$ and $g(x) = f(x)$ for all $1 < \beta \le 2$. Since α and β are arbitrary, we infer that $g(x) = 0$ for all $0 \le x \le 1$ and $g(x) = f(x)$ for all $1 < x \le 2$. Now notice that from $g(0) + g(2) = 1 \ne 0 = g(1)$ we get $g \notin L$, which is impossible. Hence, $\sup\{f, 0\}$ does not exist in L.

We shall show next that L has the Riesz decomposition property. For this, according to Theorem 1.54, it suffices to show that L satisfies the interpolation property. To this end, assume that two subsets $A = \{f_1, f_2\}$ and $B = \{g_1, g_2\}$ of L satisfy $A \le B$. We must prove that there exists a function of L between A and B.

Without loss of generality, we can assume that

$$f_2(1) \le f_1(1) \le g_1(1) \le g_2(1)\,. \tag{$\dagger$}$$

We claim that

$$f_1(1) \le \max\{f_1(0), f_2(0)\} + \max\{f_1(2), f_2(2)\} \le g_1(1)\,. \tag{$\dagger\dagger$}$$

The left inequality follows immediately from $f_1(1) = f_1(0) + f_1(2)$. For the right inequality, assume by way of contradiction that

$$\max\{f_1(0), f_2(0)\} + \max\{f_1(2), f_2(2)\} > g_1(1)\,.$$

If $\max\{f_1(0), f_2(0)\} = f_1(0)$ and $\max\{f_1(2), f_2(2)\} = f_1(2)$, then we see that $f_1(1) = f_1(0) + f_1(2) > g_1(1)$, which contradicts $(\dagger)$. This means that, without loss of generality, we can assume

$$\max\{f_1(0), f_2(0)\} = f_1(0) \quad \text{and} \quad \max\{f_1(2), f_2(2)\} = f_2(2).$$

This in turn yields $g_1(1) = g_1(0) + g_1(2) \geq f_1(0) + f_2(2) > g_1(1)$, which is also absurd, and the validity of (††) has been established.

Now, for simplicity put

$$a = \max\{f_1(0), f_2(0)\} \quad \text{and} \quad b = \max\{f_1(2), f_2(2)\},$$

and note that (††) becomes $f_1(1) \leq a + b \leq g_1(1)$.

Next, let h be the piecewise linear function defined on $[0, 2]$ by joining the points $(0, a)$, $(1, a + b)$, and $(2, b)$ by line segments, and then consider the function

$$f = (h \wedge g_1 \wedge g_2) \vee (f_1 \vee f_2),$$

where $\vee$ and $\wedge$ are the (pointwise) lattice operations in $C[0, 2]$. Clearly, $f \in C[0, 2]$ and from (††), it follows that $f(0) + f(2) = a + b = f(1)$, so that $f \in L$. Finally, note that

$$f_1 \vee f_2 \leq f \leq (h \wedge g_1 \wedge g_2) \vee (g_1 \wedge g_2) = g_1 \wedge g_2.$$

Consequently, f is a desired intermediate function between A and B. ■

Example 1.58. The third example of an ordered vector space with the Riesz decomposition property that fails to be a Riesz space is $L = C^1[0, 1]$, the real vector space of all continuously differentiable functions on $[0, 1]$, equipped with the cone $L_+ = \{x \in C^1[0, 1]\colon\ x(t) \geq 0 \text{ for all } t \in [0, 1]\}$ of all nonnegative functions of L. As in Example 1.57, we can show that if $f \in L$ is defined by $f(t) = t - \frac{1}{2}$, then f^+ does not exist in L, and so L is not a Riesz space. Since **1** (the constant function one) is an order unit, it follows that the cone L_+ is generating. We claim that L has the Riesz decomposition property.[18]

To see this, assume that three functions $x, y_1, y_2 \in L_+$ satisfy the inequality $0 \leq x \leq y_1 + y_2$. We must show that there exist two functions $x_1, x_2 \in L$ satisfying $0 \leq x_1 \leq y_1$, $0 \leq x_2 \leq y_2$ and $x = x_1 + x_2$. We shall establish that the functions $x_1, x_2\colon [0, 1] \to \mathbb{R}$, defined by

$$x_i(t) = \begin{cases} \frac{y_i(t)x(t)}{y_1(t)+y_2(t)} & \text{if } y_1(t) + y_2(t) > 0, \\ 0 & \text{if } y_1(t) + y_2(t) = 0, \end{cases}$$

satisfy the desired properties. To this end, notice first that $0 \leq x_i \leq y_i$ for $i = 1, 2$ and $x_1 + x_2 = x$ are trivially true. So what remains to be shown is that $x_1, x_2 \in L$.

[18] The ordered vector space $C^1[0, 1]$ enjoys many other interesting properties and will be invoked several times in this work; see Examples 2.41, 2.72 and Exercises 8 and 9 of Section 2.8.

We shall prove next that $x_1 \in L$; it will then follow by symmetry that $x_2 \in L$ is also true. This will be done by establishing that x_1 is differentiable on $[0,1]$ and that

$$x_1'(t) = \begin{cases} \left[\frac{y_1'x+y_1x'}{y_1+y_2} - \frac{y_1x[y_1'+y_2']}{[y_1+y_2]^2}\right](t) & \text{if } t\in[0,1] \text{ and } y_1(t)+y_2(t)>0, \\ 0 & \text{if } t\in(0,1) \text{ and } y_1(t)+y_2(t)=0, \\ (-1)^t\frac{y_1'(t)x'(t)}{y_1'(t)+y_2'(t)} & \text{if } t\in\{0,1\}, [y_1+y_2](t)=0 \,\&\, [y_1'+y_2'](t)\neq 0, \\ 0 & \text{if } t\in\{0,1\}, [y_1+y_2](t)=0 \,\&\, [y_1'+y_2'](t)=0. \end{cases}$$

We shall check each case separately. In what follows, keep in mind that if $y_1(t)+y_2(t)>0$, then $0 \le \frac{y_i(t)}{y_1(t)+y_2(t)} \le 1$ and $0 \le \frac{x(t)}{y_1(t)+y_2(t)} \le 1$.

CASE I: $t \in [0,1]$ and $y_1(t)+y_2(t)>0$.

In this case, the formula for the derivative follows immediately from the quotient rule. The formula also shows that in this case the derivative x_1' is continuous.

CASE II: $t \in (0,1)$ and $y_1(t)+y_2(t)=0$.

Fix some $t_0 \in (0,1)$ with $y_1(t_0)+y_2(t_0) = 0$. It easily follows that $y_1(t_0) = y_2(t_0) = x(t_0) = 0$ and so y_1, y_2, and x attain their minimum at t_0. Since t_0 is an interior point of $[0,1]$, we see that $y_1'(t_0) = y_2'(t_0) = x'(t_0) = 0$. Now note that for sufficiently small $h \neq 0$ we have

$$\left|\frac{x_1(t_0+h)-x_1(t_0)}{h}\right| = \left|\frac{x_1(t_0+h)}{h}\right| \le \left|\frac{x(t_0+h)}{h}\right| \xrightarrow[h\to 0+]{} 0\,.$$

This implies that x_1 is differentiable at t_0 and that $x_1'(t_0)=0$.

Next, notice that if $y_1(t_0)+y_2(t_0)>0$, then from $0 \le \frac{y_i(t_0)x(t_0)}{[y_1(t_0)+y_2(t_0)]^2} \le 1$, $0 \le \frac{y_i(t_0)y_j(t_0)}{[y_1(t_0)+y_2(t_0)]^2} \le 1$ and the formula for $x_1'(t_0)$, it follows that for each $t \in (0,1)$ sufficiently close to t_0 we have

$$|x_1'(t)-x_1'(t_0)| = |x_1'(t)| \le |y_1'(t)| + 2|x'(t)| + |y_2'(t)| \xrightarrow[t\to t_0]{} 0\,. \qquad (\star)$$

This shows that x' is continuous at t_0.

CASE III: $t \in \{0,1\}$, $y_1(t)+y_2(t)=0$ and $y_1'(t)+y_2'(t)\neq 0$.

Consider the case $t=0$ and note that $y_1'(t)+y_2'(t)>0$ must be true. (The case $t=1$ can be obtained from this case by replacing x, y_1, and y_2 by the functions v, u_1, and u_2 defined by $v(t)=x(1-t)$ and $u_i(t)=y_i(1-t)$.) In this case we have $y_1(0)=y_2(0)=x(0)=0$ and moreover $y_1(h)+y_2(h)>0$ for all $h>0$ sufficiently close to 0. The latter implies that for all $h>0$ sufficiently close to 0 we have

$$x_1'(h) = \frac{y_1'(h)x(h)+y_1(h)x'(h)}{y_1(h)+y_2(h)} - \frac{y_1(h)x(h)[y_1'(h)+y_2'(h)]}{[y_1(h)+y_2(h)]^2}\,. \qquad (\star\star)$$

Now note that

$$\frac{x_1(0+h)-x_1(0)}{h}=\frac{x_1(h)}{h}=\frac{[y_1(h)/h][x(h)/h]}{[y_1(h)/h]+[y_2(h)/h]}\xrightarrow[h\to 0+]{}\tfrac{y_1'(0)x'(0)}{y_1'(0)+y_2'(0)}\,.$$

So, in this case we have $x'(0)=\frac{y_1'(0)x'(0)}{y_1'(0)+y_2'(0)}$. Next, according to Taylor's formula, for sufficiently small $h>0$ we can write

$$y_i(h)=y_i'(0)h+o(h)\quad\text{and}\quad x(h)=x'(0)h+o(h)\,.$$

It follows that

$$\begin{aligned}\frac{y_1'(h)x(h)+y_1(h)x'(h)}{y_1(h)+y_2(h)} &= \frac{y_1'(h)[x'(0)+\frac{o(h)}{h}]+[y_1'(0)+\frac{o(h)}{h}]x'(h)}{y_1(h)/h+y_2(h)/h}\\ &\xrightarrow[h\to 0+]{} \frac{2y_1'(0)x'(0)}{y_1'(0)+y_2'(0)}\end{aligned}$$

and that

$$\begin{aligned}\frac{y_1(h)x(h)[y_1'(h)+y_2'(h)]}{[y_1(h)+y_2(h)]^2} &= \frac{[y_1'(0)+\frac{o(h)}{h}][x'(0)+\frac{o(h)}{h}][y_1'(h)+y_2'(h)]}{[y_1(h)/h+y_2(h)/h]^2}\\ &\xrightarrow[h\to 0+]{} \frac{y_1'(0)x'(0)}{y_1'(0)+y_2'(0)}\,.\end{aligned}$$

So, from $(\star\star)$, we see that $\lim_{h\to 0+}x_1'(h)=\frac{y_1'(0)x'(0)}{y_1'(0)+y_2'(0)}=x_1'(0)$, which proves that in this case x_1' is continuous at 0.

CASE IV: $t\in\{0,1\}$, $y_1(t)+y_2(t)=0$ and $y_1'(t)+y_2'(t)=0$.

Again, we let $t=0$ and notice that in this case $y_1(0)=y_2(0)=x(0)=0$. So, for each $0<h<1$ we have

$$0\le\tfrac{x(h)}{h}=\tfrac{x(h)-x(0)}{h}\le\tfrac{y_1(h)}{h}+\tfrac{y_2(h)}{h}=\tfrac{y_1(h)-y_1(0)}{h}+\tfrac{y_2(h)-y_2(0)}{h}\xrightarrow[h\to 0+]{}0\,,$$

and from this we see that x is differentiable at 0 and that $x'(0)=0$. Now use $(\star)$ to conclude that $\lim_{t\to 0+}x'(t)=x'(0)$ is also true in this case and the proof of the Riesz decomposition property of L has been completed. ■

For more about the Riesz decomposition property see Theorems 2.47 and 3.9. See also the works of L. Fuchs [**44, 45, 46**].

Exercises

1. For every nonempty set S of a partially ordered set, let $\mathcal{U}(S)$ denote the (possibly empty) collection of all upper bounds of the set S. For an ordered vector space L show that the following statements are equivalent.
   (a) L has the Riesz decomposition property.
   (b) For every nonempty finite subset F of L the collection of all upper bounds $\mathcal{U}(F)$ of F is directed downward.

(c) For every pair of nonempty finite subsets F and G of L we have $\mathcal{U}(F)+\mathcal{U}(G)=\mathcal{U}(F+G)$.

2. Consider the cone $\mathcal{K}=\{(x,y)\in\mathbb{R}^2\colon\ x>0 \text{ and } y>0\}\cup\{(0,0)\}$ of $\mathbb{R}^2$. Show that $\mathcal{K}$ is not a lattice cone and that it satisfies the Riesz decomposition property. [*Hint*: Show that the cone $\mathcal{K}$ has the interpolation property.]

3. If in an ordered vector space with the Riesz decomposition property three vectors x,y,z satisfy $x\wedge y=x\wedge z=0$, then show that $x\wedge(y+z)=0$. [*Hint*: If $u\le\{x,y+z\}$, then $\{0,u-z\}\le\{x,y\}$.]

4. Establish that a cone $\mathcal{K}$ of a vector space X has the Riesz decomposition property (in X) if and only if the cone $\mathcal{K}$ as a cone of the vector subspace $Y=\mathcal{K}-\mathcal{K}$ (= the linear span of $\mathcal{K}$) also has the Riesz decomposition property in Y.

In particular, conclude from this that if $\mathcal{K}-\mathcal{K}$ is a lattice-subspace of $(X,\mathcal{K})$, then $\mathcal{K}$ has the Riesz decomposition property.

5. For the product ordered vector space $\prod_{i\in I}L_i$ of a family $\{L_i\}_{i\in I}$ of ordered vector spaces establish the following.

(a) L is a Riesz space if and only if each factor L_i is a Riesz space.

(b) L has the Riesz decomposition property if and only if each factor L_i has the Riesz decomposition property.

6. Establish the following property that was used in the proof of Lemma 1.53. Let A and B be two nonempty subsets of an ordered vector space L such that A has m vectors and B has n vectors. If $n\neq m$, then there exist two subsets A' and B' of L each of which has $\max\{m,n\}$ vectors such that $A\subseteq A'$, $B\subseteq B'$ and $A'\le B'$.

7. Show that a generating cone $\mathcal{K}$ of a vector space X satisfies the interpolation property if and only if $\{a_1,a_2\}\ge\{b_1,b_2\}>0$ implies the existence of some $x\in X$ satisfying $a_i\ge x\ge b_j$ for all $i,j=1,2$. [*Hint*: For the "only if" part assume $\{a_1,a_2\}\ge\{b_1,b_2\}$. Since $\mathcal{K}$ is generating, for each $i=1,2$ there exists some $u_i>0$ such that $u_i\ge -b_i$. If $u=u_1+u_2$, then $\{a_1+u,a_2+u\}\ge\{b_1+u,b_2+u\}>0$. If $x\in X$ satisfies $a_i+u\ge x\ge b_j+u$ for all i and j, then $a_i\ge x-u\ge b_j$ for all $i,j=1,2$.]

8. Let L be an ordered vector space with a generating cone and the Riesz decomposition property. Show that if $y\le\sum_{i=1}^n x_i$ holds in L, then there exist vectors $y_1,\dots,y_n$ in L satisfying $y_i\le x_i$ for each i and $y=\sum_{i=1}^n y_i$. [*Hint*: Pick a vector $z\in L_+$ satisfying $z+y\ge 0$ and $z+x_i\ge 0$ for each i and consider the inequality $0\le y+nz\le\sum_{i=1}^n(x_i+z)$.]

9. Show that the ordered vector spaces of all polynomials and all rational functions on $[0,1]$ are not Riesz spaces. [*Hint*: We provide a hint for the rational functions—the same arguments work for the polynomials. Assume that f,g,h are rational functions satisfying $h>f$ and $h>g$. Pick $c>0$ so that the functions $\mathbf{1}-c(h-f)$ and $\mathbf{1}-c(h-g)$ are strictly positive on $[0,1]$. Now consider the rational function $\pi=h-c(h-f)(h-g)$ and note that $\pi\ge f$, $\pi\ge g$ and $\pi<h$.]

10. (Fuchs [**46**]) This exercise presents another proof of the fact that the ordered vector space L of all rational functions on $[0,1]$ satisfies the Riesz decomposition property (see Example 1.56) by establishing that L has the interpolation property. Prove the following.

(a) Assume that four real numbers a_1, a_2, b_1, b_2 that are not all equal satisfy $\{a_1, a_2\} \le \{b_1, b_2\}$. If $c = \frac{b_1 b_2 - a_1 a_2}{(b_1+b_2)-(a_1+a_2)}$, then for all i and j we have

$$c - a_i = \frac{(b_1-a_i)(b_2-a_i)}{(b_1+b_2)-(a_1+a_2)} \quad \text{and} \quad b_j - c = \frac{(b_j-a_1)(b_j-a_2)}{(b_1+b_2)-(a_1+a_2)} .$$

In particular, $\{a_1, a_2\} \le c \le \{b_1, b_2\}$ and if $\{a_1, a_2\} < \{b_1, b_2\}$, then $\{a_1, a_2\} < c < \{b_1, b_2\}$.

(b) L has the interpolation property if and only if whenever four polynomials p_1, p_2, q_1, q_2 satisfy $\{p_1, p_2\} \le \{q_1, q_2\}$ (on $[0,1]$), there exists a rational function f such that $\{p_1, p_2\} \le f \le \{q_1, q_2\}$.

(c) If four distinct polynomials p_1, p_2, q_1, q_2 satisfy $\{p_1, p_2\} \le \{q_1, q_2\}$ (on $[0,1]$), then the function $c\colon [0,1] \to \mathbb{R}$, defined by

$$c(t) = \begin{cases} \frac{q_1(t)q_2(t)-p_1(t)p_2(t)}{[q_1(t)+q_2(t)]-[p_1(t)+p_2(t)]} & \text{if } q_1(t)+q_2(t) \ne p_1(t)+p_2(t), \\ p_1(t) & \text{if } q_1(t)+q_2(t) = p_1(t)+p_2(t) , \end{cases}$$

is a rational function satisfying $\{p_1, p_2\} \le c \le \{q_1, q_2\}$.

[*Hint*: For part (b), notice that $\{p_1(t), p_2(t)\} \le c(t) \le \{q_1(t), q_2(t)\}$ for all $t \in [0,1]$. Moreover, observe that $(q_1+q_2)(t) - (p_1+p_2)(t) = 0$ if and only if $q_1(t) = q_2(t) = p_1(t) = p_2(t)$. Let

$$N = \{t_1, \dots, t_k\} = \{t \in [0,1]\colon\ (q_1+q_2)(t) - (p_1+p_2)(t) = 0\}$$

and write $(q_1+q_2)(t) - (p_1+p_2)(t) = (t-t_1)^{r_1} \cdots (t-t_k)^{r_k}\pi(t)$, where the r_i are natural numbers and π is a polynomial satisfying $\pi(t) \ne 0$ for all $t \in [0,1]$.]

11. Consider the positive rational functions x, y_1, y_2 defined on $[0,1]$ by

$$x(t) = \tfrac{1+2t}{4}, \quad y_1(t) = \tfrac{t}{1+t^2} \text{ and } y_2(t) = \tfrac{1}{2-t} ,$$

and note that $0 \le x \le y_1 + y_2$. Find two rational functions x_1 and x_2 on $[0,1]$ satisfying $x = x_1 + x_2$ and $0 \le x_i \le y_i$ for $i = 1, 2$.

12. Let us say that a function $f\colon [0,1] \to \mathbb{R}$ is **strictly positive** if $f(t) > 0$ holds true for all $t \in [0,1]$. Now let L denote the partially ordered vector space of all polynomials on $[0,1]$ and let $\mathcal{K}$ be the collection of all strictly positive polynomials of L together with the zero polynomial. Establish the following.

(a) The set $\mathcal{K}$ is a non-Archimedean cone of L.

(b) $\mathcal{K}$ is not a lattice cone of L.

(c) $\mathcal{K}$ satisfies the Riesz decomposition property.

[*Hint*: For (c) use Exercise 10 in conjunction with the classical Weierstrass approximation theorem.]

13. Let P_n denote the ordered vector space of all polynomials on $[0,1]$ of degree less than or equal to a given natural number n. (The vector ordering is the pointwise ordering.) Establish the following.

(a) P_1 is a Riesz space (and so it has the Riesz decomposition property).

(b) If $n \geq 2$, then P_n does not have the Riesz decomposition property.

14. Give a direct proof of Corollary 1.55. That is, show that if in a vector lattice we have $0 \leq x \leq u + v$ with $u \geq 0$ and $v \geq 0$, then there exist $0 \leq x_1 \leq u$ and $0 \leq x_2 \leq v$ such that $x = x_1 + x_2$. [*Hint*: Let $x_1 = x \wedge u$ and show that the vector $x_2 = x - x_1$ satisfies $0 \leq x_2 \leq v$.]

15. Show that every Riesz space satisfies the following general Riesz Decomposition Property: *If $|x| \leq |y_1 + \cdots + y_n|$ holds true in a Riesz space, then there exist vectors $x_1, \ldots, x_n$ satisfying $x = x_1 + \cdots + x_n$ and $|x_i| \leq |y_i|$ for each $i = 1, \ldots, n$.* [*Hint*: Consider the case $n = 2$ and put $x_1 = [x \vee (-|y_1|)] \wedge |y_1|$ and $x_2 = x - x_1$.]

16. Let us say that an ordered vector space has the **infinite interpolation property** if for every pair of nonempty subsets A and B satisfying $A \leq B$ there exists some x such that $A \leq x \leq B$.

Show that a vector lattice L is order complete (i.e., every nonempty subset of L that is order bounded from above has a supremum) if and only if L has the infinite interpolation property. [*Hint*: If A is an arbitrary nonempty subset of the vector lattice L that is order bounded from above, let $B = \{b \in L\colon\ a \leq b \text{ for all } a \in A\}$.]

17. Let us say that an ordered vector space has the **strong interpolation property** if for every pair of nonempty subsets A and B one of which is finite satisfying $A \leq B$ there exists some x such that $A \leq x \leq B$.

Show that an ordered vector space with a generating cone has the strong interpolation property if and only if it is a vector lattice.

1.9. The Riesz–Kantorovich formulas

Although (as we have seen in Section 1.8) an ordered vector space with the Riesz decomposition property need not be a Riesz space, we have the following basic result that is due to the founders of Riesz space theory, F. Riesz [**80**] and L. V. Kantorovich [**57, 58**].

Theorem 1.59 (Riesz–Kantorovich). *If an ordered vector space L with a generating cone has the Riesz decomposition property, then for every Dedekind complete Riesz space M the ordered vector space $\mathcal{L}_b(L, M)$ is also a Dedekind complete Riesz space—and therefore, in this case, we have $\mathcal{L}_r(L, M) = \mathcal{L}_b(L, M)$.*

Moreover, for any $S, T \in \mathcal{L}_b(L, M)$ and any $x \in L_+$ the lattice operations of $\mathcal{L}_b(L, M)$ satisfy:

$$\begin{aligned} [S \vee T](x) &= \sup\{S(y) + T(z)\colon\ y, z \in L_+ \text{ and } y + z = x\}, \\ [S \wedge T](x) &= \inf\{S(y) + T(z)\colon\ y, z \in L_+ \text{ and } y + z = x\}, \\ |S|(x) &= \sup\{S(y)\colon\ -x \leq y \leq x\}. \end{aligned}$$

In addition, $T_\alpha \uparrow T$ *holds in* $\mathcal{L}_b(L, M)$ *if and only if* $T_\alpha(x) \uparrow T(x)$ *holds in* M *for each* $x \in L_+$.

Proof. Assume that M is a Dedekind complete Riesz space. Start by observing that since L_+ is generating, it follows that the collection of all positive operators from L to M is a cone and so $\mathcal{L}_b(L, M)$ is an ordered vector space. From the lattice identities

$$S \vee T = (S - T)^+ + T\,, \quad S \wedge T = -[(-S) \vee (-T)]\,, \quad \text{and} \quad |S| = S \vee (-S)\,,$$

it suffices to prove the validity of the first formula, when $T = 0$. That is, it is enough to show that $S^+(x) = [S \vee 0](x) = \sup\{S(y)\colon\ 0 \le y \le x\}$ holds true for all $S \in \mathcal{L}_b(L, M)$ and all $x \in L_+$.

To this end, fix an operator $S \in \mathcal{L}_b(L, M)$, and then define the mapping $R\colon L_+ \to M_+$ for each $x \in L_+$ by

$$R(x) = \sup\{S(y)\colon\ 0 \le y \le x\} = \sup S([0, x])\,.$$

Notice that since $S \in \mathcal{L}_b(L, M)$, we indeed have $R(x) \in M_+$ for each $x \in L_+$. We claim that S is additive. To establish this claim, let $u, v \in L_+$. If $0 \le x \le u$ and $0 \le y \le v$, then $0 \le x + y \le u + v$ and from

$$S(x) + S(y) = S(x + y) \le R(u + v)\,,$$

it follows that $R(u) + R(v) \le R(u+v)$. On the other hand, if $0 \le x \le u+v$, then by the Riesz decomposition property there exist vectors $0 \le x_1 \le u$ and $0 \le x_2 \le v$ satisfying $x_1 + x_2 = x$, and so we have

$$S(x) = S(x_1 + x_2) = S(x_1) + S(x_2) \le R(u) + R(v)\,.$$

This implies $R(u + v) \le R(u) + R(v)$, and thus $R(u + v) = R(u) + R(v)$, i.e., the mapping $R\colon L_+ \to M_+$ is additive.

Since M is Archimedean (see Exercise 6 of Section 1.3), it follows from part (3) of Lemma 1.26 that the additive mapping R has a unique positive linear extension to all of L (which we shall denote by R again). Clearly, if $x \in L_+$, then $R(x) \ge S(x)$ and $R(x) \ge 0$, i.e., $R \ge S$ and $R \ge 0$. On the other hand, if some $\Phi \in \mathcal{L}_b(L, M)$ satisfies $\Phi \ge S$ and $\Phi \ge 0$, then for each $0 \le y \le x$ we have $\Phi(x) \ge \Phi(y) \ge S(y)$, from which it follows that $\Phi(x) \ge R(x)$ for all $x \in L_+$. Consequently, $R = S \vee 0$ holds in $\mathcal{L}_b(L, M)$, as desired.

The Dedekind completeness of $\mathcal{L}_b(L, M)$ (as well as the last part of the theorem) is left as an exercise for the reader; see also Exercise 2 at the end of this section. ■

For a glimpse of the usefulness of the Riesz–Kantorovich theorem we establish that a positive operator whose domain is a Riesz subspace extends to a positive operator if and only if it is dominated by a monotone sublinear mapping. As usual, a mapping $f\colon L \to M$ between two ordered vector spaces is called **monotone** whenever $x \le y$ in L implies $f(x) \le f(y)$ in M.

Theorem 1.60. *Let L and M be Riesz spaces with M Dedekind complete. If G is a Riesz subspace of L and $T\colon G \to M$ is a positive operator, then the following statements are equivalent.*

1. *T extends to a positive operator from L to M.*
2. *T extends to an order bounded operator from L to M.*
3. *There exists a monotone sublinear mapping $p\colon L \to M$ satisfying $T(x) \le p(x)$ for all $x \in G$.*

Proof. (1) $\Longrightarrow$ (2) Obvious.

(2) $\Longrightarrow$ (3) Let $S \in \mathcal{L}_{\rm b}(E,F)$ satisfy $S(x) = T(x)$ for all $x \in G$. Then the mapping $p\colon L \to M$ defined by $p(x) = |S|(x^+)$ is monotone, sublinear and satisfies

$$T(x) \le T(x^+) = S(x^+) \le |S|(x^+) = p(x)$$

for all $x \in G$.

(3) $\Longrightarrow$ (1) Let $p\colon L \to M$ be a monotone sublinear mapping satisfying $T(x) \le p(x)$ for all $x \in G$. Then the formula $q(x) = p(x^+)$ defines a sublinear mapping from L to M such that

$$T(x) \le T(x^+) \le p(x^+) = q(x)$$

holds for all $x \in G$. Thus, by the Hahn–Banach Extension Theorem 1.27 there exists an extension $R \in \mathcal{L}(E,F)$ of T satisfying $R(x) \le q(x)$ for all $x \in L$. In particular, if $x \in L^+$, then the relation

$$-R(x) = R(-x) \le q(-x) = p\big((-x)^+\big) = p(0) = 0$$

implies $R(x) \ge 0$. That is, R is a positive linear extension of T to all of L, and the proof is finished. ■

When $M = \mathbb{R}$, Theorem 1.59 specializes to the following important case that will be used extensively in this monograph.

Theorem 1.61 (Riesz–Kantorovich)**.** *If an ordered vector space L with a generating cone has the Riesz decomposition property, then its order dual $L^\sim$ is a Dedekind complete Riesz space and coincides with its regular dual, that is, $L^\sim = L^r$. Moreover, for any $f, g \in L^\sim$ and any $x \in L_+$ the lattice operations satisfy*

$$\begin{aligned}
[f \vee g](x) &= \sup\big\{f(y) + g(z)\colon\ y, z \in L_+ \text{ and } y + z = x\big\},\\
[f \wedge g](x) &= \inf\big\{f(y) + g(z)\colon\ y, z \in L_+ \text{ and } y + z = x\big\},\\
|f|(x) &= \sup\big\{f(y)\colon\ -x \le y \le x\big\}.
\end{aligned}$$

The formulas in Theorem 1.59 describing the lattice operations in $\mathcal{L}_b(L, M)$ are known as the **Riesz–Kantorovich formulas**. The formula

$$[S \vee T](x) = \sup\bigl\{S(y) + T(z)\colon\ y, z \in L_+ \text{ and } y + z = x\bigr\}$$

will especially be referred to as the **Riesz–Kantorovich formula** of the linear functionals f and g and it will play a fundamental role in this monograph. Its general formulation is as follows: If $S_1, \ldots, S_m$ belong to $\mathcal{L}_b(L, M)$, then their supremum for each $x \in L_+$ is given by the Riesz–Kantorovich formula

$$\Bigl[\bigvee_{i=1}^m S_i\Bigr](x) = \sup\Bigl\{\sum_{i=1}^m S_i(x_i)\colon\ x_i \in L_+ \text{ for each } i \text{ and } \sum_{i=1}^m x_i = x\Bigr\}.$$

The Riesz–Kantorovich formula will be the subject of an extensive investigation in Sections 5.2, 6.2, and 6.3. The absolute value $|T|$ of an operator $T \in \mathcal{L}_b(L, M)$ is also known as the **modulus of the operator** T.

In terms of directed sets the lattice operations of $\mathcal{L}_b(L, M)$ are described as follows.

Theorem 1.62. *Assume that L is an ordered vector space with a generating cone and the Riesz decomposition property and let M be a Dedekind complete Riesz space. Then for all $S, T \in \mathcal{L}_b(L, M)$ and each $x \in L^+$ we have*

$$(1)\quad \Bigl\{\sum_{i=1}^n S(x_i) \vee T(x_i)\colon\ x_i \in L^+ \text{ and } \sum_{i=1}^n x_i = x\Bigr\} \uparrow [S \vee T](x),$$

$$(2)\quad \Bigl\{\sum_{i=1}^n S(x_i) \wedge T(x_i)\colon\ x_i \in L^+ \text{ and } \sum_{i=1}^n x_i = x\Bigr\} \downarrow [S \wedge T](x),$$

$$(3)\quad \Bigl\{\sum_{i=1}^n |T(x_i)|\colon\ x_i \in L^+ \text{ and } \sum_{i=1}^n x_i = x\Bigr\} \uparrow |T|(x).$$

Proof. (1) Consider the set

$$D = \Bigl\{\sum_{i=1}^n S(x_i) \vee T(x_i)\colon\ x_i \in L^+ \text{ for each } i \text{ and } \sum_{i=1}^n x_i = x\Bigr\}.$$

Since $\sum_{i=1}^n x_i = x$ with each $x_i \in L^+$ implies

$$\sum_{i=1}^n S(x_i) \vee T(x_i) \le \sum_{i=1}^n \bigl[(S \vee T)x_i\bigr] \vee \bigl[(S \vee T)x_i\bigr] = [S \vee T](x),$$

we see that $D \le [S \vee T](x)$. On the other hand, if $D \le u$ holds, then for each $y, z \in L^+$ with $y + z = x$ we have

$$S(y) + T(z) \le S(y) \vee T(y) + S(z) \vee T(z) \le u,$$

and consequently

$$[S \vee T](x) = \sup\bigl\{S(y) + T(z)\colon\ y, z \in L^+ \text{ and } y + z = x\bigr\} \le u.$$

Thus, $\sup D = [S \vee T](x)$, and it remains to be shown that D is directed upward.

To this end, let $x = \sum_{i=1}^n x_i = \sum_{j=1}^m y_j$ with all the x_i and y_j in L^+. By Theorem 1.54 there exists a finite collection $\{z_{ij}\colon\ i = 1, \dots, n;\ j = 1, \dots, m\}$ of positive vectors such that

$$x_i = \sum_{j=1}^m z_{ij}\,, \quad \text{for each } i = 1, \dots, n\,,$$

and

$$y_j = \sum_{i=1}^n z_{ij}\,, \quad \text{for each } j = 1, \dots, m\,.$$

In particular, we have $\sum_{i=1}^n \sum_{j=1}^m z_{ij} = x$. On the other hand, using the lattice identity $x \vee y = \frac{1}{2}(x + y + |x - y|)$, we see that

$$\begin{aligned}
\sum_{i=1}^n & S(x_i) \vee T(x_i) \\
&= \tfrac{1}{2} \sum_{i=1}^n \big[S(x_i) + T(x_i) + |S(x_i) - T(x_i)|\big] \\
&= \tfrac{1}{2} \sum_{i=1}^n \Big[\sum_{j=1}^m S(z_{ij}) + \sum_{j=1}^m T(z_{ij}) + \Big|\sum_{j=1}^m \{S(z_{ij}) - T(z_{ij})\}\Big|\Big] \\
&\le \tfrac{1}{2} \sum_{i=1}^n \Big[\sum_{j=1}^m \{S(z_{ij}) + T(z_{ij}) + |S(z_{ij}) - T(z_{ij})|\}\Big] \\
&= \sum_{i=1}^n \sum_{j=1}^m S(z_{ij}) \vee T(z_{ij})\,.
\end{aligned}$$

Similarly,

$$\sum_{j=1}^m S(y_j) \vee T(y_j) \le \sum_{i=1}^n \sum_{j=1}^m S(z_{ij}) \vee T(z_{ij})$$

holds, and so D is directed upward.

(2) Use (1) in conjunction with the identity $T \wedge S = -\big[(-S) \vee (-T)\big]$.

(3) Use (1) and the identity $|T| = T \vee (-T)$. ■

Regarding the Riesz decomposition property and the Riesz–Kantorovich formula we have the following open problems.

(1) *Is the converse of the Riesz–Kantorovich Theorem* 1.61 *true? That is, does an ordered vector space whose cone is generating and whose order dual is a Riesz space have the Riesz decomposition property?*

(2) *Assume that L is an ordered vector space with a generating cone and that two linear functions $f, g \in L^\sim$ have a supremum (in the lattice sense) in $L^\sim$. Is $f \vee g$ given by the Riesz–Kantorovich formula? That is, if $f \vee g$ exists in $L^\sim$, is it then true that*

$$[f \vee g](x) = \sup\{f(y) + g(x - y)\colon\ 0 \le y \le x\}$$

for each $x \in L_+$?

There are several papers in the literature that deal with these open problems; see, for instance, [**2, 7, 8, 9, 10, 21, 83**]. These questions will be discussed in detail in Chapter 5 (and in particular in Sections 6.2 and 6.3).

Exercises

1. If L is an ordered vector space with the Riesz decomposition property and a generating cone and M is a Dedekind complete Riesz space, then show that for all $S \in \mathcal{L}_b(L, M)$ and all $x \in L_+$ we have $|S(x)| \le |S|(x)$.

2. Establish the Dedekind completeness of $\mathcal{L}_b(L, M)$ in Theorem 1.59. [*Hint*: Assume $0 \le T_\alpha \uparrow \le T$ in $\mathcal{L}_b(L, M)$. For each vector $x \in L_+$ let $S(x) = \sup\{T_\alpha(x)\}$ in M. Show that the mapping $S\colon L_+ \to M_+$ is additive and conclude (by using part (3) of Lemma 1.26) that it extends uniquely to a positive operator from L to M, which we denote by S again. Now verify that $S = \sup\{T_\alpha\}$ holds in $\mathcal{L}_b(L, M)$.]

3. Verify that the Riesz space $L_1[0, 1]$ is Dedekind complete. Now assume that $L = M = L_1[0, 1]$ and consider the positive operators $S, T\colon L \to M$ defined by

$$S(f) = f \quad \text{and} \quad T(f) = \left[\int_0^1 f(x)\,dx\right] \cdot \mathbf{1},$$

where $\mathbf{1}$ is the constant function one. Show that $S \wedge T = 0$ holds in $\mathcal{L}_b(L, M)$.

4. Let L be an ordered vector space with the Riesz decomposition property and a generating cone and let M be a Dedekind complete Riesz space. Use the formula

$$[S \vee T](x) = \sup\bigl\{S(y) + T(z)\colon\ y, z \in L_+ \text{ and } y + z = x\bigr\}$$

for all $S, T \in \mathcal{L}_b(L, M)$ and all $x \in L_+$ to prove that for all $S \in \mathcal{L}_b(L, M)$ and all $x \in L_+$ we have

$$|S|(x) = \sup\bigl\{S(y)\colon\ -x \le y \le x\bigr\}.$$

[*Hint*: Recall the definition of the absolute value: $|S| = S \vee (-S)$. Now note that if $0 \le y \le x$, then the vector $z = 2y - x$ satisfies $-x \le z \le x$. On the other hand, if $-x \le z \le x$, then the vector $y = \frac{z+x}{2}$ satisfies $0 \le y \le x$ and $2y - x = z$.]

5. Let L be an ordered vector space with a generating cone and let M be an arbitrary ordered vector space. Assume that for two operators $S, T \in \mathcal{L}_b(L, M)$ their supremum $S \vee T$ exists in $\mathcal{L}_b(L, M)$ (in the lattice sense). Show that for each $x \in L_+$ we have

$$\sup\{S(y) + T(x - y)\colon\ 0 \le y \le x\} \le [S \vee T](x)\,.$$

6. Let $T\colon L \to M$ be a positive operator between two Riesz spaces with M Dedekind complete. If $x, y \in L$, then show that
(a) $T(x \vee y) = \max\{Rx + Sy\colon\ R, S \in \mathcal{L}_b^+(L, M) \text{ and } R + S = T\}$,
(b) $T(x \wedge y) = \min\{Rx + Sy\colon\ R, S \in \mathcal{L}_b^+(L, M) \text{ and } R + S = T\}$.

7. Consider the continuous function $g\colon [0, 1] \to [0, 1]$ defined by

$$g(x) = \begin{cases} x & \text{if } 0 \le x \le \frac{1}{2}, \\ \frac{1}{2} & \text{if } \frac{1}{2} < x \le 1\,. \end{cases}$$

Next consider the operator $T\colon C[0, 1] \to C[0, 1]$ defined via the formula $[Tf](x) = f\big(g(x)\big) - f\big(\frac{1}{2}\big)$.

Show that T is a regular operator whose modulus does not exist in $\mathcal{L}_b(C[0, 1], C[0, 1])$.

8. Let $T\colon C[0, 1] \to C[0, 1]$ be the regular operator defined by

$$[Tf](x) = f(\sin x) - f(\cos x)\,.$$

Show that T^+ and T^- both exist and that

$$[T^+ f](x) = f(\sin x) \quad \text{and} \quad [T^- f](x) = f(\cos x)\,.$$

Chapter 2

Cones in topological vector spaces

The purpose of this chapter is to make a thorough investigation of the interplay between the order and topological structures of a vector space. As we shall see, several useful interrelationships are valid once one assumes that the cone is (topologically) closed or whenever its interior is nonempty. However, the most important connection between the order and topological structures of a vector space is achieved when we impose a "smallness" condition, known as *normality*, on the cone. A cone is said to be normal (with respect to a given linear topology) if the topology has a base at zero consisting of full neighborhoods; where, as before, a set is called *full* if it contains all order intervals whose endpoints lie in the set.

Section 2.1 presents a variety of order and topological properties of cones that are based on some elementary connections between order and topology. For instance, in this section, we shall see that closed cones can "live" only in Hausdorff topological vector spaces (Lemma 2.3) and that the order units of a cone are precisely its interior points (Theorem 2.8). Duality properties of cones and wedges are discussed in Section 2.2, where we shall establish that the strictly positive vectors of a cone are its interior points (Lemma 2.17).

Section 2.3 introduces normal cones and presents some of their basic characterizations. Here we establish the basic result that the dual wedge of a normal cone in a locally convex space is generating (Theorem 2.26). Moreover, we show that for normal cones the topological convergence and weak convergence of monotone nets coincide (Lemma 2.27).

When we specialize the topological properties of cones to Banach spaces, we obtain much sharper results. This task is taken in Sections 2.4 and 2.5. Quite often positive operators between ordered Banach spaces are continuous. In Section 2.4 we study the continuity of positive operators and obtain some general results that guarantee the continuity of positive operators; see Theorem 2.32 and Corollary 2.33.

In Section 2.5 we present a complete study of ordered normed and Banach spaces. Specifically, we characterize closed, generating, and normal cones as well as several combinations of these properties in ordered Banach spaces. The results here are impressive. In a Banach space a closed cone is norm normal if and only if its order intervals are norm bounded (Theorem 2.40). The norm dual cone of a closed and generating cone of a Banach space is norm normal (Theorem 2.42). If an ordered Banach space L has a closed and generating cone, then every positive operator from L into an ordered Banach space with a closed cone is continuous (Theorem 2.32). We also prove Andô's duality theorem (Theorem 2.47) asserting that an ordered Banach space with a closed, generating and normal cone has the Riesz decomposition property if and only if its norm dual is a Riesz space.

In Section 2.6 we discuss "ice cream cones" in normed spaces and present their geometrical structure (Theorem 2.52). The objective of Section 2.7 is to introduce ideals in ordered vector spaces and to study them in conjunction with the notion of uniform convergence. Their norm structure is included in Theorem 2.55, and in Theorem 2.60 the norm normality of a closed cone of a Banach space is characterized in terms of order ideals.

The chapter culminates with a comprehensive study in Section 2.8 of the order topology generated by cone. If the order dual $L^\sim$ of an ordered vector space L separates the points of L, then the order topology coincides with the Mackey topology $\tau(L, L^\sim)$ (Corollary 2.70). On the other hand, if there exists an order unit, then the order topology is normable (Theorem 2.71).

2.1. Ordered topological vector spaces

We are now ready to mix up order and topology. This will be done with the introduction of the notion of an ordered topological vector space.

Definition 2.1. *An* ***ordered topological vector space*** *is simply an ordered vector space equipped with a linear topology.*[1]

[1] Some authors define an ordered topological vector space to be an ordered vector space L equipped with a linear topology τ for which the positive cone L_+ is τ-closed. We emphasize that in our definition here we do not impose such a condition.

As usual, the **topological dual** of a topological vector space (X, τ) will be denoted by X', that is, X' is the vector subspace of the algebraic dual X^* of X consisting of all τ-continuous linear functionals.

Definition 2.2. *If (L, τ) is an ordered topological vector space, then the wedge of all positive τ-continuous linear functionals on L will be denoted by L'_+, that is,*

$$L'_+ = L' \cap L^*_+ = \{f \in L' \colon\ f \text{ is positive}\}.$$

The first simple connections between order and topology are included in the following result.

Lemma 2.3. *If an ordered vector space L admits a linear topology τ for which its cone L_+ is τ-closed, then*

(1) *the topology τ is Hausdorff,*

(2) *the cone L_+ is Archimedean,*

(3) *the order intervals of L are τ-closed, and*

(4) *$x_\alpha\uparrow$ (resp. $x_\alpha\downarrow$) with $x_\alpha \xrightarrow{\tau} x$ implies $x_\alpha \uparrow x$ (resp. $x_\alpha \downarrow x$).*

Proof. Let τ be a linear topology on an ordered vector space L such that L_+ is τ-closed.

(1) Since $\{0\} = L_+ \cap (-L_+)$, it follows that $\{0\}$ is τ-closed and from this it easily follows that the linear topology τ is Hausdorff.

(2) Assume $nx \le y$ for all n and some $x, y \in L$. Then $\frac{1}{n}y - x \ge 0$ (that is, $\frac{1}{n}y - x \in L_+$ for each n) and $\frac{1}{n}y - x \xrightarrow{\tau} -x$ imply $-x \ge 0$ or $x \le 0$.

(3) This follows immediately from $[x, y] = (x + L_+) \cap (y - L_+)$.

(4) Suppose that $x_\alpha\uparrow$ and $x_\alpha \xrightarrow{\tau} x$. From $x_\beta - x_\alpha \in L_+$ for each $\beta \succeq \alpha$, $x_\beta - x_\alpha \xrightarrow[\beta \succeq \alpha]{\tau} x - x_\alpha$ and the τ-closedness of L_+, we get $x - x_\alpha \in L_+$ or $x \ge x_\alpha$ for all α. To see that x is the least upper bound of the net $\{x_\alpha\}$, assume that some $y \in L$ satisfies $x_\alpha \le y$ or $y - x_\alpha \in L_+$ for all α. From $y - x_\alpha \xrightarrow{\tau} y - x$ and the τ-closedness of L_+ we get $y - x \in L_+$ or $y \ge x$. This shows that $x = \sup\{x_\alpha\}$ holds. ∎

For a cone with a nonempty interior the converse of statement (2) of the preceding result is also valid.

Lemma 2.4. *Let (L, τ) be an ordered Hausdorff topological vector space whose cone L_+ has a nonempty interior. Then the cone L_+ is Archimedean if and only if it is τ-closed.*

Proof. Assume that L_+ is Archimedean and let $y \in \operatorname{Int}(L_+)$. Fix $x \in \overline{L}_+$, the τ-closure of L_+. By a well-known result of topological vector spaces, we know that $\alpha y+(1-\alpha)x \in \text{Int}(L_+)$ for each $0<\alpha\leq 1$; see for instance [**12**, Lemma 5.28, p. 182]. In particular, we have $\frac{1}{n+1}y+\left(1-\frac{1}{n+1}\right)x \in \text{Int}(L_+)$ or $y+nx \in \text{Int}(L_+)$ for each n. This implies $n(-x)\leq y$ for each n, and from the Archimedean property of L_+ we get $-x\leq 0$ or $x\in L_+$. Therefore, $\overline{L_+}\subseteq L_+$ so that L_+ is τ-closed. ■

The vectors that belong to the interior of the cone of an ordered topological vector space are characterized as follows.

Lemma 2.5. *In an ordered topological vector space L, a positive vector u is an interior point of the cone L_+ if and only if the order interval $[-u,u]$ is a neighborhood of zero.*

In particular, if L is an ordered topological vector space and u is an interior point of L_+, then

(1) *u is an order unit and*

(2) *$L^\sim \subseteq L'$.*

Proof. Let L be an ordered topological vector space and let $u\in L_+$. If $[-u,u]$ is a neighborhood of zero, then from $u+[-u,u]=[0,2u]\subseteq L_+$, we see that u is an interior point of L_+.

For the converse, assume that u is an interior point of L_+. Pick a circled neighborhood V of zero such that $u+V\subseteq L_+$. We claim that $V\subseteq[-u,u]$. Indeed, since for each $x\in V$ we have $-x\in V$, it follows from $u\pm x\in L_+$ that $u+x\geq 0$ and $u-x\geq 0$ or $-u\leq x\leq u$. Thus, $x\in[-u,u]$ for each $x\in V$ or $V\subseteq[-u,u]$. This shows that $[-u,u]$ is a neighborhood of zero.

That u is an order unit follows immediately from the fact that every neighborhood of zero is an absorbing set.

For part (2), note that each $f\in L^\sim$ is bounded on the neighborhood of zero $[-u,u]$. Hence, every $f\in L^\sim$ is continuous,[2] and so $L^\sim\subseteq L'$. ■

The interior points of a cone with a base also have another nice characterization.

Lemma 2.6. *Let $\mathcal{K}$ be a cone of a topological vector space L such that $\mathcal{K}$ has a base $\mathcal{B}=\{x\in\mathcal{K}\colon\ f(x)=1\}$, where $f\in L'$. Put $\mathcal{F}=\{x\in L\colon\ f(x)=1\}$ and consider $\mathcal{F}$ as a topological space under the induced topology from L. Then for a vector $u\in\mathcal{F}$ the following statements are equivalent.*

(1) *The vector u is an interior point of the cone $\mathcal{K}$.*

[2] This is a consequence of the following general result in topological vector spaces: *let (L,τ) be a topological vector space and let $f\in L^*$. Then f is τ-continuous if and only if it is bounded on a τ-neighborhood of zero.* [Indeed, if $|f(x)|\leq M$ holds for all x in a τ-neighborhood of zero, then for each $\epsilon>0$ and each $x\in\frac{\epsilon}{M}V$ we have $|f(x)|\leq\epsilon$.]

(2) *The vector u is an interior point of $\mathcal{B}$ (as a subset of the topological space $\mathcal{F}$).*

Proof. Clearly (1) $\Longrightarrow$ (2). So, we need to show only (2) $\Longrightarrow$ (1). To see this, assume that u is an interior point of $\mathcal{B}$ and that u is not an interior point of $\mathcal{K}$. It follows that u must be a boundary point of $\mathcal{K}$. Pick a net $\{x_\alpha\}$ of L satisfying $x_\alpha \to u$ and $x_\alpha \notin \mathcal{K}$ for each α. Since f is a continuous linear functional and $f(u) = 1$, we can suppose without loss of generality that $f(x_\alpha) > 0$ holds for each α. Also, replacing each x_α by $\frac{x_\alpha}{f(x_\alpha)}$, we can assume that $x_\alpha \in \mathcal{F}$ for all α. But then $x_\alpha \to u$ and $x_\alpha \notin \mathcal{B}$ for each α shows that u is not an interior point of $\mathcal{B}$, which is impossible. ∎

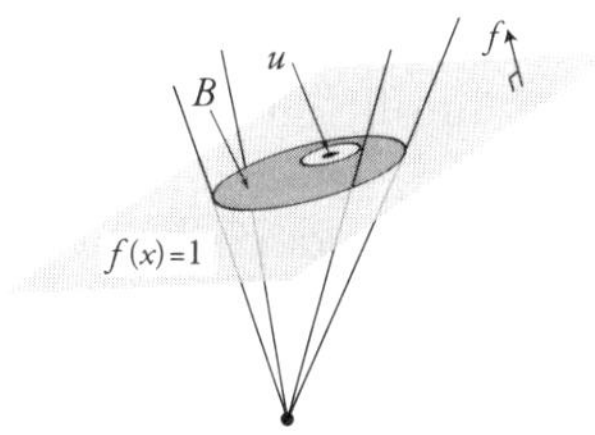

The extremal vectors of cones in topological vector spaces are always boundary points of the cones.

Lemma 2.7. *No interior point of a cone of a topological vector space of dimension more than one can be an extremal vector. In other words, in a topological vector space of dimension more than one the extremal points of a cone are always boundary points of the cone.*

Proof. If $e \in \operatorname{Int}(\mathcal{K})$, then there exists a circled neighborhood V of zero such that $e + V \subseteq \mathcal{K}$. Pick a vector $v \in V$ that is linearly independent from e and note that the vectors $e+v$ and $e-v$ are likewise linearly independent (why?). However, the identity $e = \frac{1}{2}(e+v)+\frac{1}{2}(e-v)$ contradicts Lemma 1.43 and so e cannot be an extremal vector of $\mathcal{K}$. ∎

From Lemma 1.7 we know that the order units of an ordered vector space are precisely the internal points of its cone. When the ordered vector space is also a completely metrizable topological vector space with a closed cone, the order units are precisely the interior points of the cone.

Theorem 2.8. *For a nonzero positive vector u of a completely metrizable ordered topological vector space L (in particular, this is the case if L is finite dimensional) with a closed cone the following statements are equivalent.*

(1) *u is an order unit.*

(2) *u is an internal point of the positive cone L_+.*

(3) *u is an interior point of the positive cone L_+.*

Proof. From Lemma 1.7 we know that (1) and (2) are equivalent. Also Lemma 2.5 implies that every interior point of L_+ is an order unit. So, what needs to be proven is that (1) $\Longrightarrow$ (3). Thus, assume that u is an order unit. This implies that $L = \bigcup_{n=1}^{\infty} n[-u,u]$. The closedness of the cone guarantees that $[-u,u]$ is a closed set and from the Baire category theorem we infer that $[-u,u]$ has an interior point, say v. Pick a circled neighborhood V of zero such that $v + V \subseteq [-u,u]$.

We shall finish the proof by showing that $u + \frac{1}{2}V \subseteq L_+$. To see this, note that if $x \in V$, then $u + x \geq v + x \geq -u$ and so $2u + x \geq 0$ or $u + \frac{1}{2}x \geq 0$. This implies $u + \frac{1}{2}V \subseteq L_+$ so that u is an interior point of L_+. ■

It should be noticed that the preceding result does not assert that a completely metrizable ordered topological vector space with a closed (and even with a closed and generating) cone has order units. For instance, if $L = \mathbb{R}^3$ and $L_+ = \{(x,y,0)\colon\ x \geq 0 \text{ and } y \geq 0\}$, then L (with its Euclidean topology) is an ordered Banach space and L_+ is closed but L does not have order units. On the other hand, if $L = \ell_2$ and L_+ consists of all non-negative sequences, then L is an ordered Banach space, L_+ is norm closed and generating, and L does not have order units.

The existence of a continuous strictly positive linear functional guarantees an "abundance" of continuous strictly positive linear functionals.

Lemma 2.9. *If an ordered topological vector space (L,τ) admits a τ-continuous strictly positive linear functional, then the collection of all τ-continuous strictly positive linear functionals on L is dense in L'_+ for any linear topology on L'.*

Proof. If $f \in L'_+$ is a strictly positive linear functional and $g \in L'_+$, then for each $0 < \alpha < 1$ the linear functional $h_\alpha = \alpha f + (1-\alpha)g \in L'_+$ is strictly positive. Now if τ' is any linear topology on L', then $h_\alpha \xrightarrow[\alpha\downarrow 0]{\tau'} g$. ■

We shall close the section with an important result (due to T. Andô [**23**]) regarding completely metrizable ordered topological vector spaces.

Theorem 2.10 (Andô). *Let (L,τ) be a completely metrizable ordered topological vector space such that L_+ is τ-closed and generating. If $\{U_n\colon n \in \mathbb{N}\}$ is a countable τ-neighborhood base of zero consisting of τ-closed circled sets such that $U_{n+1} + U_{n+1} \subseteq U_n$ holds for each n, then the countable collection $\{V_n\colon\ n \in \mathbb{N}\}$, where*

$$V_n = U_n \cap L_+ - U_n \cap L_+\,,$$

is also a τ-neighborhood base of zero satisfying $V_{n+1} + V_{n+1} \subseteq V_n$ for all n.

Proof. We shall prove first by steps that the countable collection of sets $\{V_n\colon\ n \in \mathbb{N}\}$ is a base for a metrizable linear topology on L.

(**1**) *Each V_n is a circled set.*

If $x = x_1 - x_2$ with $x_1, x_2 \in U_n \cap L_+$ and $\lambda \in \mathbb{R}$ satisfies $|\lambda| \leq 1$, then we have $\lambda x_i \in U_n \cap L_+$ for $0 \leq \lambda \leq 1$ and $-\lambda x_i \in U_n \cap L_+$ if $-1 \leq \lambda < 0$. From these conclusions, we easily infer that $\lambda x \in V_n$ for all $|\lambda| \leq 1$.

(**2**) *Each V_n is an absorbing set.*

Fix $x \in L$. Since the cone L_+ is generating, there exist $x_1, x_2 \in L_+$ such that $x = x_1 - x_2$. Since U_n is absorbing, there exists some $\alpha_0 > 0$ such that $\alpha x_i \in U_n$ for all $|\alpha| \leq \alpha_0$ and each $i = 1, 2$. From this and arguing as in the previous part, we see that $\alpha x \in V_n$ for all $|\alpha| \leq \alpha_0$.

(**3**) *For each n we have $V_{n+1} + V_{n+1} \subseteq V_n$ and $V_{n+1} \subseteq U_n$.*

These follow immediately from $U_{n+1} + U_{n+1} \subseteq U_n$.

(**4**) $\bigcap_{n=1}^{\infty} V_n = \{0\}$.

This is a direct consequence of the inclusion $V_{n+1} \subseteq U_n$.

(**5**) *If τ_1 is the metrizable topology generated by $\{V_n\colon\ n \in \mathbb{N}\}$, then $\tau \subseteq \tau_1$ and (L, τ_1) is a completely metrizable topological vector space.*

The inclusion $\tau \subseteq \tau_1$ is a direct consequence of the fact that $V_{n+1} \subseteq U_n$ holds true for each n. For the rest of the proof note first that if a sequence $\{y_n\} \subseteq L_+$ satisfies $y_n \in U_n$ for each n, then the series $\sum_{n=1}^{\infty} y_n$ is necessarily τ-convergent in L. To see this, let $x_n = \sum_{i=1}^{n} y_i$ and observe that the relation

$$0 \leq x_{n+p} - x_n = \sum_{i=n+1}^{n+p} y_i \in U_{n+1} + U_{n+2} + \cdots + U_{n+p} \subseteq U_n \qquad (\star)$$

implies that $\{x_n\}$ is a τ-Cauchy sequence. If y is its τ-limit in L, then it follows from $(\star)$ and the τ-closedness of U_n and L_+ that for each n we have

$$y - \sum_{i=1}^{n} y_i = y - x_n \in U_n \cap L_+ \,. \qquad (\star\star)$$

Now let $\{z_n\}$ be a τ_1-Cauchy sequence of L. By passing to a subsequence, we can assume that $z_{n+1} - z_n \in V_n$ holds for each n. Therefore, for each $n \in \mathbb{N}$ there exist $u_n, v_n \in U_n \cap L_+$ satisfying $z_{n+1} - z_n = u_n - v_n$. By the preceding discussion, both series $u = \sum_{n=1}^{\infty} u_n$ and $v = \sum_{n=1}^{\infty} v_n$ are τ-convergent in L. In addition, from $(\star\star)$ we have

$$u - \sum_{i=1}^{n} u_i \in U_n \cap L_+ \quad \text{and} \quad v - \sum_{i=1}^{n} v_i \in U_n \cap L_+$$

for each n. Now let $z = u - v - z_1$ and note that for each $n \in \mathbb{N}$ we have

$$\begin{aligned} z - z_{n+1} &= u - v - \sum_{i=1}^{n}(z_{i+1} - z_i) \\ &= \Big[u - \sum_{i=1}^{n} u_i\Big] - \Big[v - \sum_{i=1}^{n} v_i\Big] \in U_n \cap L_+ - U_n \cap L_+ = V_n\,. \end{aligned}$$

This shows that $z_n \xrightarrow{\tau_1} z$, and so (L, τ_1) is complete.

Finally, let us complete the proof of the theorem. Consider the identity operator $I\colon (L,\tau_1) \to (L,\tau)$ between the two completely metrizable topological vector spaces (L,τ_1) and (L,τ), and note that I is a continuous operator. But then, according to Corollary 8.37, I must be a homeomorphism. This implies $\tau_1 = \tau$, and the proof is finished. ■

Two useful consequences of the preceding result follow.

Corollary 2.11. *If (L,τ) is a completely metrizable ordered topological vector space with a τ-closed and generating cone, then for each τ-neighborhood V of zero the set $V \cap L_+ - V \cap L_+$ is likewise a τ-neighborhood of zero.*

Corollary 2.12. *If the cone of an ordered Banach space L is closed and generating and $U = \{x \in L\colon\ \|x\| \le 1\}$ (the closed unit ball of L), then the set $U \cap L_+ - U \cap L_+$ is a neighborhood of zero.*

Exercises

1. Establish the following fact that was used in the proof of part (1) of Lemma 2.3: *A topological vector space is Hausdorff if and only if the singleton $\{0\}$ is a closed set.*
2. Prove that a linear topology τ on a nontrivial vector space X is Hausdorff if and only if X admits a τ-closed cone. [*Hint*: If τ is a Hausdorff topology on the vector space X, pick a nonzero vector $x \in X$ and note that the cone $\mathcal{K} = \{\lambda x\colon\ \lambda \ge 0\}$ is τ-closed.]
3. In a Hausdorff topological vector space, is a cone with the Riesz decomposition property necessarily closed? [*Hint*: See Exercise 2 of Section 1.8.]
4. For a cone $\mathcal{K}$ of a topological vector space and a vector u establish the following.
   (a) The vector u is an interior point of $\mathcal{K}$ if and only if λu is likewise an interior point of $\mathcal{K}$ for each $\lambda > 0$.
   (b) The vector u is a boundary point of $\mathcal{K}$ if and only if λu is a boundary point of $\mathcal{K}$ for each $\lambda \ge 0$.
5. Show that the standard cones of $C[0,1]$ and $L_\infty[0,1]$ have interior points.

6. Verify that for each $0 \le p < \infty$ the standard cone of $L_p[0,1]$ has no interior points.

7. Show that in a topological vector space every proper face of a cone consists of boundary points of the cone.

8. Let $\mathcal{W}$ be a closed wedge in an ordered topological vector space. Show that the wedge $\mathcal{W}$ contains a straight line, i.e., there exist two distinct vectors $x, y \in W$ such that $\alpha x + (1-\alpha)y \in \mathcal{W}$ holds for all $\alpha \in \mathbb{R}$, if and only if $\mathcal{W} \cap (-\mathcal{W}) \neq \{0\}$ (in which case $\mathcal{W}$ contains a line passing through the origin). Can a nonclosed cone in an ordered Hausdorff topological vector space contain a straight line? [*Hint*: Assume that the closed wedge $\mathcal{W}$ contains the straight line passing through two distinct vectors x, y of $\mathcal{W}$ and that $\mathcal{W} \cap (-\mathcal{W}) = \{0\}$ holds true. Then $\mathcal{W}$ is a closed cone. By Lemma 2.3 the cone $\mathcal{W}$ is also Archimedean. Now note that from $y \pm n(x-y) \in \mathcal{W}$ we get $\pm n(x-y) \le y$ for each $n \in \mathbb{N}$, and from this we see that $x = y$, which is impossible.]

9. For a closed cone $\mathcal{K}$ of a topological vector space X show the following.

(a) For each $x \in \mathcal{K}$ and each $y \neq 0$ there exists a (not necessary unique) real number α such that $x + \alpha y \in \partial\mathcal{K}$, the boundary of $\mathcal{K}$.

(b) If $\dim X > 2$ and $u \in \mathrm{Int}(\mathcal{K})$, then there exists some $v \in \partial\mathcal{K}$ satisfying $v \notin \mathbb{R}u$.

[*Hint*: If $x \in \partial\mathcal{K}$, then $\alpha = 0$ is the desired scalar. So, we can assume that x lies in the interior of $\mathcal{K}$. Put

$$\beta = \inf\{\alpha \in \mathbb{R}\colon\ x + \alpha y \in \mathcal{K}\}$$

and

$$\gamma = \sup\{\alpha \in \mathbb{R}\colon\ x + \alpha y \in \mathcal{K}\}.$$

Since $x \in \mathrm{Int}(\mathcal{K})$, we see that $\beta < 0 < \gamma$. We claim that either β or γ is finite. Indeed, if $\beta = -\infty$ and $\gamma = \infty$, then (by the convexity of $\mathcal{K}$) it follows that $x + \alpha y \in \mathcal{K}$ for all $\alpha \in \mathbb{R}$. In particular, we have $\pm ny \le x$ for each n. Since $\mathcal{K}$ is Archimedean, we infer that $\pm y \le 0$ or $y = 0$, which is a contradiction. Now note that if $\beta \in \mathbb{R}$, then $x + \beta y \in \partial\mathcal{K}$; and similarly if $\gamma \in \mathbb{R}$, then $x + \gamma y \in \partial\mathcal{K}$. For part (b) pick some $x \notin \mathcal{K}$ and $x \notin \mathbb{R}u$ and then choose $\lambda \in \mathbb{R}$ such that the vector $v = u + \lambda x$ belongs to $\partial\mathcal{K}$.]

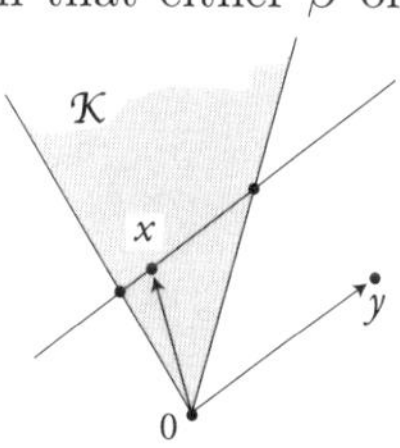

10. Let $\mathcal{K}$ be a cone of a topological vector space L such that $\mathcal{K}$ has a base $\mathcal{B} = \{x \in \mathcal{K}\colon\ f(x) = 1\}$, where $f \in L'$. Put $\mathcal{F} = \{x \in L\colon\ f(x) = 1\}$ and consider $\mathcal{F}$ as a topological space under the induced topology from L.

Show that a vector $u \in \mathcal{F}$ is a boundary point of the cone $\mathcal{K}$ if and only if u is a boundary point of $\mathcal{B}$ (where $\mathcal{B}$ is considered as a subset of the topological space $\mathcal{F}$). [*Hint*: Repeat the proof of Lemma 2.6.]

11. Prove Corollaries 2.11 and 2.12.

12. Show that in a two-dimensional vector space every closed and generating cone is a lattice cone. In particular, deduce from this that if a partially ordered topological vector space X has a closed cone, then every two-dimensional subspace M of X whose induced cone $M^+ = M \cap X^+$ is generating is a lattice-subspace. [*Hint*: Let $\mathcal{K}$ be a closed and generating cone in $\mathbb{R}^2$ and let $x_1, x_2 \in \mathcal{K}$ be two unit vectors satisfying

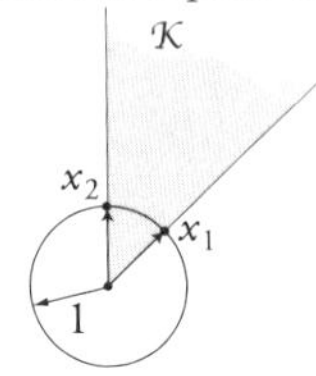

$$x_1 \cdot x_2 = \min\{x \cdot y\colon\ x, y \in \mathcal{K} \text{ and } \|x\| = \|y\| = 1\}.$$

Now establish that $\mathcal{K} = \{\lambda_1 x_1 + \lambda_2 x_2\colon\ \lambda_1, \lambda_2 \in \mathbb{R}_+\}$. (See also Exercise 7 of Section 1.7.)]

13. Let (L, τ) be a locally convex ordered topological space. Show that there exists a nonzero positive continuous linear functional on L if and only if the cone L_+ is not τ-dense in L.

14. In the proof of Theorem 2.10 we established the validity of property $(\star)$ by using the following claim: *If in a topological vector space (X, τ) a sequence $\{U_n\}$ of τ-neighborhoods of zero satisfies $U_{n+1} + U_{n+1} \subseteq U_n$ for all n, then for each n and p we have $U_{n+1} + U_{n+2} + \cdots + U_{n+p} \subseteq U_n$.* Prove this claim.

2.2. Wedge duality

We now turn our attention to dual systems. Let $\langle L, L' \rangle$ be a dual system under the duality $\langle \cdot, \cdot \rangle$, and let $\mathcal{W}$ be a wedge of L. The $\mathcal{W}$-positive linear functionals of L' will be denoted by $\mathcal{W}'$, that is,

$$\mathcal{W}' = \{x' \in L'\colon\ \langle x, x' \rangle \geq 0 \text{ for all } x \in \mathcal{W}\}.$$

We also let

$$\mathcal{W}'' = \{x \in L\colon\ \langle x, x' \rangle \geq 0 \text{ for all } x' \in \mathcal{W}'\}.$$

It should be clear that $\mathcal{W}'$ is a wedge of L', called the **dual wedge** of $\mathcal{W}$, and that $\mathcal{W}''$ is a wedge of L (called the **double dual wedge** of $\mathcal{W}$) satisfying $\mathcal{W} \subseteq \mathcal{W}''$. If $\mathcal{W}'$ is a cone, then $\mathcal{W}'$ is called the **dual cone** of $\mathcal{W}$. Similar definitions can be given for a wedge $\mathcal{W}$ of L'.

In geometric terms the dual wedge consists of all linear functionals of L' that support the wedge $\mathcal{W}$ at the origin; see the figure below, where for clarity the wedges $\mathcal{W}$ and $-\mathcal{W}'$ are shown.

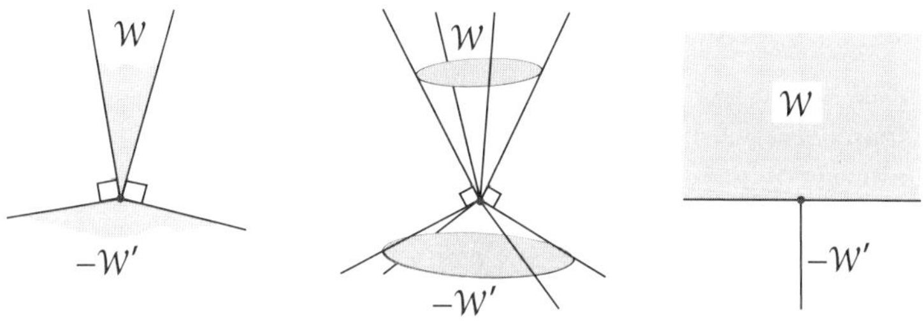

The basic properties of the dual wedges are included in the following result.

Theorem 2.13. *For a dual system $\langle L, L'\rangle$ and a wedge $\mathcal{W}$ of L we have the following.*

(1) *The dual wedge $\mathcal{W}'$ is always $\sigma(L', L)$-closed.*

(2) *The dual wedge $\mathcal{W}'$ of $\mathcal{W}$ is a cone of L' if and only if $\mathcal{W} - \mathcal{W}$ is $\sigma(L, L')$-dense in L.*

(3) *The double dual wedge $\mathcal{W}''$ of $\mathcal{W}$ is the $\sigma(L, L')$-closure of $\mathcal{W}$ in L (and so $\mathcal{W}$ is $\sigma(L, L')$-closed if and only if $\mathcal{W} = \mathcal{W}''$).*

(4) *If $\mathcal{W}$ is $\sigma(L, L')$-closed, then a vector $x \in L$ is positive (i.e., $x \in \mathcal{W}$) if and only if $\langle x, x'\rangle \geq 0$ holds for all $x' \in \mathcal{W}'$.*

(5) *If $\mathcal{W}$ is a $\sigma(L, L')$-closed cone, then the vector subspace $\mathcal{W}' - \mathcal{W}'$ of L' is $\sigma(L', L)$-dense in L'.*

Proof. (1) It follows from $\mathcal{W}' = \bigcap_{x\in\mathcal{W}}\{x' \in L' \colon \langle x, x'\rangle \geq 0\}$.

(2) Assume first that $\mathcal{W} - \mathcal{W}$ is $\sigma(L, L')$-dense in L, and then fix some $x' \in \mathcal{W}' \cap (-\mathcal{W}')$. It follows that $\langle x, x'\rangle = 0$ for all $x \in \mathcal{W} - \mathcal{W}$, and the $\sigma(L, L')$-continuity of x' coupled with the $\sigma(L, L')$-density of $\mathcal{W} - \mathcal{W}$ guarantees that $x' = 0$. Hence $\mathcal{W}'$ is a cone of L'.

For the converse, suppose that $\mathcal{W}'$ is a cone and assume by way of contradiction that $\mathcal{W} - \mathcal{W}$ is not $\sigma(L, L')$-dense in L. The separation theorem guarantees the existence of some nonzero $x' \in L'$ such that $\langle x, x'\rangle = 0$ for all $x \in \mathcal{W} - \mathcal{W}$. This implies that $x' \in \mathcal{W}' \cap (-\mathcal{W}') = \{0\}$, which is impossible. Hence, $\mathcal{W} - \mathcal{W}$ must be $\sigma(L, L')$-dense in L.

(3) Clearly, $\mathcal{W} \subseteq \mathcal{W}''$ and $\mathcal{W}''$ is $\sigma(L, L')$-closed. Let $\overline{\mathcal{W}}$ denote the $\sigma(L, L')$-closure of $\mathcal{W}$, and note that $\overline{\mathcal{W}} \subseteq \mathcal{W}''$. Now assume by way of contradiction that there exists some $x_0 \in \mathcal{W}''$ such that $x_0 \notin \overline{\mathcal{W}}$. By the separation theorem there exist some nonzero $x' \in L'$ and some $c \in \mathbb{R}$ such that $\langle x_0, x'\rangle < c \leq \langle x, x'\rangle$ holds for all $x \in \mathcal{W}$. Since $\mathcal{W}$ is a wedge, it follows that $\langle x, x'\rangle \geq 0$ for all $x \in \mathcal{W}$ (and so $x' \in \mathcal{W}'$) and that $\langle x_0, x'\rangle < 0$. However, the latter is a contradiction since $x_0 \in \mathcal{W}''$ and $x' \in \mathcal{W}'$ imply $\langle x_0, x'\rangle \geq 0$. Hence, $\overline{\mathcal{W}} = \mathcal{W}''$ is true.

(4) In this case, notice that (3) implies $\mathcal{W} = \mathcal{W}''$, and our conclusion follows.

(5) If the vector subspace $\mathcal{W}' - \mathcal{W}'$ of L' is not $\sigma(L', L)$-dense in L', then there exists some nonzero vector $x \in L$ satisfying $\langle x, x' \rangle = 0$ for all $x' \in \mathcal{W}' - \mathcal{W}'$. This implies $\langle x, x' \rangle = 0$ for all $x' \in \mathcal{W}'$. But then, by part (4), we get $x \geq 0$ and $-x \geq 0$, or $x \in \mathcal{W} \cap (-\mathcal{W}) = \{0\}$, which is impossible. So, $\mathcal{W}' - \mathcal{W}'$ is $\sigma(L', L)$-dense in L'. ■

Corollary 2.14. *If $\langle L, L' \rangle$ is a dual system and $\mathcal{W}$ is a generating wedge of L, then the dual wedge $\mathcal{W}'$ of $\mathcal{W}$ is a cone of L'.*

If L is infinite dimensional and $\mathcal{K}$ is a $\sigma(L, L')$-closed cone of L, then the vector space $\mathcal{K} - \mathcal{K}$ generated by $\mathcal{K}$ need not be $\sigma(L, L')$-closed. The example below is due to V. G. Troitsky (oral communication). For a related example see Exercise 8 of Section 2.5.

Example 2.15. Let $L = \ell_p$, where $1 < p < \infty$, and let L' be the norm dual of L. For each n let e_n be the standard unit vector of ℓ_p, i.e., e_n is the sequence whose n^{th}-coordinate is one and every other is zero. Define the sequence of vectors $u_0, u_1, \ldots$ of ℓ_p by

$$u_0 = e_1 \quad \text{and} \quad u_n = e_1 + \tfrac{1}{n} e_n \ \text{ for } \ n = 1, 2, \ldots .$$

Next, let $\mathcal{K}_0$ be the cone generated by $\{u_0, u_1, \ldots\}$, i.e., let

$$\mathcal{K}_0 = \Big\{ x \in \ell_p \colon \ \exists\ \alpha_0, \alpha_1, \ldots, \alpha_k \in \mathbb{R}_+ \ \text{such that} \ x = \sum_{i=0}^{k} \alpha_i u_i \Big\} .$$

Clearly, $\mathcal{K}_0$ is a subcone of the standard cone ℓ_p^+ and so its norm closure $\mathcal{K} = \overline{\mathcal{K}_0}$ is a closed subcone of ℓ_p^+. We claim that $\mathcal{K} - \mathcal{K}$ is a nonclosed dense vector subspace of L.

To see that $\mathcal{K} - \mathcal{K}$ is norm dense in ℓ_p, notice that it suffices to show that $\mathcal{K} - \mathcal{K}$ contains all eventually zero sequences. This follows easily by observing that if $\alpha_0, \alpha_1, \ldots, \alpha_k \in \mathbb{R}$, then we have

$$\sum_{i=0}^{k} \alpha_i u_i = \big(\alpha_0 + \alpha_1 + \cdots + \alpha_k, \alpha_1, \tfrac{\alpha_2}{2}, \ldots, \tfrac{\alpha_k}{k}, 0, 0, \ldots \big) .$$

We shall establish that $\mathcal{K} - \mathcal{K} \neq \ell_p$ holds true by proving that the vector $x = \big(0, 1, \frac{1}{2}, \frac{1}{3}, \ldots\big) \in \ell_p$ does not belong to $\mathcal{K} - \mathcal{K}$. To this end, suppose by way of contradiction that $x \in \mathcal{K} - \mathcal{K}$. This implies the existence of some vector $y = (y_0, y_1, y_2, \ldots) \in \mathcal{K}$ with $y \geq x$. Since $\mathcal{K}$ is a subcone of ℓ_p^+, we see that $y_n \geq \frac{1}{n}$ for each $n = 1, 2, \ldots$. Fix a natural number k. Since $y \in \mathcal{K}$, there exists some

$$a = \sum_{i=0}^{m} \alpha_i u_i = \left(\alpha_0 + \alpha_1 + \cdots + \alpha_m, \alpha_1, \tfrac{\alpha_2}{2}, \ldots, \tfrac{\alpha_m}{m}, 0, 0, \ldots\right),$$

where $\alpha_i \geq 0$ for each $i = 1, \ldots, m$, such that $\|y - a\| < \frac{1}{k}$. Without loss of generality, we can assume that $m \geq k$. It follows that

$$\left[y_0 - (\alpha_0 + \alpha_1 + \cdots + \alpha_m)\right]^2 + (y_1 - \alpha_1)^2 + (y_2 - \tfrac{\alpha_2}{2})^2 + \cdots + (y_m - \tfrac{\alpha_m}{m})^2 < \tfrac{1}{k^2},$$

and from this we see that $\left|y_i - \frac{\alpha_i}{i}\right| < \frac{1}{k}$ for each $i = 1, \ldots, m$. This implies

$$\alpha_i \geq i y_i - \tfrac{i}{k} \geq 1 - \tfrac{i}{k} = \tfrac{k-i}{k}$$

for each $i = 1, \ldots, m$. Since $\alpha_i \geq 0$ for each $i = 0, 1, \ldots, m$, it follows that

$$\alpha_0 + \alpha_1 + \cdots + \alpha_m \geq \tfrac{k-1}{k} + \tfrac{k-2}{k} + \cdots + \tfrac{1}{k} = \tfrac{k-1}{2}.$$

Since $|y_0 - (\alpha_0 + \alpha_1 + \cdots + \alpha_m)| \leq \frac{1}{k}$, we infer that $y_0 > \frac{k-1}{2} - \frac{1}{k}$. Since k is arbitrary, the latter is a contradiction. Hence $x \notin \mathcal{K} - \mathcal{K}$. ∎

We continue with the introduction of the strictly positive vectors.

Definition 2.16. *Let $\langle L, L' \rangle$ be a dual system and let $\mathcal{W}$ be a wedge of L. A nonzero vector $x \in \mathcal{W}$ is called $\mathcal{W}$-**strictly positive** (in symbols $x \gg_{\mathcal{W}} 0$) if for every nonzero linear functional $x' \in \mathcal{W}'$ we have $\langle x, x' \rangle > 0$.*

In other words, a vector $x \in \mathcal{W}$ is $\mathcal{W}$-strictly positive if and only if x acting as a linear functional on L' is $\mathcal{W}'$-strictly positive. When $\mathcal{W}$ has a nonempty interior, the strictly positive linear functionals are simply the interior points of $\mathcal{W}$.

Lemma 2.17. *Let $\langle L, L' \rangle$ be a dual system and let $\mathcal{W}$ be a wedge of L that has a nonempty interior* $\mathrm{Int}(\mathcal{W})$ *with respect to some consistent locally convex topology τ on L.*[3] *Then the $\mathcal{W}$-strictly positive vectors of L are precisely the vectors in* $\mathrm{Int}(\mathcal{W})$.

Proof. Let x be a nonzero vector of L. Assume first that $x \in \mathrm{Int}(\mathcal{W})$ and fix $x' \in \mathcal{W}'$ with $x' \neq 0$. Suppose also by way of contradiction that $\langle x, x' \rangle = 0$. Pick a circled τ-neighborhood V of zero with $x + V \subseteq \mathcal{W}$ and note that $\pm\langle v, x' \rangle = \langle x \pm v, x' \rangle \geq 0$ holds for all $v \in V$. This implies $\langle v, x' \rangle = 0$ for all $v \in V$ and so $x' = 0$, a contradiction. Therefore, $\langle x, x' \rangle > 0$ is true and so x is $\mathcal{W}$-strictly positive.

Now assume that x is $\mathcal{W}$-strictly positive. If $x \notin \mathrm{Int}(\mathcal{W})$, then there exist (by the Separation Theorem 8.16) some nonzero $x' \in L'$ and some constant c satisfying $\langle x, x' \rangle \leq c \leq \langle w, x' \rangle$ for all $w \in \mathcal{W}$. Since $\mathcal{W}$ is a wedge, it

[3] Recall that a locally convex topology τ on L is said to be **consistent**—or **compatible**—if the topological dual of (L, τ) coincides with L'; see Section 8.2.

follows that $c \leq 0$ and $\langle w, x' \rangle \geq 0$ for all $w \in \mathcal{W}$. This implies $x' \in \mathcal{W}'$ and so $\langle x, x' \rangle > 0$, contrary to $\langle x, x' \rangle \leq c \leq 0$. Hence, $x \in \text{Int}(\mathcal{W})$, and the proof is finished. ■

The existence of a $\mathcal{W}$-strictly positive linear functional does not imply that the wedge $\mathcal{W}$ has a nonempty interior. For instance, consider the dual system $\langle \ell_2, \ell_2 \rangle$ (under the standard duality $\langle x, y \rangle = \sum_{n=1}^{\infty} x_n y_n$) and let $\mathcal{W}$ be the standard cone of ℓ_2. It is not difficult to verify that $\mathcal{W}' = \mathcal{W}$ and from this we see that a nonzero vector $x = (x_1, x_2, \ldots) \in \mathcal{W}$ is $\mathcal{W}$-strictly positive if and only if $x_n > 0$ holds for each $n \in \mathbb{N}$. However, it can be shown that $\mathcal{W}$ has no interior point with respect to any consistent topology on ℓ_2.

For a dual system and a generating closed cone of L without strictly positive vectors see Exercise 10 at the end of the section. (See also Exercise 4 of Section 3.2.)

Exercises

1. Let $\langle L, L' \rangle$ be a dual system and let $\mathcal{W}$ be a wedge of L such that $\mathcal{W} - \mathcal{W}$ is $\sigma(L, L')$-dense in L. Show that the dual wedge $\mathcal{W}'$ is an Archimedean $\sigma(L', L)$-closed cone of L'.

2. Let $L = \mathbb{R}^{\Omega}$, where Ω is a nonempty set. Also let L' consist of all linear functionals $x' : L \to \mathbb{R}$ for which there exist "weights" $w_1, \ldots, w_n \in \mathbb{R}$ and points $\omega_1, \ldots, \omega_n \in \Omega$ (all depending on x') such that for each $x \in \mathbb{R}^{\Omega}$ we have
$$x'(x) = \sum_{i=1}^{n} w_i x(\omega_i) .$$
Establish the following.
   (a) The pair $\langle L, L' \rangle$ under the duality $\langle x, x' \rangle = \sum_{i=1}^{n} w_i x(\omega_i)$ is a dual system.
   (b) The vector space L' is the topological dual of $\mathbb{R}^{\Omega}$ (equipped with its product topology).
   (c) If $\mathcal{W} = \mathbb{R}^{\Omega}_{+}$, then $\mathcal{W}'$ consists of all $x' \in L'$ whose weights are nonnegative.
   (d) The strictly positive vectors of L consist of all functions $x \in \mathbb{R}^{\Omega}$ satisfying $x(\omega) > 0$ for each $\omega \in \Omega$.

3. In $\langle \mathbb{R}^3, \mathbb{R}^3 \rangle$ find the dual wedges of the following wedges:
   (a) $\mathcal{W} = \{(x, y, 0) \in \mathbb{R}^3 \colon\ x \geq 0 \text{ and } y \geq 0\}$,
   (b) $\mathcal{W} = \{\alpha(1, 1, 1) \colon\ \alpha \geq 0\}$,
   (c) $\mathcal{W} = \{(x, y, z) \in \mathbb{R}^3 \colon\ x + y + z \geq 0\}$.

4. Consider the dual system $\langle \mathbb{R}^3, \mathbb{R}^3 \rangle$ and the ice cream cone of $\mathbb{R}^3$ given by $\mathcal{K} = \{(x, y, z) \colon\ z \geq \sqrt{x^2 + y^2}\}$. Find the dual cone $\mathcal{K}'$.

5. Consider the dual system $\langle \ell_2, \ell_2 \rangle$ and consider the ice cream cone given by $\mathcal{K} = \{(x_1, x_2, \ldots) \in \ell_2 \colon x_1 \geq \sqrt{\sum_{n=2}^{\infty} x_n^2}\}$. Find the dual cone $\mathcal{K}'$.

6. Let Ω by a Hausdorff compact topological vector space. We shall denote by $\mathrm{ca}(\Omega)$ the vector space $\mathrm{ca}(\Sigma)$, where Σ is the σ-algebra of all **Borel sets** of Ω, i.e., Σ is the smallest σ-algebra of subsets of Ω that contains all open sets. (For the definition of $\mathrm{ca}(\Sigma)$ see Exercise 17 of Section 1.3.) Establish the following.
  (a) Under the bilinear function $\langle x, \mu \rangle = \int x \, d\mu$ the pair $\langle C(\Omega), \mathrm{ca}(\Omega) \rangle$ is a dual system.
  (b) If $\mathcal{K} = [C(\Omega)]^+$, the standard cone of $C(\Omega)$, then $\mathcal{K}'$ consists of all finite measures on the σ-algebra Σ.
  (c) A function $x \in C(\Omega)$ is strictly positive if and only if $x(\omega) > 0$ holds for each $\omega \in \Omega$.

7. Consider the dual system $\langle \ell_\infty, \ell_1 \rangle$, where as usual the duality is defined by $\langle x, y \rangle = \sum_{n=1}^{\infty} x_n y_n$ for all $x = (x_1, x_2, \ldots) \in \ell_\infty$ and $y = (y_1, y_2, \ldots) \in \ell_1$. Find the strictly positive vectors of ℓ_∞ and ℓ_1.

8. Let $\langle L, L' \rangle$ be a dual system and let $\mathcal{W}$ be a wedge of L such that its $\sigma(\mathcal{W}, \mathcal{W}')$-interior is nonempty. Then show that the interior of $\mathcal{W}$ is the same for all consistent locally convex topologies on L. [*Hint*: Use Lemma 2.17.]

9. This exercise clarifies the conclusion of Lemma 2.17 by exhibiting an example showing that $\mathcal{W}$-strictly positive vectors might exist without the wedge $\mathcal{W}$ having a nonempty interior. Consider the dual system $\langle \ell_2, \ell_2 \rangle$ under the standard duality $\langle x, y \rangle = \sum_{n=1}^{\infty} x_n y_n$ and let $\mathcal{W} = \ell_2^+$, the standard cone of ℓ_2. Establish the following.
  (a) The interior of $\mathcal{W}$ is empty with respect to any consistent topology on ℓ_2.
  (b) A nonzero vector $x = (x_1, x_2, \ldots) \in \mathcal{W}$ is $\mathcal{W}$-strictly positive if and only if $x_n > 0$ holds for all $n \in \mathbb{N}$.

10. This exercise presents an example of a cone without strictly positive vectors. Let Ω be an uncountable set and let $L = \ell_2(\Omega)$. That is, L consists of all functions $x \colon \Omega \to \mathbb{R}$ for which there exists an at most countable subset Ω_x of Ω such that $x(\omega) = 0$ for all $\omega \notin \Omega_x$ and with $\sum_{\omega \in \Omega_x} |x(\omega)|^2 < \infty$. Following the standard convention, we shall write $\sum_{\omega \in \Omega} |x(\omega)|^2$ instead of $\sum_{\omega \in \Omega_x} |x(\omega)|^2$. Establish the following.
  (a) The function $(x, y) = \sum_{\omega \in \Omega} x(\omega) y(\omega)$ defines an inner product on L under which L is a Hilbert space—that is, L under the norm $\|x\| = \sqrt{(x, x)} = \sum_{\omega \in \Omega} [x(\omega)]^2$ is a Banach space.
  (b) The norm dual of L coincides with L, i.e., every linear functional $f \in L'$ is of the form $f(x) = (x, y)$ for a unique $y \in L$ and $\|f\| = \|y\|$. In particular, $\langle L, L' \rangle$ is a dual system.
  (c) Let $\mathcal{W}$ denote the standard cone of ℓ_2. Then (with appropriate identifications) we have $\mathcal{W}' = \mathcal{W}$.
  (d) The cone $\mathcal{W}$ is closed and generating.
  (e) The vector space L has no $\mathcal{W}$-strictly positive vectors.

11. Show that in a separable Banach space[4] every closed and generating cone has a strictly positive vector. [*Hint*: Let $\mathcal{K}$ be a generating cone of a separable Banach space X. The separability of X implies the separability of $\mathcal{K}$ (why?). Let $\{x_1, x_2, \ldots\}$ be a dense subset of $\mathcal{K}$ consisting of nonzero vectors and let $x = \sum_{n=1}^{\infty} \frac{1}{2^n} \frac{x_n}{\|x_n\|} \in \mathcal{K}$. Verify that x is a $\mathcal{K}$-strictly positive vector—with respect, of course, to the dual system $\langle X, X' \rangle$, where X' is the norm dual of X.]

2.3. Normal cones

In order to create a useful theory of ordered topological vector spaces, we need to introduce some extra connection between order and topology. The most useful such connection is achieved by the notion of a normal cone that was introduced first by M. G. Krein [**62**] for Banach spaces. (Recall that a subset A of an ordered vector space is said to be *full* if for each pair $x, y \in A$ we have $[x, y] \subseteq A$.)

Definition 2.18 (M. G. Krein). *A cone $\mathcal{K}$ of a topological vector space (X, τ) is said to be τ-**normal** (or simply **normal** if there is no ambiguity about the topology) whenever τ has a base at zero consisting of $\mathcal{K}$-full sets.*

*We shall also say that the cone of an ordered topological space L is **normal** if L_+ is a normal cone.*

The proof of the following simple but useful result should be immediate from the preceding definition.

Lemma 2.19. *Every subcone of a normal cone is itself normal.*

The objective of this section is to present the basic properties of the normal cones. Before doing so, we need the following result.

Lemma 2.20. *In an ordered topological vector space the interior of a full set is likewise a full set.*

Proof. Let A be a full subset of an ordered topological vector space and let $x \leq y \leq z$ with $x, z \in \mathrm{Int}(A)$. Pick two neighborhoods V_x and V_z of zero such that $x + V_x \subseteq A$ and $z + V_z \subseteq A$. Let $V_y = V_x \cap V_z$. If $w \in V_y$, then from $x + w \in A$, $z + w \in A$ and $x + w \leq y + w \leq z + w$, we get $y + w \in A$. Hence, $y + V_y \subseteq A$, and this proves that $y \in \mathrm{Int}(A)$ or $[x, y] \subseteq \mathrm{Int}(A)$. ■

Since the full hull of a circled set is circled (why?), the interior of a circled neighborhood of zero is circled (why?), and the full hull of a convex set is convex (why?), we have the following consequence of the preceding lemma.

[4] Recall that a topological space is said to be **separable** if it has a countable dense subset.

Corollary 2.21. *If the cone of an ordered topological vector space is normal, then the topology has a base at zero consisting of open, circled, and full neighborhoods.*

Moreover, if the topology is also locally convex, then the topology has a base at zero consisting of open, circled, convex, and full neighborhoods.

An immediate consequence of the definition of a normal cone is that order bounded sets are topologically bounded.

Lemma 2.22. *If the cone of an ordered topological vector space is normal, then every order bounded set is topologically bounded.*[5]

Proof. It suffices to show that every order interval $[x, y]$ is τ-bounded. To see this, let V be a full neighborhood of zero and then pick some $\lambda > 0$ such that $\lambda x, \lambda y \in V$. Now notice that the fact that V is a full set implies $\lambda[x, y] = [\lambda x, \lambda y] \subseteq V$. ∎

The next result presents a useful characterization of normal cones.

Theorem 2.23. *For a cone $\mathcal{K}$ of a topological vector space (X, τ) the following statements are equivalent.*

(1) *The cone $\mathcal{K}$ is τ-normal.*

(2) *If two nets $\{y_\alpha\}$ and $\{x_\alpha\}$ of X (with the same index set) satisfy $0 \leq y_\alpha \leq x_\alpha$ for each α and $x_\alpha \xrightarrow{\tau} 0$, then $y_\alpha \xrightarrow{\tau} 0$.*[6]

Proof. Let $\mathcal{K}$ be a cone of a topological vector space (X, τ). It should be clear that (1) $\Longrightarrow$ (2). Next, we shall prove that (2) $\Longrightarrow$ (1). So, assume by way of contradiction that (2) is true and that $\mathcal{K}$ is not a τ-normal cone.

This means that there exists a neighborhood U of zero such that for each neighborhood V we have $[V] \nsubseteq U$, where $[V]$ is the full hull of the neighborhood V. Therefore for each neighborhood V of zero there exist $x_V, y_V \in V$ and some $z_V \in X$ satisfying $x_V \leq z_V \leq y_V$ and $z_V \notin U$.

It follows that $0 \leq z_V - x_V \leq y_V - x_V$ for each $V \in \mathcal{N}_0$, where $\mathcal{N}_0$ is the set consisting of all τ-neighborhoods of zero. If we consider $\mathcal{N}_0$ as a directed set under the direction $V \succeq W$ whenever $W \subseteq V$, then clearly the nets $\{x_V\}_{V \in \mathcal{N}_0}$ and $\{y_V\}_{V \in \mathcal{N}_0}$ satisfy $x_V \xrightarrow{\tau} 0$ and $y_V \xrightarrow{\tau} 0$. This implies $y_V - x_V \xrightarrow{\tau} 0$ and consequently from our hypothesis we get $z_V - x_V \xrightarrow{\tau} 0$. But then we have $z_V = (z_V - x_V) + x_V \xrightarrow{\tau} 0$, contrary to $z_V \notin U$ for each $V \in \mathcal{N}_0$. This contradiction establishes that the cone $\mathcal{K}$ is τ-normal. ∎

---

[5] Recall that a subset A of a topological vector space (X, τ) is said to be **topologically bounded** (or simply **τ-bounded**) if for each τ-neighborhood V of zero there exists some $\lambda > 0$ such that $\lambda A \subseteq V$.

[6] If τ is metrizable, then clearly the nets can be replaced by sequences.

Using the preceding theorem, one can establish the following result. (For a complete discussion see Exercise 9 at the end of the section.)

Corollary 2.24. *The closure of a normal cone in a Hausdorff topological vector space is likewise a normal cone.*

Recall that a mapping $f\colon L^+ \to M$, where L and M are ordered vector spaces, is said to be **monotone** if $0 \le x \le y$ in L implies $f(x) \le f(y)$ in M.

Theorem 2.25. *The cone of an ordered locally convex space (L, τ) is normal if and only if the locally convex topology τ is generated by a family of monotone τ-continuous seminorms.*

Proof. If the topology τ is generated by a family of monotone seminorms, then it is easy to see that statement (2) of Theorem 2.23 is true, and so L_+ is a τ-normal cone.

For the converse, assume that L_+ is τ-normal. Let $\{V_i\}_{i\in I}$ be a base at zero for τ consisting of convex, circled, and full neighborhoods of zero. If p_i is the Minkowski functional of V_i, i.e.,

$$p_i(x) = \inf\bigl\{\lambda > 0\colon\ x \in \lambda V_i\bigr\},$$

then it is easy to check that p_i is a monotone τ-continuous seminorm and that the family of seminorms $\{p_i\}_{i\in I}$ generates the topology τ. ■

The dual wedge of a normal cone is always generating.

Theorem 2.26. *If a cone $\mathcal{K}$ in a locally convex space L is normal, then its dual wedge $\mathcal{K}'$ is generating in L', i.e., $L' = \mathcal{K}' - \mathcal{K}'$.*

Proof. Let (L, τ) be an ordered locally convex topology with a τ-normal cone and let $f \in L'$. Pick a circled, convex, and full neighborhood V of zero such that $|f(x)| \le 1$ holds for all $x \in V$. Now let p denote the Minkowski functional of V, i.e.,

$$p(x) = \inf\bigl\{\lambda > 0\colon\ x \in \lambda V\bigr\}.$$

Since V is a full τ-neighborhood of zero, it follows that p is a monotone τ-continuous seminorm satisfying $|f(x)| \le p(x)$ for all $x \in L$. Let τ_1 denote the locally convex topology generated by p and note that $\tau_1 \subseteq \tau$.

Next consider the locally convex ordered vector space

$$M = (L, \tau_1) \times (L, \tau_1) \times \mathbb{R}$$

and its vector subspace $X = \bigl\{(x, -x, -f(x))\colon\ x \in L\bigr\}$. We claim that the convex set $A = V \times V \times (-1, 1)$ of M satisfies

$$\bigl[X + A - (0, 0, 5)\bigr] \cap \bigl(L_+ \times L_+ \times \mathbb{R}_+\bigr) = \emptyset\,. \tag{$\star$}$$

To see this, assume that $\big[X + A - (0,0,5)\big] \cap \big(L_+ \times L_+ \times \mathbb{R}_+\big) \neq \emptyset$. This means that there exist $x \in L$, and $y, z \in V$ and $-1 < \alpha < 1$ such that $x + y \geq 0$, $-x + z \geq 0$ (and so $y + z \geq y + x \geq 0$) and $-f(x) + \alpha - 5 \geq 0$. Using the monotonicity of the seminorm p, we obtain

$$\begin{aligned} p(x) &= p\big((x+y) - y\big) \leq p(x+y) + p(y) \\ &\leq p(y+z) + p(y) \leq p(y) + p(z) + p(y) \leq 3\,, \end{aligned}$$

and so $-f(x) \leq |f(x)| \leq p(x) \leq 3$. Hence,

$$0 \leq -f(x) + \alpha - 5 \leq 3 + 1 - 5 = -1 < 0\,,$$

which is impossible. This contradiction establishes the validity of ($\star$).

Since the point $(0,0,0)$ is an interior point of the set A (as a subset of M), it follows from the Separation Theorem 8.16 that there exists some nonzero $\phi \in M'$ such that $\phi \geq 0$ on $L_+ \times L_+ \times \mathbb{R}_+$ and $\phi \leq 0$ on $X + A - (0,0,5)$. Since X is a vector subspace, we easily infer that $\phi = 0$ on X or

$$\phi(x,0,0) - \phi(0,x,0) - \phi(0,0,1)f(x) = 0$$

for each $x \in L$. Notice that $\lambda = \phi(0,0,1) > 0$. To see this, note first that $\phi \geq 0$ on $L_+ \times L_+ \times \mathbb{R}_+$ implies $\lambda \geq 0$. If $\phi(0,0,1) = 0$, then it follows from $\phi(x,y,0) \leq 0$ for all $x, y \in V$ that $\phi(x,y,0) = 0$ for all $x, y \in V$, and from this we get that $\phi = 0$, a contradiction. Hence, $\phi(0,0,1) > 0$.

Finally, observe that $g(x) = \phi(x,0,0)$ and $h(x) = \phi(0,x,0)$ define positive p-continuous (and therefore positive τ-continuous) linear functionals on L. To finish the proof, notice that $f = \frac{1}{\lambda}g - \frac{1}{\lambda}h$. ■

From the preceding theorem and Lemma 2.5 we get the following.

Corollary 2.27. *If the cone of a locally convex ordered topological vector space L is normal and has an interior point, then $L^r = L^\sim = L'$.*

We shall close the section with two more useful properties of normal cones.

Lemma 2.28. *Assume that a Hausdorff ordered locally convex space (L,τ) has a τ-normal cone and that a net $\{x_\alpha\} \subseteq L_+$ is decreasing, i.e., $x_\alpha \downarrow$. Then $x_\alpha \xrightarrow{\tau} 0$ if and only if $x_\alpha \xrightarrow{\sigma(L,L')} 0$.*

Proof. Assume $x_\alpha \downarrow$ and that $x_\alpha \xrightarrow{\sigma(L,L')} 0$ holds in L. Also let V be a convex full τ-neighborhood of zero. Notice that 0 is in the $\sigma(L,L')$-closure of the set $\{x_\alpha\}$ and so 0 belongs to the $\sigma(L,L')$-closure of the convex set $\operatorname{co}\{x_\alpha\}$. Since the $\sigma(L,L')$- and τ-closures of any convex set coincide, it follows that 0 belongs also to the τ-closure of $\operatorname{co}\{x_\alpha\}$. In particular, we have $V \cap \operatorname{co}\{x_\alpha\} \neq \emptyset$. Pick indices $\alpha_1, \ldots, \alpha_n$ and positive constants $\lambda_1, \ldots, \lambda_n$ satisfying $\sum_{i=1}^n \lambda_i = 1$ and $\sum_{i=1}^n \lambda_i x_{\alpha_i} \in V$. Now note that if $\alpha \succeq \alpha_i$ holds for each i, then we have $0 \leq x_\alpha = \sum_{i=1}^n \lambda_i x_\alpha \leq \sum_{i=1}^n \lambda_i x_{\alpha_i} \in V$. Thus, if we pick an index α_0 satisfying $\alpha_0 \succeq \alpha_i$ for each $1 \leq i \leq n$, then $x_\alpha \in V$ for all $\alpha \succeq \alpha_0$ so that $x_\alpha \xrightarrow{\tau} 0$. ∎

Lemma 2.29. *Let $\langle L, L' \rangle$ be a dual system and let $\mathcal{K}$ be a cone of L. Then the dual wedge $\mathcal{K}'$ of $\mathcal{K}$ is generating in L' if and only if the cone $\mathcal{K}$ is weakly normal, i.e., $\sigma(L, L')$-normal.*

Proof. If $\mathcal{K}$ is $\sigma(L, L')$-normal, then (by Theorem 2.26) $\mathcal{K}'$ is generating. For the converse, assume that $\mathcal{K}'$ is generating and assume that two nets $\{x_\alpha\}$ and $\{y_\alpha\}$ of L satisfy $0 \leq_{\mathcal{K}} y_\alpha \leq_{\mathcal{K}} x_\alpha$ for each α and $x_\alpha \xrightarrow{\sigma(L,L')} 0$. From $0 \leq f(y_\alpha) \leq f(x_\alpha)$ and $f(x_\alpha) \to 0$ for each $f \in \mathcal{K}'$, we get $f(y_\alpha) \to 0$ for each $f \in \mathcal{K}'$. Since $L' = \mathcal{K}' - \mathcal{K}'$, we obtain $f(y_\alpha) \to 0$ for each $f \in L'$ so that $y_\alpha \xrightarrow{\sigma(L,L')} 0$. By Theorem 2.23 the cone $\mathcal{K}$ is $\sigma(L, L')$-normal. ∎

Exercises

1. Prove Lemma 2.19 and Corollary 2.21.
2. Let $\{(L_i, \tau_i)\}_{i \in I}$ be a family of ordered topological vectors spaces. If the cone L_i^+ is τ_i-normal for each $i \in I$, then show that the product cone $\prod_{i \in I} L_i^+$ is a normal cone of the product ordered topological vector space $\left(\prod_{i \in I} L_i, \prod_{i \in I} \tau_i\right)$.
3. Show that the standard cone of $C[0, 1]$ with respect to the sup norm is normal. Likewise show that standard cones of the $L_p(\mu)$-spaces are all normal.
4. Establish that the property of being a "normal cone" is a hereditary property. That is, a cone $\mathcal{K}$ of a topological vector space (X, τ) that lies in a vector subspace Y of X is a τ-normal cone of X if and only if it is a τ-normal cone of Y.
5. Show that for a cone $\mathcal{K}$ of a topological vector space (L, τ) the following statements are equivalent.
   (a) The cone $\mathcal{K}$ is τ-normal.
   (b) The collection of neighborhoods of zero
   $$\left\{V \in \mathcal{N}_0 \colon\ [0, x] \subseteq V \text{ for all } x \in V\right\},$$
   where $\mathcal{N}_0$ denotes the τ-neighborhood system at zero, is a neighborhood base for $\mathcal{N}_0$.
   (c) For each τ-neighborhood U of zero there is another τ-neighborhood V of zero such that for each $x \in V$ we have $[0, x] \subseteq U$.
6. If the cone of an ordered topological vector space is normal, then show that the full hull of a topologically bounded set is topologically bounded.
7. Assume that the cone of an ordered topological vector space (L, τ) is normal and that a net $\{x_\alpha\}$ of L satisfies $x_\alpha \uparrow$ in L. Show that the net

$\{x_\alpha\}$ is τ-convergent (i.e., there exists some $x \in L$ with $x_\alpha \xrightarrow[\alpha]{\tau} x$) if and only if there exists a subnet $\{x_{\alpha_\lambda}\}$ of the net $\{x_\alpha\}$ satisfying $x_{\alpha_\lambda} \xrightarrow[\lambda]{\tau} x$.

8. Let A be a full, convex, and absorbing subset of an arbitrary ordered vector space L and consider its Minkowski functional $p_A \colon L \to \mathbb{R}$ defined by $p_A(x) = \inf\{\lambda > 0\colon\ x \in \lambda A\}$. Show that p_A is a monotone sublinear functional. (This property was used in the proof of Theorem 2.25.)

9. Prove that the closure of a normal cone in a Hausdorff topological vector space is likewise a normal cone. [*Hint*: Assume that $\mathcal{K}$ is a normal cone in a Hausdorff topological vector space (X, τ) and let $\overline{\mathcal{K}}$ denote the τ-closure of $\mathcal{K}$. Clearly, $\overline{\mathcal{K}}$ is a wedge. To see that $\overline{\mathcal{K}}$ is a cone, let $x \in \overline{\mathcal{K}} \cap (-\overline{\mathcal{K}})$. Pick two nets $\{x_\alpha\}$ and $\{y_\lambda\}$ of K such that $x_\alpha \xrightarrow{\tau} x$ and $y_\lambda \xrightarrow{\tau} -x$. So, $x_\alpha + y_\lambda \xrightarrow[\alpha,\lambda]{\tau} x - x = 0$ and from $0 \leq x_\alpha \leq x_\alpha + y_\lambda$ and the τ-normality of the cone $\mathcal{K}$, we see that $x_\alpha \xrightarrow{\tau} 0$. Since τ is Hausdorff, we get $x = 0$ so that $\overline{\mathcal{K}}$ is a cone.

Now assume that two nets $\{y_\alpha\}$ and $\{x_\alpha\}$ of X (with the same index set) satisfy $0 \leq y_\alpha \leq x_\alpha$ for each α with respect to the cone $\overline{\mathcal{K}}$ and $x_\alpha \xrightarrow{\tau} 0$. Fix some τ-closed neighborhood U of zero and then select a $\mathcal{K}$-full τ-neighborhood V of zero satisfying $V \subseteq U$. By Lemma 2.20 the τ-interior $\operatorname{Int}(V)$ of V is an open full τ-neighborhood of zero. Pick some α_0 such that $x_\alpha \in \operatorname{Int}(V)$ for all $\alpha \succeq \alpha_0$. Next, fix some $\alpha \succeq \alpha_0$ and then pick two nets $\{b_\lambda\}$ and $\{c_\mu\}$ of $\mathcal{K}$ such that $b_\lambda \xrightarrow{\tau} y_\alpha$ and $c_\mu \xrightarrow{\tau} x_\alpha - y_\alpha$. From $b_\lambda + c_\mu \xrightarrow[\lambda,\mu]{\tau} x_\alpha$, it follows that there exist indices λ_0 and μ_0 such that $\lambda \succeq \lambda_0$ and $\mu \succeq \mu_0$ imply $b_\lambda + c_\mu \in \operatorname{Int}(V)$. Now since $0 \leq_{\mathcal{K}} b_\lambda \leq_{\mathcal{K}} b_\lambda + c_\mu$ holds for all λ and μ and from the fact that $\operatorname{Int}(V)$ is $\mathcal{K}$-full, we see that $b_\lambda \in \operatorname{Int}(V) \subseteq U$ for all $\lambda \succeq \lambda_0$. This implies $y_\alpha \in U$ for all $\alpha \succeq \alpha_0$. Thus, $y_\alpha \xrightarrow{\tau} 0$, and so (by Theorem 2.23) the cone $\overline{\mathcal{K}}$ is τ-normal.]

10. Let L be an ordered Hausdorff locally convex space and let $\{x_1, \ldots, x_n\}$ be a collection of linearly independent vectors of L. Consider the set

$$\mathcal{K} = \Big\{\sum_{i=1}^{n} \lambda_i x_i\colon\ \lambda_i \geq 0 \text{ for each } i = 1, \ldots, n\Big\}.$$

(a) Show that $\mathcal{K}$ is a normal cone. (See also Exercise 7 of Section 1.7.)

(b) According to Theorem 2.26 the wedge $\mathcal{K}'$ is generating in L'. If $f \in L'$, construct two linear functionals $g, h \in \mathcal{K}'$ such that $f = g - h$. [*Hint*: Assume that $f \in L'$. Let X be the linear span of $\{x_1, \ldots, x_n\}$ and then pick a closed vector subspace Y of L such that $L = X \oplus Y$. Write $f|_X = f_1 - f_2$, where $f_1, f_2 \in X_+^*$, and then consider the linear functionals $g, h \in L^*$ defined by $g = f_1$ on X and $g = f$ on Y, $h = f_2$ on X and $h = 0$ on Y. Now note that g and h are $\mathcal{K}$-positive continuous linear functionals on L satisfying $f = g - h$.]

11. Show that if the cone of a Hausdorff ordered topological vector space (L, τ) is τ-normal with a nonempty interior, then the topology τ is normable. [*Hint*: If u is an interior point of L_+, then the order interval $[-u, u]$ is a circled convex neighborhood of zero that is also τ-bounded since the cone L_+ is τ-normal.]

2.4. Positivity and continuity

It is remarkable that quite often positive operators are automatically continuous. This was first proven by M. G. Krein for positive linear functionals [**63**] and later generalized in several contexts by various authors; see for instance [**26, 70, 71, 89, 85**]. In this section, we shall present some general conditions according to which every positive operator is continuous.

We start our discussion with a useful result.

Lemma 2.30. *Let (L, τ) be an ordered topological vector space such that either*

(a) *its positive cone L_+ has a τ-interior point or else*

(b) *the positive cone L_+ is generating and τ-closed and the linear topology τ is completely metrizable.*

If $x_n \xrightarrow{\tau} x$ holds in L, then there exist a subsequence $\{u_n\}$ of $\{x_n\}$ and some $u \in L_+$ such that for each n we have

$$-\tfrac{1}{n}u \leq u_n - x \leq \tfrac{1}{n}u\,.$$

In particular, in this case, every τ-convergent sequence of L has an order bounded subsequence.

Proof. Assume $x_n \xrightarrow{\tau} 0$ in L. Suppose first that e is an interior point of L_+. In this case the conclusion follows from the fact that for each n the interval $\left[-\frac{1}{n}e, \frac{1}{n}e\right]$ is a neighborhood of zero; see Lemma 2.5.

Next, suppose that τ is a metrizable and complete linear topology. Choose a countable neighborhood base $\{U_n\}$ of zero consisting of τ-closed circled sets satisfying $U_{n+1} + U_{n+1} \subseteq U_n$ for each n. Let

$$V_n = U_n \cap L_+ - U_n \cap L_+\,.$$

By Theorem 2.10, the sets $\{V_n\}$ form a τ-neighborhood base at zero. So, there exists a subsequence $\{u_n\}$ of $\{x_n\}$ such that $u_n \in \frac{1}{n}V_n$ (or $nu_n \in V_n$) holds for each n. For each n pick $y_n, z_n \in U_n \cap L_+$ such that $nu_n = y_n - z_n$.

Now let $w_n = \sum_{i=1}^n y_i \in L_+$ and $v_n = \sum_{i=1}^n z_i \in L_+$. Clearly,

$$w_{n+p} - w_n \in U_{n+1} + U_{n+2} + \cdots + U_{n+p} \subseteq U_n$$

holds for each n. This implies that $\{w_n\}$ is a τ-Cauchy sequence and hence $\{w_n\}$ converges to some w. Since L_+ is closed, $w \in L_+$. Letting $p \to \infty$ in $w_{n+p} - w_n \in L_+$, we see that $w - w_n \in L_+$, and so $w \geq w_n \geq y_n$ for each n. Similarly, there exists some $v \in L_+$ such that $v_n \xrightarrow{\tau} v$ and $z_n \leq v$ hold for each n. If we let $u = w + v \in L_+$, then we have

$$-u \leq -v \leq -z_n \leq y_n - z_n = nu_n \leq y_n \leq w \leq u\,,$$

and so $-\frac{1}{n}u \leq u_n \leq \frac{1}{n}u$ holds for each n, as desired. ■

Corollary 2.31. *Let (L,τ) be a metrizable ordered topological vector space such that either*

(a) *the positive cone L_+ has a τ-interior point or else*

(b) *the positive cone L_+ is generating and τ-closed and the linear topology τ is completely metrizable.*

Then an operator $T\colon L \to X$, where X is an arbitrary topological vector space, is continuous if and only if $T\colon L_+ \to X$ is continuous at zero.

Proof. Assume that $T\colon L \to X$ is an operator such that $T\colon L_+ \to X$ is continuous at zero and let $x_n \to 0$ in L. It suffices to show that $\{x_n\}$ has a subsequence $\{u_n\}$ such that $Tu_n \to 0$ in X (why?).

So let $x_n \to 0$ in L. By Lemma 2.30 there exist a subsequence $\{u_n\}$ of $\{x_n\}$ and some $u \in L_+$ such that $-\frac{1}{n}u \le u_n \le \frac{1}{n}u$ holds for all n. Now let $v_n = u_n + \frac{1}{n}u$ and $w_n = \frac{1}{n}u$, and note that $u_n = v_n - w_n$. Clearly, $v_n, w_n \in L_+$ holds for each n, $v_n \to 0$ and $w_n \to 0$ in L. By our hypothesis we have $Tv_n \to 0$ and $Tw_n \to 0$ in X. But then $Tu_n = Tv_n - Tw_n \to 0$ also holds in X, as desired. ∎

The next result presents general conditions under which a positive operator is continuous.

Theorem 2.32. *Let L and M be two completely metrizable ordered topological vector spaces with closed cones. If the cone of L is generating, then every positive operator from L to M is continuous.*

Proof. Let $T\colon L \to M$ be a positive operator. To establish that T is continuous, it suffices to show that the operator T has a closed graph; see Theorem 8.38. So, assume $x_n \to 0$ in L and $Tx_n \to y$ in M. We must show that $y = 0$.

According to Lemma 2.30, there exist some $u \in L_+$ and a subsequence of $\{x_n\}$ which we shall denote by $\{x_n\}$ such that $-\frac{1}{n}u \le x_n \le \frac{1}{n}u$ holds for each n. But then the positivity of T implies $-\frac{1}{n}Tu \le Tx_n \le \frac{1}{n}Tu$ for each n. Using that M_+ is also closed, we conclude that $0 \le y \le 0$. That is, $y = 0$ and the proof is finished. ∎

Corollary 2.33. *Let L and M be two ordered Banach spaces*[7] *with closed cones. If the cone of L is also generating, then every positive operator $T\colon L \to M$ is continuous.*

In particular, positive operators between Banach lattices are continuous.

[7] As the name suggests and as we shall define in the next section, an *ordered Banach space* is a Banach space equipped with a cone.

Theorem 2.32 specialized to linear functionals yields the following result due to G. Ya. Lozanovsky [**89**].

Corollary 2.34 (Lozanovsky). *If the cone of a completely metrizable ordered vector space (in particular, of an ordered Banach space) L is closed and generating, then every positive linear functional on L is continuous. In particular, we have $L^r \subseteq L'$.*

Exercises

1. Let (L, τ) be a metrizable ordered topological vector space such that either
   (a) its positive cone L_+ has a τ-interior point or else
   (b) the positive cone L_+ is generating and τ-closed and the linear topology τ is completely metrizable.

   Show that every positive operator from L to any ordered topological vector space with a normal cone is continuous. [*Hint*: Take a look at the last part of the proof of Theorem 2.32.]

2. Show that for any given Riesz space L there is at most one (up to equivalence) lattice norm that makes L a Banach lattice. [*Hint*: Assume that a Riesz space L is a Banach lattice under the two lattice norms $\|\cdot\|_1$ and $\|\cdot\|_2$. Then, by Theorem 2.32, the identity operator $I\colon (L, \|\cdot\|_1) \to (L, \|\cdot\|_2)$ is a homeomorphism.]

3. Let L be the Riesz space of all eventually zero real sequences (with, of course, the pointwise ordering). Establish the following.
   (a) The sup norm on L, i.e., the norm $\|x\| = \sup_{n\in\mathbb{N}} |x_n|$, is a lattice norm—and so L is a normed Riesz space.
   (b) The cone of L is (norm) closed and generating.
   (c) If we define the function $f\colon L \to \mathbb{R}$ via the formula

   $$f(x_1, x_2, \ldots) = \sum_{n=1}^{\infty} n x_n ,$$

   then f is a strictly positive linear functional that is not norm continuous.
   (d) Why does the conclusion of part (c) not contradict Corollary 2.34?

4. Let L be the ordered vector space of all continuous real functions on $\mathbb{R}$ with compact support. That is, L is ordered pointwise and a continuous function $x\colon \mathbb{R} \to \mathbb{R}$ belongs to L if and only if there exists a closed interval $[a, b]$ such that $x(t) = 0$ for all $t \notin [a, b]$. Establish the following.
   (a) The ordered vector space L is a Riesz space without order units.
   (b) The sup norm $\|x\| = \sup_{t\in\mathbb{R}} |x(t)|$ is a lattice norm on L (and so L equipped with the sup norm is a normed Riesz space and has a closed and generating cone).

(c) The function $f\colon L \to \mathbb{R}$, defined by $f(x) = \int_{-\infty}^{\infty} x(t)\,dt$, is a strictly positive linear functional that fails to be norm continuous. (See also Exercise 10 of Section 1.5.)

(d) Why does the conclusion of part (c) not contradict Corollary 2.34?

2.5. Ordered Banach spaces

We start with the definition of an ordered Banach space.

Definition 2.35. *An ordered topological vector space whose topology is generated by a complete norm is called an* **ordered Banach space**. *That is, an ordered Banach space is simply a Banach space equipped with a cone.*[8]

Similarly, an **ordered normed space** is an ordered topological vector space whose topology is generated by a norm. In other words, an ordered normed space is a normed space equipped with a cone.

Definition 2.36. *A norm $\|\cdot\|$ on a vector lattice is said to be a* **lattice norm** *if $|x| \le |y|$ implies $\|x\| \le \|y\|$.*

A **normed vector lattice** (or a **normed Riesz space**) is a vector lattice equipped with a lattice norm. A complete normed vector lattice is called a **Banach lattice**. Thus, the Banach lattices are special examples of ordered Banach spaces. Examples of Banach lattices are provided by the $C(\Omega)$-spaces with their sup norms (where Ω are compact topological spaces) and the classical $L_p(\mu)$-spaces (where $1 \le p \le \infty$) with their L_p-norms.

We start our discussion by characterizing the closed cones of Banach spaces that are generating.

Theorem 2.37 (M. G. Krein–Šmulian)**.** *For a closed cone $\mathcal{K}$ of a Banach space X the following statements are equivalent.*

(1) *The cone $\mathcal{K}$ is generating.*

(2) *If U is the closed unit ball of X, i.e., $U = \{x \in X\colon \|x\| \le 1\}$, then the convex set $U \cap \mathcal{K} - U \cap \mathcal{K}$ is a neighborhood of zero. That is, there exists some $\alpha > 0$ such that $\alpha U \subseteq U \cap \mathcal{K} - U \cap \mathcal{K}$.*

(3) *There exists some constant $\rho > 0$ such that for each $x \in X$ there exist vectors $x_1, x_2 \in \mathcal{K}$ satisfying*

$$x = x_1 - x_2, \quad \|x_1\| \le \rho\|x\|, \quad \textit{and} \quad \|x_2\| \le \rho\|x\|\,.$$

Proof. (1) $\Longrightarrow$ (2) This follows immediately from Corollary 2.12.

[8] As mentioned before, some authors define an ordered Banach space to be a Banach space equipped with a closed cone.

(2)$\Longrightarrow$(3) Let $x \in X$ be a nonzero vector. Clearly, $\frac{\alpha}{\|x\|}x \in \alpha U$. Therefore, there exist $y_1, y_2 \in U \cap \mathcal{K}$ such that $\frac{\alpha}{\|x\|}x = y_1 - y_2$. This implies $x = \frac{1}{\alpha}\|x\|y_1 - \frac{1}{\alpha}\|x\|y_2$. Letting $x_i = \frac{1}{\alpha}\|x\|y_i \in \mathcal{K}$ for $i = 1, 2$ and $\rho = \frac{1}{\alpha} > 0$, we see that $x = x_1 - x_2$ and $\|x_i\| \leq \rho\|x\|$ for each i.

(3)$\Longrightarrow$(1) Obvious. ■

The following result characterizes normality in ordered normed spaces. (Keep in mind that the norm of an ordered normed space is said to be **monotone** if $0 \leq x \leq y$ implies $\|x\| \leq \|y\|$.)

Theorem 2.38. *For an ordered normed space L the following statements are equivalent.*

(1) *The cone L_+ is normal.*

(2) *The normed space L admits an equivalent monotone norm.*

(3) *There exists a constant $c > 0$ such that $a \leq x \leq b$ implies*
$$\|x\| \leq c \max\{\|a\|, \|b\|\}.$$

(4) *The full hull $[U] = (U + L_+) \cap (U - L_+)$ of the closed unit ball U of L is norm bounded.*

(5) *There is a constant $c > 0$ such that $0 \leq x \leq y$ implies $\|x\| \leq c\|y\|$.*

Proof. (1) $\Longrightarrow$ (2) Let U denote the closed unit ball of L and let W be a circled, convex, and full neighborhood of zero such that $W \subseteq U$. Then it is easy to see that the Minkowski functional $\|\cdot\|_W$ of W is a monotone norm that is equivalent to the norm of L.

(2) $\Longrightarrow$ (3) Let $|||\cdot|||$ be an equivalent monotone norm. Pick two constants $c_1 > 0$ and $c_2 > 0$ such that $c_1\|x\| \leq |||x||| \leq c_2\|x\|$ holds for all $x \in L$. Now assume that $a \leq x \leq b$ and put $m = \max\{\|a\|, \|b\|\}$. From $0 \leq x - a \leq b - a$ and the monotonicity of the norm $|||\cdot|||$ we get
$$\begin{aligned} |||x||| - |||a||| &\leq |||x - a||| \leq |||b - a||| \leq |||b||| + |||a||| \\ &\leq c_2[\|a\| + \|b\|] \leq 2c_2 m. \end{aligned}$$
Consequently, $c_1\|x\| \leq |||x||| \leq |||a||| + 2c_2 m \leq 3c_2 m$, from which it follows that $\|x\| \leq \frac{3c_2}{c_1}m = \frac{3c_2}{c_1}\max\{\|a\|, \|b\|\}$.

(3) $\Longrightarrow$ (4) If $x \in [U] = (U + L_+) \cap (U - L_+)$, pick $u_1, u_2 \in U$ and $y_1, y_2 \in L_+$ such that $x = u_1 + y_1 = u_2 - y_2$ and note that $u_1 \leq x \leq u_2$. This implies $\|x\| \leq c \max\{\|u_1\|, \|u_2\|\} \leq c$ so that $[U]$ is norm bounded.

(4) $\Longrightarrow$ (5) Pick some $c > 0$ such that $\|x\| \leq c$ for all $x \in [U]$. Now let $0 < x \leq y$. Clearly, $\frac{x}{\|y\|} = \frac{y}{\|y\|} - \left(\frac{y}{\|y\|} - \frac{x}{\|y\|}\right) \in U - L_+$ and moreover $\frac{x}{\|y\|} = 0 + \frac{x}{\|y\|} \in U + L_+$. Hence $\frac{x}{\|y\|} \in [U]$ and so $\left\|\frac{x}{\|y\|}\right\| \leq c$ or $\|x\| \leq c\|y\|$.

(5) $\Longrightarrow$ (1) If $0 \le x_n \le y_n$ holds in L with $y_n \to 0$, then it follows from $\|x_n\| \le c\|y_n\| \to 0$ that $x_n \to 0$. So, by Theorem 2.23, the cone L_+ is (norm) normal. ■

The preceding result in connection with Exercise 7 at the end of the section yields the following.

Lemma 2.39. *The cone of any normed Riesz space is normal, closed, and generating.*

If an ordered Banach space has a closed cone, then the normality of the cone is equivalent to the norm boundedness of its order intervals.

Theorem 2.40. *For an ordered Banach space L with a closed cone the following statements are equivalent.*

(1) *The cone L_+ is (norm) normal.*

(2) *The dual wedge L'_+ is generating in L'.*

(3) *The order intervals of L are weakly bounded.*

(4) *The order intervals of L are norm bounded.*

Proof. (1) $\Longrightarrow$ (2) If the cone L_+ is normal, then it follows from Theorem 2.26 that the dual wedge L'_+ is generating.

(2) $\Longrightarrow$ (3) Let $[0, x]$ be an order interval of L and let $f \in L'_+$. It follows from $f([0, x]) \subseteq [0, f(x)]$ that $f([0, x])$ is a bounded subset of $\mathbb{R}$. From this and the fact that the wedge L'_+ is generating, it should be clear that every order interval of L is weakly bounded.

(3) $\Longrightarrow$ (4) This is a consequence of the following well-known result: *A subset of L is norm bounded if and only if it is weakly bounded.*

(4) $\Longrightarrow$ (1) To see that L_+ is norm normal, assume by way of contradiction that there exist two sequences $\{x_n\}$ and $\{y_n\}$ of L_+ such that $0 \le x_n \le y_n$ holds for each n, $\|y_n\| \to 0$ and $\|x_n\| \not\to 0$. By passing to a subsequence, we can assume that there exists some $\epsilon > 0$ such that $\|x_n\| \ge \epsilon$ holds for each n. Also, replacing each x_n by $x_n/\|y_n\|$ and each y_n by $y_n/\|y_n\|$ (to justify this, notice that $y_n > 0$ holds true for all n), we can assume that $0 \le x_n \le y_n$ and $\|y_n\| = 1$ hold for each n and $\|x_n\| \to \infty$. By passing to one more subsequence of $\{x_n\}$, we can also suppose that $\|x_n\| \ge n2^n$ for all n.

Put $y = \sum_{n=1}^{\infty} \frac{1}{2^n} y_n$ and note that the closedness of the cone L_+ implies $y \in L_+$ and $0 \le \frac{1}{2^n} x_n \le y$ for all n. But then the inequality $\left\|\frac{1}{2^n} x_n\right\| \ge n$ for each n contradicts the norm boundedness of the order interval $[0, y]$. This contradiction establishes that L_+ must be a normal cone. ■

Surprisingly, a closed and generating cone in a Banach space need not be normal. The following example clarifies the situation.

Example 2.41. Let $L = C^1[0,1]$, the real vector space of all continuously differentiable functions on $[0,1]$, equipped with the cone L_+ of all nonnegative functions of L, i.e., $L_+ = \{x \in C^1[0,1]\colon\ x(t) \geq 0 \text{ for all } t \in [0,1]\}$. Under the norm

$$\|x\| = \|x\|_\infty + \|x'\|_\infty\,,$$

it is easy to see that L is a Banach space (where x' denotes the derivative of the function $x \in L$).

It should be clear that L_+ is a closed cone and that (since L_+ has order units) it is also generating—and it also satisfies (according to Example 1.58) the Riesz decomposition property. However, the cone L_+ is not normal. To see this, assume by way of contradiction that L_+ is (norm) normal. So, according to Theorem 2.38, there exists a constant $c > 0$ such that $0 \leq x \leq y$ implies $\|x\| \leq c\|y\|$. Now consider the functions $x_n \in C^1[0,1]$ defined by $x_n(t) = t^n$. Clearly, $0 \leq x_n \leq \mathbf{1}$ holds for all n and so $\|x_n\| \leq c$ for each n. Now note that $\|x_n\| = n+1$ and consequently $n+1 \leq c$ holds for each n, which is impossible. (For more about this example see Examples 1.58 and 2.72 and Exercises 8 and 9 of Section 2.8.) ■

Although a closed and generating cone in a Banach space need not be normal, T. Andô [**23**] has shown that its dual cone is always normal.

Theorem 2.42 (Andô). *If the cone of an ordered Banach space is closed and generating, then its dual cone is norm normal.*

Proof. Let L be an ordered Banach space whose cone L_+ is closed and generating. By Theorem 2.37 there exists some $\rho > 0$ such that for each $x \in L$ there exist $y, z \in L_+$ satisfying $x = y - z$, $\|y\| \leq \rho\|x\|$ and $\|z\| \leq \rho\|x\|$.

We shall finish the proof by establishing that

$$0 \leq y' \leq x' \ \text{ in } \ L' \implies \|y'\| \leq \rho\|x'\|\,.$$

To see this, assume that $0 \leq y' \leq x'$ holds in L' and fix $x \in L$ with $\|x\| \leq 1$. Write $x = y - z$, where $y, z \in L_+$ satisfy $\|y\| \leq \rho$ and $\|z\| \leq \rho$, and note that

$$y'(x) = y'(y) - y'(z) \leq y'(y) \leq x'(y) = \rho x'\big(\tfrac{y}{\rho}\big) \leq \rho\|x'\|\,.$$

Similarly, $-y'(x) = y'(-x) \leq \rho\|x'\|$ and so $|y'(x)| \leq \rho\|x'\|$ for all $\|x\| \leq 1$. Consequently, $\|y'\| = \sup_{\|x\|\leq 1} |y'(x)| \leq \rho\|x'\|$, as desired. ■

As a consequence of the preceding result we have the following companion to Theorem 2.40 that is due to M. G. Krein [**62**].

Corollary 2.43 (M. G. Krein). *For an ordered norm space L whose cone L_+ satisfies $\overline{L_+ - L_+} = L$ we have the following.*

(a) *The cone L_+ is norm normal if and only if the dual cone L'_+ is generating in L'.*

(b) *If L is also a reflexive Banach space, then the cone L_+ is generating in L if and only if the dual cone L'_+ is norm normal.*

Proof. Notice that the condition $\overline{L_+ - L_+} = L$ is equivalent to asserting that the (norm closed) wedge L'_+ is indeed a cone of the norm dual L'.

(a) If the cone L_+ is normal, then (in view of Theorem 2.26) the (norm) closed cone L'_+ is generating. For the converse, assume that the closed cone L'_+ is generating. So, by Theorem 2.42, the cone L''_+ of the double dual L'' is a normal cone. Now the result follows from the fact that L_+ can be viewed as a subcone of L''_+; see also Lemma 2.19.

(b) In this case the natural embedding of L to L'' is a surjective norm preserving order isomorphism. By part (a), the cone L'_+ is norm normal if and only if its norm dual cone $L''_+ = L_+$ is generating in $L'' = L$. ■

There is another class of normal cones that is associated with certain "order continuity" properties. Recall that the classical Beppo Levi theorem from the theory of integration states the following.

- If a sequence of functions $\{f_n\} \subseteq L_1(\mu)$ satisfies $0 \le f_n \le f_{n+1}$ for each n and $\sup_n \|f_n\|_1 < \infty$, then there exists a function $f \in L_1(\mu)$ such that $\|f_n - f\|_1 \to 0$ (and, of course, $f_n \uparrow f$ also holds in $L_1(\mu)$).

Inspired from this result, we give the following definition.

Definition 2.44. *If $\mathcal{K}$ is a closed cone in a Banach space L, then we say*

(1) *the cone $\mathcal{K}$ satisfies the* ***strong Levi property*** *(or the ordered Banach space $(L, \mathcal{K})$ has the* ***strong Levi property****) if it follows from $0 \le x_n \uparrow$ in L and $\sup_n \|x_n\| < \infty$ that the sequence $\{x_n\}$ is norm convergent in L and*

(2) *the cone $\mathcal{K}$ satisfies the* ***Levi property*** *(or the ordered Banach space $(L, \mathcal{K})$ has the* ***Levi property****) if it follows from $0 \le x_n \uparrow \le x$ in L that the sequence $\{x_n\}$ is norm convergent in L.*

The next result presents a connection between the Levi properties and normal cones.

Theorem 2.45. *For a closed cone $\mathcal{K}$ of a Banach space L consider the following statements.*

(1) *The cone $\mathcal{K}$ has the strong Levi property.*

(2) *The cone $\mathcal{K}$ has the Levi property.*

(3) *The cone $\mathcal{K}$ is norm normal.*

Then (1) $\Longrightarrow$ (2) $\Longrightarrow$ (3) *and no other implication is true.*

Moreover, if L is reflexive, then the three statements are equivalent.

Proof. We shall show first that (1) $\Longrightarrow$ (3) and (2) $\Longrightarrow$ (3) simultaneously. To this end, assume that the cone $\mathcal{K}$ is not norm normal. This implies the existence of two sequences $\{x_n\}$ and $\{y_n\}$ of L satisfying $0 \le x_n \le y_n$ and $\|x_n\| > 2^n\|y_n\|$ for each n. Letting $u_n = \frac{x_n}{\|x_n\|}$ and $v_n = \frac{y_n}{2^n\|y_n\|}$, we see that $0 \le u_n \le v_n$ holds for each n and that $\sum_{n=1}^{\infty}\|v_n\| = 1$. Now let $v = \sum_{n=1}^{\infty} v_n$ and define the sequence $\{w_n\}$ by letting $w_1 = v_1$ and

$$w_n = \begin{cases} v_1 + v_2 + \cdots + v_{2m} & \text{if } n = 2m\,, \\ v_1 + v_2 + \cdots + v_{2m} + u_{2m+1} & \text{if } n = 2m+1\,. \end{cases}$$

It should be clear that $\|w_n\| \le 2$ and $\|w_{2n+1} - w_{2n}\| = \|u_{2n+1}\| = 1$ for each n and that $0 \le w_n \uparrow \le v$. These show that neither (1) nor (2) is true, and the desired implications follows.

Next, we shall establish that (1) $\Longrightarrow$ (2). To this end, let $0 \le x_n \uparrow \le x$. Since (as shown above) the validity of (1) implies the norm normality of the cone, which in turn implies the norm boundedness of the order intervals, it follows that $\{x_n\}$ is a norm bounded sequence. Consequently, $\{x_n\}$ is norm convergent.

The Banach lattice c_0 of all real sequences that converge to zero (with the pointwise ordering and the sup norm) has the Levi property and a normal cone but fails to satisfy the strong Levi property; see Exercise 13 at the end of the section. On the other hand, $C[0,1]$ (as a Banach lattice) has a normal cone but it does not satisfy the Levi property—note that if $x_n(t) = 1 - t^n$, then $0 \le x_n \uparrow \le \mathbf{1}$ while $\{x_n\}$ is not norm convergent.

Now suppose that L is reflexive and that the cone $\mathcal{K}$ is norm normal. Also, assume that a sequence $\{x_n\}$ of L satisfies $0 \le x_n \uparrow$ and $C = \sup_n \|x_n\| < \infty$. If $f \in L'_+$, then it follows from

$$f(x_n) \le f(x_{n+1}) \le C\|f\|$$

that the limit $\lim f(x_n)$ exists in $\mathbb{R}$. Since the normality of the cone implies $L' = L'_+ - L'_+$, we easily infer that $F(f) = \lim f(x_n)$ exists in $\mathbb{R}$ for each $f \in L'$. Clearly, F defines a linear functional on L', and from $|F(f)| \le C\|f\|$ we see that $F \in L''$. Since L is reflexive, this means that there exists some $x \in L$ such that $F(f) = f(x)$ for all $f \in L'$. But then we have $f(x_n) \to f(x)$ for all $f \in L'$, i.e., $x_n \xrightarrow{w} x$. It follows that $x_n \uparrow x$ in L and a glance at Lemma 2.28 of Section 2.3 yields $\|x_n - x\| \to 0$. ■

In other words, for closed cones in reflexive Banach lattices the Levi properties characterize their (norm) normality. For more characterizations of normal cones in ordered Banach spaces see Theorems 2.60 and 2.63.

We shall close the section with two characterizations (both due to T. Andô [**23**]) of the Riesz decomposition property in ordered Banach spaces. The first one is in terms of an approximation property.

Theorem 2.46 (Andô). *A closed and normal cone $\mathcal{K}$ in a Banach space X has the Riesz decomposition property if and only if it satisfies the following approximate interpolation property:*

- ($\bullet$) *if two nonempty subsets A and B of X each of which has at most three vectors satisfy $A \geq_{\mathcal{K}} B$, then for each $\epsilon > 0$ there exist $x_\epsilon \in X$ and $y_\epsilon \in \mathcal{K}$ satisfying $\|y_\epsilon\| < \epsilon$, $A \geq_{\mathcal{K}} x_\epsilon - y_\epsilon$ and $x_\epsilon \geq_{\mathcal{K}} B$.*

If $\mathcal{K}$ is also generating, then the Riesz decomposition property of $\mathcal{K}$ is equivalent to the statement:

- ($\bullet\bullet$) *if two nonempty subsets A and B of X each of which has at most three vectors satisfy $A \geq_{\mathcal{K}} B >_{\mathcal{K}} 0$, then for each $\epsilon > 0$ there exist $x_\epsilon \in X$ and $y_\epsilon \in \mathcal{K}$ satisfying $\|y_\epsilon\| < \epsilon$, $A \geq_{\mathcal{K}} x_\epsilon - y_\epsilon$ and $x_\epsilon \geq_{\mathcal{K}} B$.*

Proof. Let $\mathcal{K}$ be a closed and normal cone in a Banach space X. We shall denote the order induced by the cone $\mathcal{K}$ by $\geq$, i.e., $\geq = \geq_{\mathcal{K}}$. By Theorem 2.38 there exists some $c > 0$ such that $a \leq x \leq b$ implies $\|x\| \leq c \max\{\|a\|, \|b\|\}$.

If $\mathcal{K}$ has the Riesz decomposition property and two finite nonempty sets satisfy $A \geq B$, then pick some $x \in X$ with $A \geq x \geq B$ and for each $\epsilon > 0$ let $x_\epsilon = x$ and $y_\epsilon = 0$. For the converse, assume that condition ($\bullet$) is satisfied and suppose that $A = \{a_1, a_2\} \geq B = \{b_1, b_2\}$ holds in X. For each n let $\epsilon_n = \frac{1}{2^n}$.

Start by choosing $x_1 \in X$ and $y_1 \in \mathcal{K}$ satisfying $\|y_1\| < \epsilon_1$, $A \geq x_1 - y_1$ and $x_1 \geq B$. Clearly, $A \cup \{x_1\} \geq B \cup \{x_1 - y_1\}$. By our condition, there exist $x_2 \in X$ and $y_2 \in \mathcal{K}$ satisfying $\|y_2\| < \epsilon_2$, $A \cup \{x_1\} \geq x_2 - y_2$ and $x_2 \geq B \cup \{x_1 - y_1\}$. Again, note that $A \cup \{x_2\} \geq B \cup \{x_2 - y_2\}$. Now using an inductive argument, we see that there exists a sequence $\{x_n\}$ of X and a sequence $\{y_n\}$ of $\mathcal{K}$ such that for each n we have

$$\|y_n\| < \epsilon_n, \quad A \cup \{x_n\} \geq x_{n+1} - y_{n+1}, \quad \text{and} \quad x_{n+1} \geq B \cup \{x_n - y_n\}.$$

In particular, we have $-y_n \leq x_{n+1} - x_n \leq y_{n+1}$ and from this, it follows that $\|x_{n+1} - x_n\| \leq c \max\{\|y_n\|, \|y_{n+1}\|\} < c\epsilon_n = \frac{c}{2^n}$ holds for each n. The latter implies that $\{x_n\}$ is a Cauchy sequence, and so $x_n \to x$ holds in X. Given that we have $y_n \to 0$, $A \geq x_n - y_n$ and $x_n \geq B$ for each n, it follows from the closedness of the cone $\mathcal{K}$ that $A \geq x \geq B$. This (combined with

Lemma 1.53) shows that the cone $\mathcal{K}$ has the interpolation property and hence it also has the Riesz decomposition property.

The second part follows immediately from the first part and Exercise 7 of Section 1.8. ■

The second characterization of the Riesz decomposition property is in terms of duality properties.

Theorem 2.47 (Andô). *For an ordered Banach space L with a closed, generating and norm normal cone, the following statements are equivalent.*

(1) *L has the Riesz decomposition property.*

(2) *L' is a Riesz space.*

(3) *L' has the Riesz decomposition property.*

Proof. As usual, we shall denote by U and U' the closed unit balls of L and L', respectively, i.e.,

$$U = \{x \in L\colon\ \|x\| \le 1\} \quad \text{and} \quad U' = \{x' \in L'\colon\ \|x'\| \le 1\}.$$

Since the cone L_+ is closed, generating and normal, the dual wedge L'_+ is also a norm closed, generating and normal cone; see Theorems 2.26 and 2.42. So, by Theorems 2.37 and 2.38, there exists a constant $\kappa > 0$ such that the following hold.

($\star$) *For every $u \in L$ there exist $u_1, u_2 \in L_+$ satisfying $u = u_1 - u_2$ and $\|u_i\| \le \kappa\|u\|$ for $i = 1, 2$.*

($\star\star$) *If $0 \le u \le v$ holds in L, then $\|u\| \le \kappa\|v\|$.*

(1) $\Longrightarrow$ (2) By Theorem 1.59 we know that $L^\sim$ is a Riesz space, and so $L^\sim = L^r$. Also, by Corollary 2.34 we have $L^r \subseteq L'$. Now the normality of the cone L_+ implies that the order intervals of L are norm bounded. This implies that every continuous linear functional is order bounded, i.e., $L' \subseteq L^\sim$. Therefore, in this case, $L' = L^\sim = L^r$, and so L' is a Riesz space.

(2) $\Longrightarrow$ (3) Obvious.

(3) $\Longrightarrow$ (1) Start by observing that since L'_+ is a norm closed, generating and normal cone, the proof of the implication (1) $\Longrightarrow$ (2) above shows that $L'' = (L')^\sim = (L')^r$ and that L'' is a Riesz space. In particular, L'' has the interpolation property.

For our proof here, we need to introduce some notation. For each non-empty subset S of L we let

$$\begin{aligned} S^{\ge 1} &= \{x' \in L'_+\colon\ x'(s) \ge 1 \text{ for all } s \in S\} \text{ and} \\ S^{\le 1} &= \{x' \in L'_+\colon\ x'(s) \le 1 \text{ for all } s \in S\}. \end{aligned}$$

Clearly, the (possibly empty) sets $S^{\geq 1}$ and $S^{\leq 1}$ are both $\sigma(L', L)$-closed and convex subsets of L'. Moreover, if S is any finite subset of $L_+ \setminus \{0\}$, then $S^{\geq 1}$ and $S^{\leq 1}$ are both nonempty.

Since the cone L_+ is generating, in order to establish that L has the Riesz decomposition property, it suffices (according to Theorem 2.46) to show that if two subsets $\{a_1, a_2, a_3\}$ and $\{b_1, b_2, b_3\}$ of L satisfy $a_i \geq b_j > 0$ for all i and j, then for each $\epsilon > 0$ there exist two vectors $x \in L$ and $y \in L_+$ satisfying $\|y\| < \epsilon$, $a_i \geq x - y$ and $x \geq b_j$ for all i and j. So, assume that $a_i \geq b_j > 0$ holds for all $i, j = 1, 2, 3$.

Fix $\eta > 0$ and $\gamma > 0$ and define the two nonempty convex sets

$$A = \operatorname{co} \bigcup_{i=1}^{3} \left[(a_i - L_+)^{\leq 1} \cap \gamma U' \right] \quad \text{and} \quad B = (1+\eta) \operatorname{co} \bigcup_{i=1}^{3} \{b_i\}^{\geq 1}.$$

We claim that A and B are disjoint, i.e., $A \cap B = \emptyset$. To see this, assume by way of contradiction that there exists some $x' \in L'$ such that $x' \in A \cap B$. Consequently, we can write $x' = \sum_{i=1}^{3} \alpha_i f_i = (1+\eta) \sum_{i=1}^{3} \beta_i g_i$, where we have $f_i \in (a_i - L_+)^{\leq 1}$, $g_i \in \{b_i\}^{\geq 1}$, $0 \leq \alpha_i \leq 1$, and $0 \leq \beta_i \leq 1$ for all $i = 1, 2, 3$ and $\sum_{i=1}^{3} \alpha_i = \sum_{i=1}^{3} \beta_i = 1$. Next, use the interpolation property of L'' to pick some $x'' \in L''$ satisfying $b_i \leq x'' \leq a_j$ in L'' for all i and j, and note that

$$\begin{aligned} 1+\eta &\leq (1+\eta) \sum_{i=1}^{3} \beta_i g_i(b_i) \leq (1+\eta) \sum_{i=1}^{3} \beta_i g_i(x'') \\ &= x''\Big((1+\eta) \sum_{i=1}^{3} \beta_i g_i \Big) = x''\Big(\sum_{i=1}^{3} \alpha_i f_i \Big) \\ &= \sum_{i=1}^{3} \alpha_i x''(f_i) \leq \sum_{i=1}^{3} \alpha_i a_i(f_i) \leq \sum_{i=1}^{3} \alpha_i = 1, \end{aligned}$$

which is impossible. This contradiction shows that $A \cap B = \emptyset$.

Since U' is $\sigma(L', L)$-compact and each $(a_i - L_+)^{\leq 1}$ is $\sigma(L', L)$-closed, it follows that the nonempty convex set $(a_i - L_+)^{\leq 1} \cap \gamma U'$ is $\sigma(L', L)$-compact for each i. This implies that A is a nonempty $\sigma(L', L)$-compact and convex subset of L'.

Next, we claim that B is a $\sigma(L', L)$-closed set. For this, it suffices to show that $\left[\operatorname{co} \bigcup_{i=1}^{3} \{b_i\}^{\geq 1}\right] \cap \rho U'$ is $\sigma(L', L)$-closed for each $\rho > 0$. (See Corollary 8.47.) Therefore, assume that a net $\{h_\lambda\}$, where $h_\lambda = \sum_{i=1}^{3} \alpha_\lambda^i f_\lambda^i$, $0 \leq \alpha_\lambda^i \leq 1$ and f_λ^i is in $\{b_i\}^{\geq 1}$ for each i and λ and $\sum_{j=1}^{3} \alpha_\lambda^j = 1$ for each λ, satisfies $\|h_\lambda\| \leq \rho$ and $h_\lambda \xrightarrow{\sigma(L', L)} h$ in L'. Clearly, $h \in \rho U'$. By property $(\star\star)$ at the beginning of the proof, for each i the net $\{\alpha_\lambda^i f_\lambda^i\}$ is norm bounded. Thus, by passing to a subnet, we can assume that for each i we have $\alpha_\lambda^i \xrightarrow[\lambda]{} \alpha^i$

in $\mathbb{R}$ and $\alpha^i_\lambda f^i_\lambda \xrightarrow{} f^i$ in L'. Clearly, $\alpha^i \geq 0$ for each i and $\sum_{i=1}^3 \alpha^i = 1$. If $\alpha^i > 0$, then $f^i_\lambda \xrightarrow{} \frac{1}{\alpha^i} f^i = g^i \in \{b_i\}^{\geq 1}$. If each $\alpha^i > 0$, then it should be clear that $h = \sum_{i=1}^3 \alpha^i g^i \in \big[\text{co} \bigcup_{i=1}^3 \{b_i\}^{\geq 1}\big] \cap \rho U'$. If some $\alpha^i = 0$ and we let $I = \{i\colon\ \alpha^i \neq 0\}$, then $I \neq \emptyset$, $\sum_{i\in I} \alpha^i = 1$ and $h = \sum_{i \in I} \alpha^i g^i + g$, where $g = \sum_{i \notin I} f^i \geq 0$. Consequently, we have $h = \sum_{i\in I} \alpha^i (g + g^i)$ and clearly $g + g^i \in \{b_i\}^{\geq 1}$ for each $i \in I$. This implies that $h \in \big[\text{co} \bigcup_{i=1}^3 \{b_i\}^{\geq 1}\big] \cap \rho U'$ holds true in this case too and the $\sigma(L', L)$-closedness of B follows.

Now, according to the Strong Separating Hyperplane Theorem 8.17, there exists some vector $c \in L$ such that

$$g(c) < 1 \leq f(c) \ \text{ holds true for all } g \in A \text{ and all } f \in B\,. \tag{†}$$

We claim that for each $i = 1, 2, 3$ we have

$$(1+\eta)c \geq b_i\,.$$

To see this, note first that $c \in L_+$. This follows immediately from (†) and the fact that B is a comprehensive from above set; see Lemma 1.32. To see that the vector $z = (1+\eta)c$ satisfies $z \geq b_i$ for each i, assume by way of contradiction that $z \geq b_i$ is not true for some i, i.e., $z \notin b_i + L_+$ for some i. Using Theorem 8.17 once more, we see that there exist some nonzero $\theta \in L'$ and some $\alpha \in \mathbb{R}$ such that $\theta(z) < \alpha \leq \theta(b_i + x)$ holds for all $x \in L_+$. In particular, the latter implies that $\theta \in L'_+$. Notice that $\theta(b_i) > 0$. Otherwise, $\theta(b_i) = 0$ yields $\theta(z) < \alpha \leq \theta(b_i + 0) = 0$, which contradicts $z \in L_+$. Replacing θ by $\frac{\theta}{\theta(b_i)}$, we can assume that $\theta(b_i) = 1$ and thus $\theta \in \{b_i\}^{\geq 1}$. But then $(1+\eta)\theta \in B$, and so from (†) we get $1 \leq (1+\eta)\theta(c) = \theta(z) < \alpha \leq \theta(b_i + 0) = 1$, which is impossible. Hence $(1+\eta)c \geq b_i$ holds true for each $i = 1, 2, 3$.

Next, we assert that for each $i = 1, 2, 3$ we have

$$c \in \big(a_i - L_+ + \tfrac{1}{\gamma} U\big)^-,$$

where the bar denotes norm closure. To see this, assume by way of contradiction that $c \notin \big(a_i - L_+ + \frac{1}{\gamma}U\big)^-$ holds for some i. Using the Separation Theorem 8.17 once more, we see that there exist some nonzero $g \in L'$ and some $\alpha \in \mathbb{R}$ satisfying

$$g\big(a_i - x + \tfrac{1}{\gamma} u\big) \leq \alpha < g(c) \ \text{ for all } x \in L_+ \text{ and all } u \in U\,.$$

This implies $g \in L'_+$, and letting $x = a_i$, we get $\frac{1}{\gamma} g(u) \leq \alpha$ for all $u \in U$. This implies $\alpha > 0$ and so replacing g by $\frac{g}{\alpha}$, we can assume that $\alpha = 1$. From $g(\frac{1}{\gamma}u) \leq 1$ for all $u \in U$, it follows that $\|g\| = \sup_{u\in U} g(u) \leq \gamma$, and so $g \in \gamma U'$. Moreover, notice that from $g(a_i - x) = g(a_i - x + \frac{1}{\gamma}0) \leq 1$ for all $x \in L_+$, we infer that $g \in (a_i - L_+)^{\leq 1}$ and so $g \in (a_i - L_+)^{\leq 1} \cap \gamma U'$.

This implies $g \in A$ and consequently $1 < g(c) < 1$ must be the case, which is absurd. This contradiction establishes the validity of our last claim.

Now from $(c + \frac{1}{\gamma}U) \cap (a_i - L_+ + \frac{1}{\gamma}U) \neq \emptyset$, it follows that for each i there exist vectors $u_1^i, u_2^i \in U$ and $x_i \in L_+$ such that $c + \frac{1}{\gamma}u_1^i = a_i - x_i + \frac{1}{\gamma}u_2^i$. Letting $c_i = \frac{1}{\gamma}u_1^i - \frac{1}{\gamma}u_2^i$, we see that $c + c_i \le a_i$ and $\|c_i\| \le \frac{2}{\gamma}$. Using property $(\star)$ at the beginning of the proof, we can write $c_i = c_i^1 - c_i^2$ with $c_i^1, c_i^2 \in L_+$ and $\|c_i^j\| \le \kappa\|c_i\| \le \frac{2\kappa}{\gamma}$ for each $j = 1, 2$.

Next, let $x = (1+\eta)c$ and $y = \eta c + \sum_{i=1}^3 c_i^2$. Clearly, $y \in L_+$ and for each $i = 1, 2, 3$ we have (as shown above) $x \ge b_i$ and moreover

$$x - y = c - \sum_{i=1}^{3} c_i^2 \le c - c_i^2 = c + c_i - c_i^1 \le a_i - c_i^1 \le a_i\,.$$

From $c + c_1 \le a_1$, we get $0 \le c \le a_1 - c_1$, and so property $(\star\star)$ at the beginning of the proof yields $\|c\| \le \kappa\|a_1 - c_1\| \le \kappa(\|a_1\| + \frac{2}{\gamma})$. Consequently,

$$\|y\| \le \eta\|c\| + \sum_{i=1}^{3} \|c_i^2\| \le \eta\kappa\big(\|a_1\| + \tfrac{2}{\gamma}\big) + \tfrac{6\kappa}{\gamma}\,. \tag{$\dagger\dagger$}$$

Finally, if $\epsilon > 0$ is given, then we can use $(\dagger\dagger)$ to choose some $\eta > 0$ and $\gamma > 0$ such that $\|y\| < \epsilon$, and the proof is finished. ∎

We list below three interesting consequences of Theorem 2.47.

Corollary 2.48. *A closed, generating and normal cone of a reflexive Banach space has the Riesz decomposition property if and only if it is a lattice cone.*

Corollary 2.49. *Assume that an ordered Banach space L with a closed and generating cone has the Riesz decomposition property. Then the norm dual L' is a Riesz space if and only if the cone L_+ is norm normal.*

Proof. If L_+ is also a norm normal cone, then from Andô's Theorem 2.47, it follows that L' is a Riesz space. Conversely, if L' is a Riesz space, then the cone L'_+ is generating. But then Corollary 2.43 shows that the cone L_+ is norm normal. ∎

Corollary 2.50. *For an ordered Banach space L with a closed and generating (but not necessarily normal) cone that satisfies the Riesz decomposition property, we have the following.*

(1) $L^\sim = L^r = L'_+ - L'_+$.

(2) *The order dual $L^\sim$ is a lattice-subspace of L'.*

(3) *L' has the Riesz decomposition property.*

Proof. (1) and (2) According to Corollary 2.34 we have $L^r \subseteq L'$. Since L has the Riesz decomposition property and L_+ is generating, it follows from Theorem 1.59 that $L^\sim$ is a Riesz space. Moreover, according to Theorem 2.32, every positive linear functional on L is continuous. These remarks easily imply that $L^\sim = L^r = L'_+ - L'_+$ and that $L^\sim$ is a lattice-subspace of L'.

(3) The Riesz decomposition property of L' follows immediately from Exercise 4 of Section 1.8. ■

Exercises

1. Assume that L is an ordered Banach space with a closed and generating cone. If $x_n \to x$ holds in L, then show that there exist a subsequence $\{u_n\}$ of $\{x_n\}$, $x, y \in L_+$ and two sequences $\{y_n\}$ and $\{z_n\}$ of L_+ satisfying
   (a) $x = y - z$,
   (b) $y_n \to y$ and $z_n \to z$, and
   (c) $u_n = y_n - z_n$ for each n.
   [*Hint*: By Theorem 2.37 there exist some $\rho > 0$ and two sequences $\{a_n\}$ and $\{b_n\}$ of L_+ satisfying $x_n - x = a_n - b_n$, $\|a_n\| \leq \rho\|x_n - x\|$, and $\|b_n\| \leq \rho\|x_n - x\|$ for each n. Therefore, $a_n \to 0$ and $b_n \to 0$. Now write $x = y - z$ with y, $z \in L_+$, and then let $y_n = y + a_n$ and $z_n = z + b_n$. The sequences $\{y_n\}$ and $\{z_n\}$ satisfy the desired properties.]
2. Show that the $C(\Omega)$-spaces with their sup norms (where Ω are compact topological spaces) and the classical $L_p(\mu)$-spaces (where $1 \leq p \leq \infty$) with their L_p-norms are all Banach lattices.
3. Show that a cone $\mathcal{K}$ of a normed space is normal if and only if there exists some $\epsilon > 0$ such that $\|x + y\| \geq \epsilon$ holds for all $x, y \in \mathcal{K}$ with $\|x\| = 1$. [*Hint*: If the condition is true and $0 < x \leq y$, then note that $\left\|\frac{y}{\|x\|}\right\| = \left\|\left(\frac{y}{\|x\|} - \frac{x}{\|x\|}\right) + \frac{x}{\|x\|}\right\| \geq \epsilon$. This implies $\|y\| \geq \epsilon\|x\|$.]
4. Prove that two equivalent norms on a vector space have the same normal cones.
5. Let L be a normed ordered vector space and assume that L_+ is a norm normal cone. If $\widehat{L}$ denotes the norm completion of L, then show that the norm closure $\overline{L_+}$ of L_+ in $\widehat{L}$ is likewise a norm normal cone. [*Hint*: Use Exercise 9 of Section 2.3 or prove it directly by establishing the validity of statement (5) of Theorem 2.38.]
6. (T. Andô [**23**]) Let X be an arbitrary Banach space and let U denote its closed unit ball, i.e., $U = \{x \in X\colon\ \|x\| \leq 1\}$. Show that a closed cone $\mathcal{K}$ of X is generating if and only if there exists some $\alpha > 0$ such that $\alpha U \subseteq \overline{U \cap \mathcal{K} - U \cap \mathcal{K}}$. [*Hint*: Assume $\alpha U \subseteq \overline{U \cap \mathcal{K} - U \cap \mathcal{K}}$ for some $\alpha > 0$ and note that $X = \bigcup_{n=1}^{\infty} n(\overline{U \cap \mathcal{K} - U \cap \mathcal{K}})$.]
7. Prove that every normed Riesz space L has a closed cone. [*Hint*: From part (b) of Exercise 12 of Section 1.3, it follows that $|x^+ - y^+| \leq |x - y|$

for all x, y. This implies $\|x^+ - y^+\| \le \|x - y\|$ and so the function $x \mapsto x^+$ is norm continuous. Now note that $L_+ = \{x \in L\colon\ x^+ = 0\}$.]

8. Consider the Banach lattice $L = C[0,1]$ (with the pointwise ordering and the sup norm). Let $\mathcal{K}_1$ denote the collection of all polynomials (restricted to $[0,1]$) with nonnegative coefficients. Establish the following.
  (a) $\mathcal{K}_1$ is a subcone of L_+ $(= C^+[0,1])$.
  (b) If $\mathcal{K} = \overline{\mathcal{K}_1}$, then $\mathcal{K}$ is a proper subcone of L_+ consisting of increasing functions.
  (c) The cone $\mathcal{K}$ is closed and norm normal.
  (d) We have $\mathcal{K} - \mathcal{K} \neq L$ and $\overline{\mathcal{K} - \mathcal{K}} = L$.

[*Hint*: For (d) notice that $\mathcal{K} - \mathcal{K}$ consists of functions of bounded variation. Thus, not every continuous function of unbounded variation belongs to $\mathcal{K} - \mathcal{K}$. The denseness of $\mathcal{K} - \mathcal{K}$ follows from the Stone–Weierstrass theorem.]

9. Let L be an ordered Banach space with a closed and generating cone and let U denote its closed unit ball. Show that a sequence $\{x'_n\}$ of L' converges to some x' in L' (i.e., $\|x'_n - x'\| \to 0$) if and only if the sequence $\{x'_n\}$ converges uniformly to x' on $U \cap L_+$. [*Hint*: Pick some $\rho > 0$ such that each $x \in L$ can be decomposed in the form $x = y - z$, where $y, z \in L_+$, $\|y\| \le \rho\|x\|$ and $\|z\| \le \rho\|x\|$, and note that for each $f \in L'$ we have $\sup_{x \in U \cap L_+} |f(x)| \le \|f\| \le 2\rho \sup_{x \in U \cap L_+} |f(x)|$.]

10. Let L be an ordered Banach space having a normal and generating cone. Assume that for a sequence $\{x_n\}$ there exists a summable sequence $\{\epsilon_n\}$ of positive real numbers, i.e., $\sum_{n=1}^{\infty} \epsilon_n < \infty$, such that for each n we have $|x'(x_{n+1} - x_n)| \le \epsilon_n \|x'\|$ for all $x' \in L'_+$. Show that the sequence $\{x_n\}$ is norm convergent. [*Hint*: Notice that the dual cone L'_+ is closed and generating. So, according to Theorem 2.37, there exists some constant $\rho > 0$ such that for each $x' \in L'$ there exist $x'_1, x'_2 \in L'_+$ satisfying $x' = x'_1 - x'_2$, $\|x'_1\| \le \rho\|x'\|$ and $\|x'_2\| \le \rho\|x'\|$. It easily follows that $\|x_{n+1} - x_n\| = \sup_{x' \in U'} |x'(x_{n+1} - x_n)| \le 2\rho\epsilon_n$ for each n, where U' is the closed unit ball of L'. This implies that $\{x_n\}$ is a norm Cauchy sequence.]

11. For a closed cone $\mathcal{K}$ of a Banach space L establish the following.
  (a) $\mathcal{K}$ satisfies the strong Levi property if and only if $0 \le x_\alpha \uparrow$ in L with $\sup_\alpha \|x_\alpha\| < \infty$ implies that the net $\{x_\alpha\}$ is norm convergent.
  (b) $\mathcal{K}$ satisfies the Levi property if and only if $0 \le x_\alpha \uparrow \le x$ in L implies that the net $\{x_\alpha\}$ is norm convergent.

12. Show that
  (a) an ordered Banach space with a normal cone and the strong Levi property is Dedekind complete,
  (b) an ordered Banach space with the Levi property is necessarily Dedekind σ-complete.

13. Show that the Banach lattice c_0 (with the pointwise ordering and the sup norm) satisfies the Levi property but fails to satisfy the strong Levi property. [*Hint*: Let $x_n = (1, 1, \ldots, 1, 0, 0, \ldots) \in c_0$, where the "1" occupy the first n coordinates, and note that $0 \le x_n \uparrow$.]

14. Assume that L is an ordered Banach space with the Levi property and let $[a,b]$ be an order interval of L. If a function $\phi\colon [a,b] \to [a,b]$ is nondecreasing (i.e., $a \le x < y \le b$ implies $\phi(x) \le \phi(y)$) and weakly continuous, then show that ϕ has a fixed point.[9] [*Hint*: Consider the sequence $\{x_n\} \subseteq [a,b]$ defined by $x_1 = a$ and $x_{n+1} = \phi(x_n)$ for each $n = 1, 2, \ldots$.]

15. Show that an ordered normed space with a closed cone has the interpolation property if and only if whenever $A \ge B$ holds true for any pair of subsets A and B each having two vectors, there exist two sequences $\{x_n\}$ and $\{y_n\}$ such that $A \ge x_n$, $y_n \ge B$ for each n and $\lim x_n = \lim y_n = x$.

16. Prove Corollary 2.48.

17. Let L be an ordered Banach space with a closed, generating and normal cone that satisfies the Riesz decomposition property. According to Andô's Theorem 2.47 the norm dual L' of L is a Riesz space. Show that in this case, L' admits an equivalent lattice norm—and so L' is a Banach lattice. [*Hint*: We claim that the mapping $x' \mapsto |x'|$, from L' to L', is (norm) uniformly continuous. If this is proven, then L' is a locally convex-solid Riesz space. That is, it has a base at zero consisting of convex and solid[10] neighborhoods; see [**14**, Theorem 2.17, p. 55]. In particular, if V is a convex and solid neighborhood of zero satisfying $V \subseteq U'$, then its Minkowski's functional p_V is a lattice norm that is equivalent to $\|\cdot\|$.

A proof of the uniform continuity of $x' \mapsto |x'|$ goes as follows. Fix some constant $\rho > 0$ such that each $x \in L$ has a decomposition of the form $x = x_1 - x_2$, where $x_1, x_2 \in L_+$ and $\|x_i\| \le \rho\|x\|$ for $i = 1, 2$. Also, since the cone L_+ is norm normal, we can assume that every $u \in [a,b]$ satisfies $\|u\| \le \rho \max\{\|a\|, \|b\|\}$; see Theorems 2.37 and 2.38.

Fix $f \in L'$ and let $x \in U \cap L_+$. If $-x \le y \le x$, then from $\|y\| \le \rho$, it follows that $f(y) = \rho f(\frac{y}{\rho}) \le \rho\|f\|$, and the Riesz–Kantorovich formula yields $|f|(x) = \sup\{f(y)\colon\ -x \le y \le x\} \le \rho\|f\|$.

Next, let $x', y' \in L'$. If $x \in U$ is arbitrary, then (as above), we can write $x = x_1 - x_2$ with $x_i \in \rho U \cap L_+$ and note that

$$\begin{aligned}
\big(|x'| - |y'|\big)(x) &\le \big|(|x'| - |y'|)(x_1)\big| + \big|(|x'| - |y'|)(x_2)\big| \\
&\le \big||x'| - |y'|\big|(x_1) + \big||x'| - |y'|\big|(x_2) \\
&\le |x' - y'|(x_1) + |x' - y'|(x_2) \le 2\rho^2\|x' - y'\|\,.
\end{aligned}$$

This implies $\big\||x'| - |y'|\big\| \le 2\rho^2\|x' - y'\|$.]

18. Assume that L is an ordered Banach space with the decomposition property whose cone is closed, normal and generating. Also suppose that for some $x \in L_+$ the order interval $[0,x]$ is weakly compact. If two sequences $\{x_n\}$ and $\{y_n\}$ satisfy $0 \le x_n \le y_n + x$ for all n and $\|y_n\| \to 0$, then show that the sequence $\{x_n\}$ has a weak accumulation point. [*Hint*: By Theorem 2.37 there exists some $\rho > 0$ such that for each n we can write

[9] Recall that if $f\colon Y \to X$ is a function, where $Y \subseteq X$, then any point $x \in Y$ that satisfies $f(x) = x$ is called a **fixed point** of f.

[10] A subset S of a Riesz space is said to be **solid** if $|x| \le |y|$ and $y \in S$ imply $x \in S$.

$y_n = y_n^1 - y_n^2$ with $\|y_n^i\| \leq \rho\|y_n\|$ $(i = 1, 2)$ and so $0 \leq x_n \leq y_n^1 + x$ holds for each n. This shows that we can assume that $y_n \in L_+$ for each n.]

2.6. Ice cream cones in normed spaces

There is a useful class of cones in normed spaces that are referred to as "ice cream cones." The objective of this section is to describe them and present their basic properties. We start with the definition of an ice cream cone.

Definition 2.51. *Let X be a normed space, let $f \in X'$ be of norm one, and let $0 < \epsilon < 1$. Then the* ***ice cream cone*** *with parameters f and ϵ is the cone*

$$\mathcal{K}_{f,\epsilon} = \big\{x \in X\colon\ f(x) \geq \epsilon\|x\|\big\}.$$

The linear functional f is called the ***axis*** *and $\omega_\epsilon = \arccos \epsilon$ is known as the* ***angle*** *of the ice cream cone $\mathcal{K}_{f,\epsilon}$.*

The reader should verify immediately that the set $\mathcal{K}_{f,\epsilon}$ is indeed a cone. The "ice cream cone" name is justified from the geometrical shape of these cones in two or three-dimensional vector spaces. Indeed, if $X = \mathbb{R}^3$ and $\|\cdot\|$ is the Euclidean norm, then from $f(x) = \|f\| \cdot \|x\| \cos\theta$, where θ is the angle between the vectors f and x, we see that $f(x) \geq \epsilon\|x\|$ holds true if and only if $\cos\theta \geq \epsilon$ or $0 \leq \theta \leq \omega_\epsilon = \arccos\epsilon$. The assumption $0 < \epsilon < 1$ guarantees, of course, that $0 < \omega_\epsilon < 90^\circ$. This shows that $\mathcal{K}_{f,\epsilon}$ is the ice cream cone whose axis is determined by the unit vector f and which has angle ω_ϵ; see the accompanying figure.

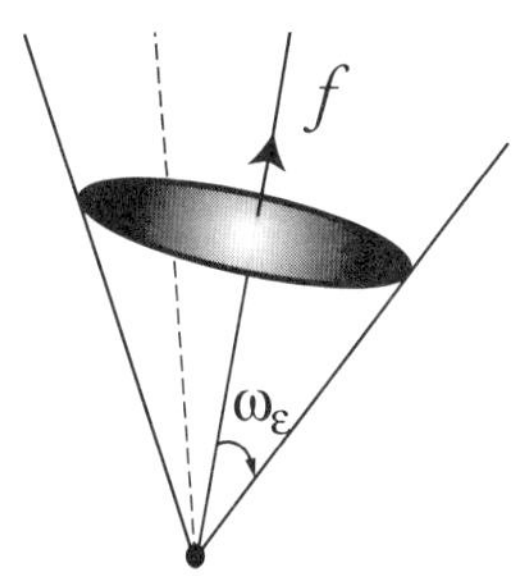

We list below the basic properties of the ice cream cones.

Theorem 2.52. *For an ice cream cone $\mathcal{K}_{f,\epsilon}$ we have the following.*

(1) *The cone $\mathcal{K}_{f,\epsilon}$ is closed and normal with a nonempty interior—and hence it is also generating. Moreover,*

$$\operatorname{Int}(\mathcal{K}_{f,\epsilon}) = \{x \in X\colon\ f(x) > \epsilon\|x\|\}.$$

(2) *The boundary of $\mathcal{K}_{f,\epsilon}$ is the set $\partial\mathcal{K}_{f,\epsilon} = \{x \in X\colon\ f(x) = \epsilon\|x\|\}$.*

(3) *The parameters of an ice cream cone are uniquely determined, i.e., two ice cream cones $\mathcal{K}_{f,\epsilon}$ and $\mathcal{K}_{g,\delta}$ satisfy $\mathcal{K}_{f,\epsilon} = \mathcal{K}_{g,\delta}$ if and only if $f = g$ and $\epsilon = \delta$.*

(4) *The linear functional f is $\mathcal{K}_{f,\epsilon}$-strictly positive—and therefore the closed convex set $\mathcal{B} = \{x \in \mathcal{K}_{f,\epsilon}\colon\ f(x) = 1\}$ is a (norm bounded) base of $\mathcal{K}_{f,\epsilon}$.*

Proof. Let $\mathcal{K}_{f,\epsilon}$ be an ice cream cone of a normed space X.

(1) The function $\phi\colon X \to \mathbb{R}$, defined by $\phi(x) = f(x) - \epsilon\|x\|$, is continuous and clearly satisfies $\mathcal{K}_{f,\epsilon} = \phi^{-1}([0,\infty))$. This shows that the cone $\mathcal{K}_{f,\epsilon}$ is closed. To see that $\mathcal{K}_{f,\epsilon}$ has interior points, fix some $x_0 \in X$ satisfying $\|x_0\| = 1$ and $f(x_0) > \epsilon$. (The existence of such a vector is guaranteed from the assumption $\|f\| = 1$.) Clearly, $x_0 \in \mathcal{K}_{f,\epsilon}$ and $\phi(x_0) > 0$. This implies that the open set $\phi^{-1}((0,\infty))$ (which is a convex subset of $\mathcal{K}_{f,\epsilon}$) is nonempty and hence $\mathcal{K}_{f,\epsilon}$ has a nonempty interior. In fact, the interior of $\mathcal{K}_{f,\epsilon}$ coincides with the open convex set $\bigl\{x \in X\colon\ f(x) > \epsilon\|x\|\bigr\} = \phi^{-1}((0,\infty))$.

To see this, notice that if some $x \in X$ satisfies $f(x) = \epsilon\|x\|$, then x cannot be an interior point of $\mathcal{K}_{f,\epsilon}$. Indeed, if any such x is an interior point of $\mathcal{K}_{f,\epsilon}$, then there exists some $\delta > 0$ such that for all $\|y\| \le \delta$ we have

$$f(x) + f(y) = f(x+y) \ge \epsilon\|x+y\| \ge \epsilon\|x\| - \epsilon\|y\| = f(x) - \epsilon\|y\|\,.$$

This implies $|f(y)| \le \epsilon\|y\|$ for all $\|y\| \le \delta$ and consequently $|f(y)| \le \epsilon\|y\|$ for all $y \in X$. So $\|f\| \le \epsilon$, which is impossible.

For the normality of the cone $\mathcal{K}_{f,\epsilon}$ assume $0 \le y_n \le x_n$ holds for all n with $\|x_n\| \to 0$. From $f(y_n) \ge \epsilon\|y_n\|$ we get $f(y_n) \ge 0$ for each n and from $f(x_n - y_n) \ge \epsilon\|x_n - y_n\|$ we see that $0 \le f(y_n) \le f(x_n)$ for all n. Consequently, $f(y_n) \to 0$. But then from $f(x_n - y_n) \ge \epsilon\|x_n - y_n\|$, it follows that $\|x_n - y_n\| \to 0$ and so $y_n = x_n - (x_n - y_n) \to 0$, proving that $\mathcal{K}_{f,\epsilon}$ is a normal cone. (For a different proof see Exercise 1 at the end of the section.)

(2) This follows from the fact that $\partial\mathcal{K}_{f,\epsilon} = \mathcal{K}_{f,\epsilon} \setminus \phi^{-1}((0,\infty))$.

(3) Assume $\mathcal{K} = \mathcal{K}_{f,\epsilon} = \mathcal{K}_{g,\delta}$. From part (2), we get $\frac{f(x)}{\epsilon} = \frac{g(x)}{\delta} = \|x\|$ for each $x \in \partial\mathcal{K}$. Now if $u \in \mathrm{Int}(\mathcal{K})$, then there exist some nonzero $v \in \partial\mathcal{K}$ and some $\alpha \in \mathbb{R}$ such that $u + \alpha v \in \partial\mathcal{K}$; see Exercise 9 of Section 2.1. It follows that $\frac{f(u+\alpha v)}{\epsilon} = \frac{g(u+\alpha v)}{\delta}$ and from this we obtain $\frac{f(u)}{\epsilon} = \frac{g(u)}{\delta}$ for all $u \in \mathrm{Int}(\mathcal{K})$. Since $\mathcal{K}$ is generating, we obtain $\frac{f(x)}{\epsilon} = \frac{g(x)}{\delta}$ for all $x \in X$, i.e., $\frac{f}{\epsilon} = \frac{g}{\delta}$. Taking norms yields $\epsilon = \delta$ and consequently $f = g$.

(4) Note that if $x > 0$, then $f(x) \ge \epsilon\|x\| > 0$, and note that $x \in \mathcal{B}$ implies $\|x\| \le \frac{1}{\epsilon}$. ∎

There is a natural way to "add an axis" to a given normed space X so that its closed unit ball $U = \{x \in X\colon\ \|x\| \le 1\}$ "becomes" a base for an appropriately constructed ice cream cone. We shall describe this construction below.

So let X be a normed space. Start by defining the norm $\|\cdot\|$ on $\mathbb{R}\times X$ by $\|(r,x)\| = \sqrt{r^2+\|x\|^2}$ and then for each $\alpha>0$ consider the cone

$$\mathcal{K}_\alpha = \big\{(r,x)\in\mathbb{R}\times X\colon\ r\geq\alpha\|x\|\big\}.$$

It should be checked that $\mathcal{K}_\alpha$ is indeed a closed cone of $\mathbb{R}\times X$. From

$$\begin{aligned} r\geq\alpha\|x\| &\iff r\geq 0 \ \text{ and } \ r^2\geq\alpha^2\|x\|^2\\ &\iff r\geq 0 \ \text{ and } \ (1+\alpha^2)r^2\geq\alpha^2(r^2+\|x\|^2)\\ &\iff r\geq\tfrac{\alpha}{\sqrt{1+\alpha^2}}\|(r,x)\|, \end{aligned}$$

we see that

$$\mathcal{K}_\alpha = \big\{(r,x)\in\mathbb{R}\times X\colon\ r\geq\tfrac{\alpha}{\sqrt{1+\alpha^2}}\|(r,x)\|\big\}.$$

Now if we consider the continuous linear functional $f\colon\mathbb{R}\times X\to\mathbb{R}$ defined by $f(r,x)=r$, then $\|f\|=1$ and moreover we have

$$\mathcal{K}_\alpha = \big\{(r,x)\in\mathbb{R}\times X\colon\ f(r,x)\geq\tfrac{\alpha}{\sqrt{1+\alpha^2}}\|(r,x)\|\big\}.$$

Consequently, $\mathcal{K}_\alpha$ is an ice cream cone of $\mathbb{R}\times X$ whose axis is the "factor" $\mathbb{R}$ (the $\mathbb{R}$-axis) and its angle is $\omega=\arccos\frac{\alpha}{\sqrt{1+\alpha^2}}=\operatorname{arccot}\alpha$.

Since f is strictly positive, it follows that the closed convex set

$$\begin{aligned} \mathcal{B} &= \big\{(r,x)\in\mathcal{K}_\alpha\colon\ f(r,x)=\alpha\big\}\\ &= \big\{(\alpha,x)\in\mathbb{R}\times X\colon\ \alpha\geq\alpha\|x\|\big\}\\ &= \big\{(\alpha,x)\in\mathbb{R}\times X\colon\ \|x\|\leq 1\big\}\\ &= \{\alpha\}\times U \end{aligned}$$

is a base for the cone $\mathcal{K}_\alpha$. That is, the closed unit ball U of X translated at the point α of the $\mathbb{R}$-axis is a base of the cone $\mathcal{K}_\alpha$.

According to the preceding discussion, the cone defined on $\mathbb{R}^3(=\mathbb{R}\times\mathbb{R}^2)$ by $z\geq\sqrt{x^2+y^2}$ is the ice cream cone having axis the z-axis and an angle of 45°. A base for this ice cream cone is the closed unit circle $x^2+y^2\leq 1$ of $\mathbb{R}^2$ translated at the point $z=1$ of the z-axis.

Likewise, the cone $y\geq 5\sqrt{2x^2+3z^2}$ is a cone of $\mathbb{R}^3$ that has axis the y-axis and a base the ellipse $2x^2+3z^2\leq 1$ of $\mathbb{R}^2$ translated at the point $y=5$ of the y-axis.

Exercises

1. Show that every ice cream cone $\mathcal{K}_{f,\epsilon}$ of a normed space is normal by proving that $0\leq x\leq y$ (where the order is the one induced by the cone $\mathcal{K}_{f,\epsilon}$) implies $\|x\|\leq\frac{1+\epsilon}{2\epsilon}\|y\|$.

2. Show that a cone $\mathcal{K}$ in a normed space L is an ice cream cone if and only if there exists a (unique) linear functional $f \in L'$ with $\|f\| > 1$ such that $\mathcal{K} = \{x \in L\colon\ f(x) \geq \|x\|\}$.

3. Recall that an extreme point c of a convex subset C of a normed space L is said to be **strongly exposed** if there exists a continuous linear f on L such that $\{c_n\} \subseteq C$ and $f(c_n) \to f(c)$ imply $\|c_n - c\| \to 0$. (The linear functional f is also called a functional that **strongly exposes** c.)

Show that if $\mathcal{K}_{f,\epsilon}$ is an ice cream cone of a normed space, then f strongly exposes the extreme point 0 of the cone $\mathcal{K}_{f,\epsilon}$.

4. Let us say that a closed cone $\mathcal{K}$ of a normed space is a **generalized ice cream cone** if there exist an equivalent norm $|||\cdot|||$ on X,[11] some $0 < \epsilon < 1$, and a nonzero linear functional $f \in X'$ of $|||\cdot|||$-norm one such that

$$\mathcal{K} = \big\{x \in X\colon\ f(x) \geq \epsilon |||x|||\big\}\,.$$

Establish the following.

(a) Every closed and generating cone in $\mathbb{R}^2$ is a generalized ice cream cone of $\mathbb{R}^2$. (See also Exercise 12 of Section 2.1.)

(b) Not every closed cone of $\mathbb{R}^3$ with nonempty interior is a generalized ice cream cone.

5. Let $\{x_1, x_2, x_3\}$ be three linearly independent vectors of $\mathbb{R}^3$ and consider the lattice cone $\mathcal{K}$ of $\mathbb{R}^3$ defined by

$$\mathcal{K} = \Big\{\lambda_1 x_1 + \lambda_2 x_2 + \lambda_3 x_3\colon\ \lambda_i \geq 0 \text{ for each } i = 1, \ldots, 3\Big\}\,.$$

(See Exercise 7 of Section 1.7.) Characterize the (Hamel) bases $\{x_1, x_2, x_3\}$ that give rise to generalized ice cream cones of $\mathbb{R}^3$. (For the definition of a generalized ice cream cone, see the preceding exercise.)

6. Show that the standard cone of $\mathbb{R}^n$ is a generalized ice cream cone. [*Hint*: Note that $\mathbb{R}^n_+ = \big\{(x_1, \ldots, x_n) \in \mathbb{R}^n\colon\ \sum_{i=1}^n |x_i| \leq \sum_{i=1}^n x_i\big\}\,.$]

7. Let $X = \mathbb{R}^n$ with $n \geq 2$ be equipped with its Euclidean norm and consider an ice cream cone $\mathcal{K}_{f,\epsilon}$ of X. Show that there exist a vector subspace Y and some $\alpha > 0$ such that (with appropriate identifications) $X = \mathbb{R} \times Y$ and the cone $\mathcal{K}_{f,\epsilon}$ can be viewed as a cone of the form $\mathcal{K}_\alpha$ with respect to the normed space Y. [*Hint*: We can assume that the linear functional f is a unit vector of X and that $f(x) = x \cdot f$ for each $x \in X$. Let $Y = \operatorname{Ker} f = \{x \in X\colon\ f(x) = x \cdot f = 0\}$ and note that $\mathbb{R}f \oplus Y = X$, where $\mathbb{R}f = \{\lambda f\colon\ \lambda \in \mathbb{R}\}$. This implies (with the obvious identification) that $\mathbb{R} \times Y = X$. Now put $\alpha = \frac{1}{\sqrt{1-\epsilon^2}}$ and note that $\mathcal{K}_{f,\epsilon} = \mathcal{K}_\alpha$ by verifying that for each $x = rf + y = (r, y)$ we have

$$\begin{aligned} x \in K_{f,\epsilon} \quad &\Longleftrightarrow \quad f(x) \geq \epsilon\|x\| \iff r \geq \epsilon\sqrt{r^2 + \|y\|^2} \\ &\Longleftrightarrow \quad r \geq \tfrac{1}{\sqrt{1-\epsilon^2}}\|y\| \iff x = (r, y) \in \mathcal{K}_\alpha\,.] \end{aligned}$$

---

[11] Recall that two norms $\|\cdot\|_1$ and $\|\cdot\|_2$ on a vector space X are said to be **equivalent** if they generate the same linear topology. Clearly, this is the case if and only if there exist constants $c_1 > 0$ and $c_2 > 0$ such that $c_1\|x\|_1 \leq \|x\|_2 \leq c_2\|x\|_1$ holds for all $x \in X$.

8. Show that every ice cream cone of a Banach space satisfies the strong Levi property—and conclude from this that every ice cream cone is normal.

9. Is the norm dual of an ice cream cone an ice cream cone?

2.7. Ideals in ordered vector spaces

In this section, we shall discuss certain vector subspaces of an ordered vector space known as ideals. We start with their definition.

Definition 2.53. *Let L be an ordered vector space and let $x \in L_+$. The (**principal**) **ideal** generated by the vector x is the vector subspace*

$$\begin{aligned} L_x &= \{y \in L\colon\ \exists\,\lambda > 0 \ \textit{such that}\ -\lambda x \le y \le \lambda x\} \\ &= \bigcup_{n=1}^{\infty}[-nx, nx] = \bigcup_{n=1}^{\infty} n[-x, x]\,. \end{aligned}$$

We immediately present three simple properties of the order ideals.

Lemma 2.54. *In any ordered vector space L the following hold.*

(a) *For each $x \in L_+$ we have $L_x \subseteq L_+ - L_+$ and $\bigcup_{x\in L_+} L_x = L_+ - L_+$.*

(b) *Any vector $x \in L_+$ is an order unit of L_x, where L_x is considered ordered by the induced cone $L_x^+ = L_x \cap L_+$ (and so a vector $x \in L_+$ is an order unit of L if and only if $L_x = L$).*

(c) *If L_+ is generating, then for each $y \in L$ there exist $x \in L_+$ and $0 < \alpha_0 < 1$ such that $0 \le \alpha \le \alpha_0$ implies $\alpha y + (1-\alpha)x \in L_+$.*

Proof. The proofs of parts (a) and (b) are straightforward. We shall prove only (c). To this end, fix $y \in L_+$. Since L_+ is generating, there exist $x_1, x_2 \in L_+$ such that $x_1 \ge y$ and $x_2 \ge -y$. Put $x = x_1 + x_2 \in L_+$ and note that $-x \le y \le x$. This implies $y \in L_x$. Now a glance at Lemma 1.7 shows that x is an internal point of the cone $L_x \cap L_+$. In particular, there exists some $0 < \alpha_0 < 1$ such that $0 \le \alpha \le \alpha_0$ implies $\alpha y + (1-\alpha)x \in L_+$, and the proof is finished. ■

On any ideal L_x we can define the Minkowski functional $\|\cdot\|_x\colon L_x \to \mathbb{R}$ by

$$\|y\|_x = \inf\{\lambda > 0\colon\ -\lambda x \le y \le \lambda x\} = \inf\{\lambda > 0\colon\ y \in \lambda[-x, x]\}\,.$$

Clearly, $\|\cdot\|_x$ is a seminorm on L_x (and, of course, it is the gauge of the order interval $[-x, x]$).

The basic properties of the ideal L_x and the seminorm $\|\cdot\|_x$ are included in the next result.

Theorem 2.55. *For a positive vector x of an ordered vector space L we have the following.*

(1) *The Minkowski functional $\|\cdot\|_x$ is a monotone seminorm on L_x (and so the cone $L_x^+ = L_x \cap L_+$ of L_x is $\|\cdot\|_x$-normal).*

(2) *If L is also Archimedean, then*
   (a) *the Minkowski functional $\|\cdot\|_x$ is a norm on L_x,*
   (b) *the closed unit ball of $\|\cdot\|_x$ coincides with the order interval $[-x, x]$, i.e., $\{y \in L_x\colon \|y\|_x \le 1\} = [-x, x]$,*
   (c) *the cone $L_x^+ = L_x \cap L_+$ of L_x is $\|\cdot\|_x$-closed.*

(3) *For any linear topology τ on L (resp. on L_x) for which the intervals are τ-bounded, the natural embedding $J\colon \big(L_x, \|\cdot\|_x\big) \to (L, \tau)$ (resp. $J\colon \big(L_x, \|\cdot\|_x\big) \to (L_x, \tau)$) is continuous.*

Proof. (1) Assume $0 \le y \le z$ in L_x. Then we have

$$\{\lambda > 0\colon \; -\lambda x \le z \le \lambda x\} \subseteq \{\lambda > 0\colon \; -\lambda x \le y \le \lambda x\},$$

and from this it easily follows that $\|y\|_x \le \|z\|_x$, i.e., $\|\cdot\|_x$ is a monotone seminorm on L_x.

(2) Fix $y \in L_x$. From the definition of $\|y\|_x$ there exists a sequence $\{\lambda_n\}$ of positive real numbers satisfying $\lambda_n \downarrow \|y\|_x$ and $-\lambda_n x \le y \le \lambda_n x$ for each n. Since L is Archimedean and clearly $y \in L_+ - L_+$, it follows from Corollary 1.13 that

$$-\|y\|_x x \le y \le \|y\|_x x \,. \tag{$\star$}$$

In other words, we have shown that if L is Archimedean, then for each $x \in L_+$ and each $y \in L_x$ we have $\|y\|_x = \min\{\lambda \ge 0\colon \; -\lambda x \le y \le \lambda x\}$.

Now assume that $\|y\|_x = 0$. Then from ($\star$) we get $0 \le y \le 0$ or $y = 0$. This proves that $\|\cdot\|_x$ is a norm on L_x. To see that the closed unit ball of $\|\cdot\|_x$ on L_x coincides with the order interval $[-x, x]$, use ($\star$) once more to notice that $\|y\|_x \le 1$ if and only if $-x \le y \le x$. For the $\|\cdot\|_x$-closedness of the cone $L_x^+ = L_x \cap L_+$, see Exercise 7 at the end of the section.

(3) Assume that $\|y_n\|_x \to 0$ holds in L_x and let V be a τ-neighborhood of zero. Choose some $\lambda > 0$ such that $\lambda[-x, x] \subseteq V$ and then note that there exists some n_0 such that $-\lambda x \le y_n \le \lambda x$ for all $n \ge n_0$. Consequently, $y_n \in V$ for all $n \ge n_0$ so that $y_n \xrightarrow{\tau} 0$. ■

When is $\big(L_x, \|\cdot\|_x\big)$ a Banach space? The rest of the discussion in this section will provide some answers to this question. But first we need a definition.

Definition 2.56. *Let $\{x_n\}$ be a sequence in an ordered vector space L and fix some vector $0 < v \in L$. The sequence $\{x_n\}$ is said to be*

(1) v-***uniformly convergent*** *to a vector* $x \in L$, *in symbols* $x_n \xrightarrow{u} x(v)$, *if for each* $\epsilon > 0$ *there exists some* n_0 *such that* $-\epsilon v \leq x_n - x \leq \epsilon v$ *holds for all* $n \geq n_0$,

(2) v-***uniformly Cauchy*** *if for each* $\epsilon > 0$ *there exists some* n_0 *such that for all* $m, n \geq n_0$ *we have* $-\epsilon v \leq x_n - x_m \leq \epsilon v$.

It should be noticed that in Archimedean ordered vector spaces the uniform limits are uniquely determined. To see this, assume that a sequence $\{x_n\}$ in an Archimedean ordered vector space L satisfies $x_n \xrightarrow{u} x\,(v)$ and $x_n \xrightarrow{u} y\,(w)$. Fix k and then choose n such that $-\frac{1}{k}v \leq x_n - x \leq \frac{1}{k}v$ and $-\frac{1}{k}w \leq y - x_n \leq \frac{1}{k}w$. Adding the inequalities yields

$$-\tfrac{1}{k}(v+w) \leq y - x \leq \tfrac{1}{k}(v+w)$$

for each k. Now invoke the Archimedean property to see that $x = y$.

Definition 2.57. *An ordered vector space* L *is said to be*

(a) v-***uniformly complete*** *if every* v*-uniformly Cauchy sequence of* L *is* v*-uniformly convergent,*

(b) ***uniformly complete*** *if* L *is* v*-uniformly complete for each* $v > 0$.

The norm completeness of $\big(L_x, \|\cdot\|_x\big)$ is characterized as follows.

Theorem 2.58. *Let* L *be an Archimedean ordered vector space and assume that* $0 < x \in L$. *Then the normed vector space* $\big(L_x, \|\cdot\|_x\big)$ *is a Banach space, i.e., norm complete, if and only if* L *is* x*-uniformly complete.*

Proof. If L is x-uniformly complete, then it should be clear from part (2) of Theorem 2.55 that $\big(L_x, \|\cdot\|_x\big)$ is a Banach space. For the converse assume that $\big(L_x, \|\cdot\|_x\big)$ is a Banach space and let $\{x_n\}$ be an x-uniformly Cauchy sequence. Pick some k such that $-x \leq x_n - x_m \leq x$ for all $n, m \geq k$.

If for each $n \geq k$ we let $z_n = x_n - x_k \in L_x$, then on one hand we have $x_n = x_k + z_n$ for each $n \geq k$ and on the other hand from $x_n - x_m = z_n - z_m$ for all $n, m \geq k$ we see that the sequence $\{z_n\colon\ n \geq k\}$ is $\|\cdot\|_x$-Cauchy. If $\|z_n - z\|_x \to 0$ in L_x, then it is easy to verify that $x_n \xrightarrow{u} x_k + z\,(x)$, and the proof is finished. ■

The next consequence of the preceding result presents a large class of uniformly complete ordered vector spaces.

Corollary 2.59. *If* (L, τ) *is a* τ*-complete ordered topological vector space with a closed and normal cone, then* L *is uniformly complete—and so for each* $x \in L_+$ *the ideal* $(L_x, \|\cdot\|_x)$ *is an ordered Banach space whose closed unit ball coincides with the order interval* $[-x, x]$.

Proof. Let $\{u_n\}$ be a v-uniformly Cauchy sequence. Fix a circled and full neighborhood V of zero and then pick some $\epsilon > 0$ such that $\epsilon v \in V$. Next, pick some n_0 such that $n, m \geq n_0$ implies

$$-\epsilon v \leq u_n - u_m \leq \epsilon v\,. \qquad (\star\star)$$

Since V is circled and full, we get $u_n - u_m \in V$ for all $n, m \geq n_0$. This shows that $\{u_n\}$ is a τ-Cauchy sequence and so there exists some $u \in L$ such that $u_n \xrightarrow{\tau} u$ in L. Since for each $\epsilon > 0$ the inequality $(\star\star)$ is true and L_+ is τ-closed, it follows that for each $\epsilon > 0$ there exists some n_0 such that $-\epsilon v \leq u_n - u \leq \epsilon v$ holds for all $n \geq n_0$. Consequently, we have shown that L is uniformly complete. The last conclusion follows immediately from Theorem 2.58. ■

The next result (due to T. Andô [**23**]) characterizes the closed normal cones in Banach spaces in terms of a "uniform completeness condition." It should be compared with Theorem 2.45.

Theorem 2.60 (Andô). *For an ordered Banach space L with a closed cone the following statements are equivalent.*

(1) *The cone L_+ is normal.*

(2) *Every increasing sequence of positive vectors which is uniformly Cauchy with respect to one of its upper bounds has a supremum.*

(3) *For each $u \in L_+$ the normed space $(L_u, \|\cdot\|_u)$ is a Banach space.*

Proof. (1)$\Longrightarrow$(2) According to condition (5) of Theorem 2.38 there exists some constant $c > 0$ such that $0 \leq x \leq y$ implies $\|x\| \leq c\|y\|$. Now suppose that a sequence $\{x_n\}$ of L satisfies $0 \leq x_n \leq x_{n+1} \leq v$ for all n and some $v \in L$ and that $\{x_n\}$ is v-uniformly Cauchy. Fix $\epsilon > 0$ and then pick some n_0 such that $-\epsilon v \leq x_n - x_m \leq \epsilon v$ holds for all $n, m \geq n_0$. This implies $\|x_n - x_m\| \leq \epsilon c\|v\|$ for all $n \geq m \geq n_0$, and from this we infer that $\{x_n\}$ is a Cauchy sequence. If $x_n \to x$ in L, then a glance at part (4) of Lemma 2.3 guarantees that $x_n \uparrow x$, i.e., $x = \sup\{x_n\}$ in L.

(2)$\Longrightarrow$(3) Fix $u \in L_+$ and let $\{x_n\}$ be a $\|\cdot\|_u$-Cauchy sequence of L_u. By passing to a subsequence, we can assume without loss of generality that $\|x_{n+1} - x_n\|_u \leq \frac{1}{2^{n+1}}$ or $-\frac{1}{2^{n+1}}u \leq x_{n+1} - x_n \leq \frac{1}{2^{n+1}}u$ holds for all n.

For each n let $y_n = x_{n+1} - x_n + \frac{1}{2^{n+1}}u$ and note that $0 \leq y_n \leq \frac{1}{2^n}u$ for all n. Next consider the sequence $\{z_n\}$ of L_u defined by $z_n = \sum_{i=1}^n y_i$. Clearly, for each n we have $0 \leq z_n \leq z_{n+1}$ and $z_n = \sum_{i=1}^n y_i \leq \sum_{i=1}^n \frac{1}{2^i}u \leq u$. Moreover, for all n and k we have

$$0 \leq z_{n+k} - z_n = \sum_{i=n+1}^{n+k} y_i \leq \sum_{i=n+1}^{n+k} \tfrac{1}{2^i}u \leq \tfrac{1}{2^n}u\,. \qquad (\dagger)$$

This clearly shows that the increasing sequence of positive vectors $\{z_n\}$ is u-uniformly Cauchy with respect to its upper bound u. By our assumption the sequence $\{z_n\}$ has a supremum in L, say z. Clearly, $z \in L_u$. From (†), it follows that for each fixed n we have $z_{n+k} \leq z_n + \frac{1}{2^n}u$ for all k. This implies $z \leq z_n + \frac{1}{2^n}u$ or $0 \leq z - z_n \leq \frac{1}{2^n}u$. Consequently, $\|z_n - z\|_u \leq \frac{1}{2^n}$ and so $z_n \to z$ in L_u.

Now let $x = z + x_1 - \frac{1}{2}u \in L_u$ and note that

$$x_{n+1} - x = \sum_{i=1}^{n}\left(x_{i+1} - x_i + \tfrac{1}{2^{i+1}}u\right) - z + \sum_{i=n+1}^{\infty} \tfrac{u}{2^{i+1}} = z_n - z + \tfrac{1}{2^{n+1}}u\,.$$

Hence $x_n = x + z_{n-1} - z + \frac{1}{2^n}u \to x$ in L_u so that the normed space $(L_u, \|\cdot\|_u)$ is a Banach space.

(3) $\Longrightarrow$ (1) Fix any vector $u \in L_+$. Since the cone $L_u^+ = L_u \cap L_+$ is $\|\cdot\|_u$-closed (see Exercise 7 at the end of the section), it easily follows from Theorem 2.32 that the identity operator $I\colon (L_u, \|\cdot\|_u) \to L$ is continuous or, equivalently, that (for each $u \in L_+$) the closed unit ball $[-u, u]$ of L_u is mapped under I to a norm bounded subset of L for each $u \in L_+$. Thus, we have shown that L has norm bounded order intervals. But then, from Theorem 2.40, we infer that the cone L_+ is norm normal. ■

Rephrasing Lemma 2.30 we also have the following useful connection between topological convergence and uniform convergence.

Theorem 2.61. *If an ordered topological vector space (L, τ) is such that either*

(a) *its positive cone L_+ has a τ-interior point or else*

(b) *the positive cone L_+ is generating and τ-closed and the linear topology τ is completely metrizable,*

then every τ-convergent sequence in L has a uniformly convergent subsequence.

The Krein spaces are introduced next.

Definition 2.62. *An ordered Banach space with a closed cone and order units is called a* ***Krein space****.*

Example 2.41 shows that the cone of a Krein space need not be normal. The proof of the next result that characterizes the Krein spaces with normal cones is left for the reader—see also Exercise 11 at the end of the section. (Keep in mind that if u is an order unit of L, then $L_u = L$.)

Theorem 2.63. *For a Krein space $(L, \|\cdot\|)$ and an order unit u of L the following statements are equivalent.*

(1) *The cone L_+ is $\|\cdot\|$-normal.*

(2) *The norms $\|\cdot\|_u$ and $\|\cdot\|$ are equivalent.*

(3) *The normed space $(L, \|\cdot\|_u)$ is a Banach space.*

Finally, we close the section with the introduction of the quasi-interior points of an ordered topological vector space.

Definition 2.64. *A positive vector u of an ordered topological vector space L is said to be a* ***quasi-interior point*** *if $\overline{L}_u = L$, i.e., if the ideal L_u generated by u is topologically dense in L.*

The quasi-interior points are always strictly positive vectors.

Lemma 2.65. *Every quasi-interior point of an ordered topological vector space is strictly positive.*

Proof. Assume $\overline{L}_u = L$ and suppose by way of contradiction that some $0 < f \in L'$ satisfies $f(u) = 0$. This implies $f(x) = 0$ for all $x \in L_u$. By continuity, $f = 0$, which is impossible. Thus $f(u) > 0$ and so u is strictly positive. ■

The converse of the preceding result is false. For instance, let $L = \mathbb{R}^3$ (with the Euclidean topology) and let $L_+ = \{(x, y, 0)\colon\ x \geq 0 \text{ and } y \geq 0\}$. If we consider the vector $u = (1, 1, 0) \in L_+$, then u is strictly positive while $\overline{L}_u = L_u = \{(x, y, 0)\colon\ x, y \in \mathbb{R}\} \neq L$.

Exercises

1. If $x_n \xrightarrow{u} x\,(v)$ and $y_n \xrightarrow{u} y\,(w)$ hold in an ordered vector space, then for each $\alpha, \beta \in \mathbb{R}$ show that $\alpha x_n + \beta_n y_n \xrightarrow{u} \alpha x + \beta y\,(|\alpha| v + |\beta| w)$.

2. Give an example of a sequence in a non-Archimedean ordered vector space that is v-uniformly convergent to two different vectors. [*Hint*: Consider the sequence $\{x_n\}$ of the lexicographic plane defined by $x_n = (\frac{1}{n}, 0)$ and note that if $v = (2, 2)$, then for every vector of the form $x = (0, \alpha)$ we have $0 \leq x_n - x \leq \frac{1}{n} v$ for each n.]

3. Prove that in an ordered vector space every v-uniformly convergent sequence is necessarily v-uniformly Cauchy.

4. Prove that a v-uniformly Cauchy sequence is v-uniformly convergent if and only if it has a v-uniformly convergent subsequence.

5. If an increasing sequence $\{x_n\}$ in an Archimedean ordered vector space is uniformly convergent to some vector x, then show that $x_n \uparrow x$.

6. Let Ω be an arbitrary topological space and let $C_b(\Omega)$ denote the ordered vector space of all (uniformly) bounded continuous real-valued functions defined on Ω. Show that $C_b(\Omega)$ is uniformly complete.

7. Show that the uniform limit of a positive sequence in an Archimedean ordered vector space is a positive vector. [*Hint*: Assume that a sequence $\{x_n\}$ of positive vectors in an Archimedean ordered vector space satisfies $x_n \xrightarrow{u} x(v)$. If $\epsilon > 0$ is given, pick some n such that $-\epsilon v \le x - x_n \le \epsilon v$ and note that $x = (x - x_n) + x_n \ge x - x_n \ge -\epsilon v$.]

8. If L is a Riesz space, then show that every ideal L_u of L is a Riesz subspace of L and that $\|\cdot\|_u$ is a lattice norm, i.e., $|x| \le |y|$ in L_u implies $\|x\|_u \le \|y\|_u$.

9. Prove Theorem 2.58.

10. Recall that an ordered vector space L is said to be **σ-order complete** (or **Dedekind σ-complete**) if every increasing order bounded sequence of L has a supremum. A cone $\mathcal{K}$ of a vector space X is called **σ-order complete** if $(X, \mathcal{K})$ is σ-order complete. Establish the following.
   (a) A Riesz space is σ-order complete if and only if every nonempty countable subset of L that is bounded from above has a supremum.
   (b) Every closed σ-order complete cone in a Banach space is normal.
[*Hint*: For (b) use Theorem 2.60.]

11. Prove Theorem 2.63. [*Hint*: Assume that u is an order unit. Clearly, $L = L_u$. From Corollary 2.59 we know that $(L, \|\cdot\|_u)$ is a Banach space. Now use part 2(c) of Theorem 2.55 in connection with Theorem 2.32 to conclude that the identity operator $I\colon (L, \|\cdot\|_u) \to (L, \|\cdot\|)$ is continuous. The Open Mapping Theorem 8.36 guarantees that I is an isomorphism.]

12. A sequence $\{x_n\}$ of an ordered vector space L is said to be **order summable** if its sequence of partial sums, i.e., the sequence $\{\sum_{i=1}^n x_i\}$, has a supremum in L. Also a sequence $\{x_n\} \subseteq L_+$ is of **ℓ_1-type** if there exist some $x \in L_+$ and a sequence $(\alpha_1, \alpha_2, \dots) \in \ell_1^+$ such that $0 \le x_n \le \alpha_n x$ holds for all n.

Assume that the cone of an ordered Banach space is closed. Show that the cone is normal if and only if every sequence of ℓ_1-type is order summable.

13. Show that in a Krein space L the collection of all order units is norm dense in L_+.

14. Let H be a complex Hilbert space and let $\mathcal{L}(H)$ denote the Banach space (in fact Banach algebra) of all bounded (linear) operators on H, where as usual the norm on $\mathcal{L}(H)$ is defined by $\|T\| = \sup_{\|x\| \le 1} \|Tx\|$. Recall that if $T \in \mathcal{L}(H)$, then its **Hilbert space adjoint operator** is the unique operator $T^* \in \mathcal{L}(H)$ that satisfies $(Tx, y) = (x, T^*y)$ for all $x, y \in H$, where $(\cdot, \cdot)$ is the inner product of H. An operator $T \in \mathcal{L}(H)$ is called **self-adjoint** if $T^* = T$. An operator $T \in \mathcal{L}(H)$ is **positive**, in symbols $T \ge 0$, if $(Tx, x) \ge 0$ holds for all $x \in H$. Establish the following.
   (a) An operator $T \in \mathcal{L}(H)$ is self-adjoint if and only if $(Tx, x) \in \mathbb{R}$ for all $x \in H$.
   (b) Every positive operator is self-adjoint.
   (c) If $T \in \mathcal{L}(H)$, then TT^* and T^*T are both positive operators.

(d) The collection of all self adjoint operators $\mathcal{J}$ is a closed vector subspace of $\mathcal{L}(H)$. Moreover, the collection of all positive operators of $\mathcal{L}(H)$ is a cone of $\mathcal{J}$ (and so $\mathcal{J}$ considered as a real vector space is an ordered Banach space).

(e) The ordered Banach space $\mathcal{J}$ is a Krein space.

[*Hint*: For part (e) note that for each $T \in \mathcal{J}$ we have $\|T\|I \geq T$, where I denotes the identity operator on H. (For more about this problem see [**66**, Chapter 8].)]

15. For an element $u > 0$ in a Krein space L show that the following statements are equivalent.

(a) The vector u is an order unit of L.

(b) The vector u is an internal point of L_+.

(c) The vector u is an interior point of L_+.

16. Let $T\colon L \to L$ be a positive operator on a Krein space. If u is an order unit, then show that the operator $T\colon (L, \|\cdot\|) \to (L, \|\cdot\|_u)$ is continuous. [*Hint*: Fix $r > 0$ such that $u + rU \subseteq L_+$ and then pick some $c > 0$ such that $T(u) \leq cu$. This implies that for every $x \in U = \{x \in L\colon\ \|x\| \leq 1\}$ we have $-\frac{c}{r}u \leq T(x) \leq \frac{c}{r}u$.]

2.8. The order topology generated by a cone

Let $\mathcal{K}$ be a cone of a vector space X. There is a finest (not necessarily Hausdorff) locally convex topology on X such that the $\mathcal{K}$-order intervals of X are all topologically bounded. This locally convex topology will be denoted by $\tau_o(\mathcal{K})$ (or simply by τ_o if there is no confusion about the cone $\mathcal{K}$ under consideration) and will be referred to as the **order topology** generated by the cone $\mathcal{K}$. Its precise definition is as follows.

Definition 2.66 (Namioka [**71**]; Schaefer [**84**]). *The* ***order topology*** *generated by a cone $\mathcal{K}$ of a vector space X is the locally convex topology $\tau_o(\mathcal{K})$ (or simply τ_o) of X that has a base at zero consisting of all nonempty convex and circled subsets of X that absorb every $\mathcal{K}$-order interval of X.*[12] *That is, a base for the τ_o-neighborhoods at zero is the filter base*

$$\mathcal{B}_o = \big\{V \subseteq X\colon\ V \text{ is convex, circled and absorbs the } \mathcal{K}\text{-order intervals}\big\}.$$

Since $X \in \mathcal{B}_o$, it should be immediate that $\mathcal{B}_o$ is indeed a filter base that generates a unique locally convex topology on X. The following result should be immediate from the preceding definition and Lemma 2.22.

Lemma 2.67. *If $\mathcal{K}$ is a cone of a vector space X, then the topology $\tau_o(\mathcal{K})$ is the finest locally convex topology on X for which every $\mathcal{K}$-order interval*

[12] Recall that a subset A of a vector space **absorbs** another set B if there exists some $\lambda_0 > 0$ such that $\lambda B \subseteq A$ for all $|\lambda| \leq \lambda_0$. A subset that absorbs the singletons is known as an **absorbing set**. From $\{x\} = [x, x]$, we see that every set that absorbs the order intervals is automatically an absorbing set.

is topologically bounded. In particular, every locally convex topology τ on X for which the cone $\mathcal{K}$ is τ-normal satisfies $\tau \subseteq \tau_0(\mathcal{K})$.

For the rest of the discussion in this section L will be an ordered vector space and τ_o will denote the order topology on L generated by the cone L_+.

Lemma 2.68. *The order topology is bornological.*[13]

Proof. Let V be a convex circled subset of L that absorbs every τ_o-bounded subset of L. In particular, V absorbs every order interval of L and is thus a τ_o-neighborhood of zero. This shows that (L, τ_o) is a bornological space. ∎

The topological dual of (L, τ_o) is precisely the order dual of L.

Theorem 2.69. *The topological dual of (L, τ_o) coincides with the order dual of L, i.e., $L^\sim = (L, \tau_0)'$. In particular, the order topology τ_0 is Hausdorff if and only if $L^\sim$ separates the points of L.*

Proof. Let $L' = (L, \tau_0)'$. It should be clear that $L' \subseteq L^\sim$. For the reverse inclusion, let $f \in L^\sim$. Put $V = f^{-1}([-1, 1])$ and note that V is a convex circled set that absorbs the order intervals of L. Hence, V is a τ_o-neighborhood of zero and since f is bounded on V, it follows that $f \in L'$. Thus, $L' = L^\sim$.

If τ_o is Hausdorff, then clearly $L^\sim$ separates the points of L. On the other hand, if $L^\sim$ separates the points of L, then $\sigma(L, L^\sim)$ is a Hausdorff locally convex topology such that every order interval of L is $\sigma(L, L^\sim)$-bounded. This implies $\sigma(L, L^\sim) \subseteq \tau_o$ and so τ_o is a Hausdorff topology. ∎

Corollary 2.70. *If $L^\sim$ separates the points of L, then the order topology coincides with the Mackey topology $\tau(L, L^\sim)$, i.e., $\tau_o = \tau(L, L^\sim)$.*

Proof. This follows immediately from the fact that τ_o is a bornological topology that is consistent with the dual system $\langle L, L^\sim \rangle$; see Exercise 5 at the end of the section. ∎

There is an important case in which the order topology is normable. It is described in the next result.

Theorem 2.71. *For an Archimedean ordered vector space L with order units, we have the following.*

(1) *If u is any order unit, then the norm $\|\cdot\|_u$ generates the order topology τ_0 (and so, in this case, τ_0 is normable and for every other order unit e the norm $\|\cdot\|_e$ is equivalent to $\|\cdot\|_u$).*

---

[13] A locally convex space (X, τ) is said to be a **bornological space** (and the topology τ is referred to as a **bornological topology** of X) if every convex circled set that absorbs the τ-bounded sets is already a τ-neighborhood of zero.

(2) *The order topology* τ_0 *is the finest linear topology on* L *for which the cone is normal.*

(3) *The cone* L_+ *is* $\|\cdot\|_u$*-closed.*

(4) *If* $L' = (L, \|\cdot\|_u)'$*, then* $L^\sim = L^r = L'_+ - L'_+ = L'$ *(and so* $L^\sim$ *separates the points of* L*).*

Proof. Fix any order unit u of L and let $\|\cdot\|_u$ denote the Minkowski functional of the convex, circled and absorbing set $[-u, u]$. According to Theorem 2.55 the seminorm $\|\cdot\|_u$ is a monotone norm whose closed unit ball coincides with the order interval $[-u, u]$.

(1) and (2) The monotonicity of $\|\cdot\|_u$ guarantees that $\|\cdot\|_u$ generates a locally convex topology for which L_+ is normal. Now let τ be another linear topology on L such that the order intervals of L are τ-bounded—in particular, let τ be a linear topology on L for which L_+ is τ-normal.

By part (3) of Theorem 2.55 the identity operator $I\colon (L, \|\cdot\|_u) \to (L, \tau)$ is continuous. This shows that the norm topology generated by $\|\cdot\|_u$ is finer than τ and the validity of (1) and (2) follows.

(3) This is part (2)(c) of Theorem 2.55.

(4) Since L_+ is $\|\cdot\|_u$-normal, it follows from Theorem 2.26 that (the cone) L'_+ is generating in L', and so $L' = L'_+ - L'_+ \subseteq L^\sim$. On the other hand, since the $\|\cdot\|_u$-closed unit ball of L coincides with the order interval $[-u, u]$, it follows that every linear functional in $L^\sim$ is $\|\cdot\|_u$-continuous, i.e., $L^\sim \subseteq L'$, and the desired identity has been established. ■

Finally, we close the section with an example that illustrates the basic results in this chapter.

Example 2.72. This is Example 2.41 revisited. Therefore, let $L = C^1[0,1]$, the real vector space of all continuously differentiable functions on $[0,1]$, equipped (as usual) with the cone L_+ of all nonnegative functions of L, i.e., $L_+ = \{x \in C^1[0,1]\colon\ x(t) \geq 0 \text{ for all } t \in [0,1]\}$. From Example 1.58 we know the following.

(i) *The cone* L_+ *is generating and has the Riesz decomposition property. In particular, the order dual* $L^\sim$ *of* L *is a Riesz space. Moreover, the lattice operations of* $L^\sim$ *are given by the Riesz–Kantorovich formulas, i.e., if* $f, g \in L^\sim$ *and* $x \in L_+$*, then*

$$\begin{aligned}
[f \vee g](x) &= \sup\{f(y) + g(z)\colon\ y, z \in L_+ \text{ and } y + z = x\},\\
[f \wedge g](x) &= \inf\{f(y) + g(z)\colon\ y, z \in L_+ \text{ and } y + z = x\},\\
|f|(x) &= \sup\{f(y)\colon\ -x \leq y \leq x\}.
\end{aligned}$$

Recall that in Example 2.41 we saw that L under the norm

$$\|x\| = \|x\|_\infty + \|x'\|_\infty\,,$$

where x' denotes the derivative of the function x, is a Banach space and that L_+ is closed—in fact, L is a Krein space. We also noticed that the cone L_+ is not normal. That is, we have shown the following.

(ii) *The cone L_+ is closed but fails to be norm normal.*

In order to study the ordered vector space L, we need to recall a few properties of the ordered vector space $M = C[0,1]$. The vector space M is ordered pointwise and is equipped with the sup norm $|||x||| = \|x\|_\infty$. Under these structures M is a Banach lattice and its norm dual $M'\,(=M^\sim)$ can be identified with the Banach lattice of all regular Borel measures on $[0,1]$ of finite total variation. The norm of M' is the total variation, i.e., $|||\mu||| = |\mu|([0,1])$, and of course the bilinearity of the dual system $\langle M, M'\rangle$ is given by $\langle x, \mu\rangle = \mu(x) = \int_{[0,1]} x\,d\mu$. For details see [**16**, Section 38].

Clearly, L is a vector subspace of M, the order of L is the one induced from M, and L majorizes M. We are now ready to exhibit a few more interesting properties of the ordered Banach space L.

(iii) *The dual wedge L'_+ is a closed and (norm) normal cone of L'. Moreover, we have $L^r = L'_+ - L'_+$.*

Since L is generating, L'_+ is clearly a cone. The norm closedness of L'_+ is obvious. The norm normality of L'_+ follows immediately from Theorem 2.42. The identity $L^r = L'_+ - L'_+$ is a simple consequence of Corollary 2.34.

(iv) *The order topology of L coincides with the norm topology generated by the sup norm. In particular, $L^\sim = (L, \|\cdot\|_\infty)' = L^r = L'_+ - L'_+$.*

To see this, notice that since $\mathbf{1}$ is an order unit of L, it follows from Theorem 2.71 that the order topology on L coincides with the norm topology generated by the norm $\|\cdot\|_{\mathbf{1}}$. Now observe that $\|x\|_{\mathbf{1}} = \|x\|_\infty$ for each $x \in L$. The description of $L^\sim$ is simply part (4) of Theorem 2.71.

(v) *The order dual $L^\sim$ is not (norm) dense in L'.*

To see that $L^\sim$ is not norm dense in L', consider the linear functional $\phi\colon L \to \mathbb{R}$ defined by $\phi(x) = x'(1)$. From $|\phi(x)| \le \|x'\|_\infty \le \|x\|$, we see that $\phi \in L'$. We claim that ϕ does not belong to the norm closure of $L^\sim$. To establish this, fix $f \in L^\sim$ and then pick some $k \ge 2$ such that whenever $x \in L$ satisfies $-\frac{1}{k}\mathbf{1} \le x \le \frac{1}{k}\mathbf{1}$ we have $|f(x)| < \frac{1}{4}$. Now consider the function $y \in L$ defined by $y(t) = \frac{1}{4k\pi}\sin(2k\pi t)$. From $y'(t) = \frac{1}{2}\cos(2k\pi t)$, we get $-\frac{1}{k}\mathbf{1} \le y \le \frac{1}{k}\mathbf{1}$, $\phi(y) = y'(1) = \frac{1}{2}$, and

$$\|y\| = \|y\|_\infty + \|y'\|_\infty \le \tfrac{1}{k} + \tfrac{1}{2} \le 1\,.$$

Consequently,

$$\|f - \phi\| \geq |f(y) - \phi(y)| \geq |\phi(y)| - |f(y)| \geq \tfrac{1}{2} - \tfrac{1}{4} = \tfrac{1}{4}\,.$$

Since $f \in L^\sim$ is arbitrary, this shows that ϕ does not belong to the norm closure of $L^\sim$. That is, $L^\sim$ is not norm dense in L'.

(vi) *Every linear positive functional on L has a unique linear positive extension to all of M.*

To see this, let $f \in L'_+$. Since L majorizes M, it follows from Theorem 1.30 that f has a positive linear extension $\widehat{f}$ to all of M. By Corollary 2.34 this extension $\widehat{f}$ must be $\|\cdot\|_\infty$-continuous. If now g is another positive linear extension of f to all of M, then g is also $\|\cdot\|_\infty$-continuous and since $\widehat{f}(p) = g(p) = f(p)$ for all polynomials p, it follows from the classical Stone–Weierstrass theorem (see for instance [**16**, Section 11]) that $g = \widehat{f}$.

Now an easy argument shows that every $f \in L^r$ extends uniquely to an order bounded linear functional $\widehat{f}$ on $M = C[0,1]$. Conversely, the restriction of every $f \in M'$ to L belongs to L^r. Thus, there exists an order isomorphism $f \mapsto \widehat{f}$, from L^r onto M', and so (under this order isomorphism) we can identify L^r with $M' = \mathrm{ca}\,[0,1]$. In other words, we have established the following property.

(vii) *The regular dual L^r $(= L^\sim)$ of L can be identified with $M' = \mathrm{ca}\,[0,1]$ in the sense that for each $f \in L^r$ there exists a unique regular signed Borel measure μ on $[0,1]$ of finite total variation such that for each $x \in L$ we have $f(x) = \int_{[0,1]} x\,d\mu$.*

Since $L^\sim = L'_+ - L'_+$ is not norm dense in L', it follows that the double wedge L''_+ is no longer a cone of the double dual L''. For more on this example see Exercises 8 and 9 at the end of the section. ■

Exercises

1. Show that a locally convex space (X, τ) is bornological if and only if every seminorm of X that is bounded on every τ-bounded subset of X is τ-continuous.
2. If X is a normed space with norm dual X', then show that the Mackey topology $\tau(X, X')$ coincides with the norm topology of X.
3. Show that every order bounded operator $T\colon L \to M$ between two ordered vector spaces is continuous when L and M are equipped with their order topologies.
4. Let $(L_i, \tau_i)_{i\in I}$ be a family of ordered topological vector spaces such that τ_i is the order topology of L_i. Show that the order topology of the product ordered vector space $\prod_{i\in I} L_i$ coincides with the product topology $\prod_{i\in I} \tau_i$.

5. Let $\langle X, X' \rangle$ be a dual system and let τ be a consistent locally convex topology on X. If τ is also bornological, then show that $\tau = \tau(X, X')$. [*Hint*: We only need to show that $\tau(X, X') \subseteq \tau$. To establish this, we must invoke the fact that all consistent locally convex topologies on X have the same bounded sets; see for instance [**12**, Thm 6.24, p. 248]. If this is the case, then since every convex circled $\tau(X, X')$-neighborhood of zero absorbs all the $\tau(X, X')$-bounded sets, it must also absorb all the τ-bounded sets and so it must be a τ-neighborhood of zero.]

6. Assume that an ordered vector space L has the Riesz decomposition property and that its order dual $L^\sim$ separates the points of L. Show that the order topology of L is the finest locally convex topology for which the cone L_+ is normal.

7. For an ordered vector space L (not necessarily Archimedean) establish the following.
   (a) For each $x \in L_+$ the order topology of L_x coincides with the topology generated by the seminorm $\|\cdot\|_x$ and is the largest locally convex topology on L_x for which the cone $L_x^+ = L_x \cap L_+$ is normal.
   (b) If L has order units, then for each order unit u the order topology of L is generated by the seminorm $\|\cdot\|_u$.
   [*Hint*: Use part (3) of Theorem 2.55.]

8. This exercise presents another way of looking at the ordered Banach space of Example 2.72. Therefore, let $L = C^1[0,1]$ be as in Example 2.72 and note that L is a Krein space. Establish the following properties.
   (a) The function $|||\cdot|||\colon L \to \mathbb{R}$, defined by $|||x||| = |x(0)| + \|x'\|_\infty$, is a norm that is equivalent to the norm $\|\cdot\|$.
   (b) The mapping $J\colon L \to \mathbb{R} \times C[0,1]$, defined by $J(x) = \big(x(0), x'\big)$, is a surjective linear homeomorphism—and so L can be identified with the Banach space $M = \mathbb{R} \times C[0,1]$.
   (c) The ordered Banach space M equipped with the cone

$$M_+ = J(L_+) = \big\{\big(x(0), x'\big)\colon\ 0 \le x \in C^1[0,1]\big\}$$

   becomes a Krein space. Moreover, we have

$$M_+ = \Big\{(\alpha, x) \in \mathbb{R} \times C[0,1]\colon\ \alpha + \int_0^t x(s)\,ds \ge 0\ \forall\, t \in [0,1]\Big\}.$$

   (d) We have $M' = \mathbb{R} \times \mathcal{M}[0,1]$, where $\mathcal{M}[0,1]$ denotes the Banach space of all signed measures on $[0,1]$ of bounded variation. Moreover, the bilinearity of the dual system $\langle M, M' \rangle$ is given by

$$\big\langle (\alpha, z), (\beta, \mu) \big\rangle = \alpha\beta + \int_0^1 z(t)\,d\mu(t)\,.$$

   (e) The dual cone M'_+ of the cone M_+ is given by

$$M'_+ = \Big\{(\beta, \mu)\colon\ \beta x(0) + \int_0^1 x'(t)\,d\mu(t) \ge 0\ \forall\, 0 \le x \in C^1[0,1]\Big\}.$$

   (f) The vector space $M'_+ - M'_+$ is not norm dense in M'.

(g) The order dual of M coincides with the order regular dual of M and is a proper vector subspace of the norm dual of M. That is, $M^\sim = M^r = M'_+ - M'_+ \subseteq M'$ and $M^\sim \neq M'$.

(h) The order dual of L coincides with the regular dual of L and we have

$$L^\sim = L^r = L'_+ - L'_+ \subseteq L' \quad \text{and} \quad L^\sim \neq L'\,.$$

9. Let $L = C^k[0,1]$, the vector space of all k-times continuously differentiable real functions defined on $[0,1]$. Clearly, $L \subseteq C[0,1]$ and so L with the pointwise ordering is an ordered vector space. Establish the following.

(a) The function $\|\cdot\|\colon L \to \mathbb{R}$ defined by

$$\|x\| = \|x\|_\infty + \|x'\|_\infty + \cdots + \|x^{(k)}\|_\infty$$

is a norm on L under which L is a Banach space and the cone L_+ is closed, generating, and fails to be (norm) normal.

(b) The ordered Banach space L satisfies the properties of Example 2.72.

(c) For each $\xi \in [0,1]$ and each $1 \leq i \leq k$ consider the linear functional $\phi_i^\xi\colon L \to \mathbb{R}$ defined by $\phi_i^\xi(x) = x^{(i)}(\xi)$. Then the following hold.

 (i) Each ϕ_i^ξ is a continuous and nonorder bounded linear functional, i.e., $\phi_i^\xi \in L' \setminus L^\sim$.

 (ii) Each ϕ_k^ξ does not belong to the norm closure of $L^\sim$ in L'.

 (iii) For each $1 \leq i \leq k-1$ (where here we assume that $k \geq 2$) the linear functional ϕ_i^ξ belongs to the norm closure of $L^\sim$ in L' (and conclude from this that $L^\sim$ is not norm closed in L').

(d) Consider $C[0,1]$ equipped with the sup norm and define the operator $T\colon L \to \mathbb{R}^k \times C[0,1]$ by

$$T(x) = \big(x(0), x'(0), \ldots, x^{(k-1)}(0), x^{(k)}\big)\,.$$

Then T is a Banach space isomorphism.

(e) Does L have the Riesz decomposition property?

[*Hint*: For (c)(iii), fix $\xi \in [0,1]$, $1 \leq i \leq k-1$, and pick a sequence $\{\xi_n\}$ of $[0,1]$ with $\xi_n \neq \xi$ for each n and $\xi_n \to \xi$. Also, let $h_n = \xi_n - \xi$ and consider the linear functional $\psi_n \in L'$ defined by $\psi_n(x) = \frac{x^{(i-1)}(\xi_n) - x^{(i-1)}(\xi)}{h_n}$. Now note that if $x \in L$, then for some τ_n between ξ and ξ_n we have

$$\begin{aligned} |\psi_n(x) - \phi_i^\xi(x)| &= \Big|\tfrac{x^{(i-1)}(\xi_n) - x^{(i-1)}(\xi)}{h_n} - x^{(i)}(\xi)\Big| \\ &= \big|x^{(i)}(\tau_n) - x^{(i)}(\xi)\big| \\ &= |\tau_n - \xi|\big|x^{(i+1)}(\tau_n)\big| \leq |h_n|\,\|x\|\,. \end{aligned}$$

This implies $\|\psi_n - \phi_i^\xi\| \leq |h_n|$ for each n. Now prove (by induction on i) that $\psi_n \in L^\sim$ holds true for each n.]

Chapter 3

Yudin and pull-back cones

The chapter unravels the properties of directional wedges and more specifically those of the Yudin cones in vector spaces. A Yudin cone is a cone generated by a linearly independent set of vectors. Here are some results regarding Yudin cones in Hausdorff topological vector spaces: Yudin cones are closed with compact bases (Theorem 3.20), and in the finite dimensional case they are precisely the Archimedean lattice cones (Theorem 3.21). Furthermore, we establish in Section 3.2 that *every vector space admits a Dedekind complete vector lattice ordering whose positive cone is a Yudin cone and has a base* (Corollary 3.18).

The conspicuous idea in this chapter is the identification of closed finite dimensional cones as the cones generated by finite dimensional subspaces of the vector lattice $C[0,1]$. The pull-back cone of a one-to-one operator $T\colon X \to L$, from a vector space to an ordered vector space, is the cone $T^{-1}(L_+)$. The pull-back cones of operators are discussed in Section 3.5. We obtain the following useful result (Theorem 3.51): *Every closed cone of a finite dimensional vector space X is the pull-back cone of a one-to-one operator from X to the vector lattice $C[0,1]$.*

The chapter takes the opportunity to discuss the properties of cones in finite dimensional vector spaces—thus cones of finite dimensional subspaces of $C[0,1]$. As we shall see, the results are much sharper than their infinite dimensional counterparts. For instance, we shall establish that the order units of cones in finite dimensional vector spaces are precisely their interior points and that closed cones are automatically normal.

Section 3.1 discusses the properties of cones in finite dimensional spaces. The results obtained here include the following: A cone is generating if and only if it has an interior point (Lemma 3.2); a cone is closed if and only if it is Archimedean (Corollary 3.4); every closed cone has a compact base (Corollary 3.8); and a closed and generating cone has the Riesz decomposition property if and only if it is a lattice cone (Theorem 3.9). Details about directional wedges and the special class of cones known as Yudin cones can be found in Section 3.2.

Sections 3.3 and 3.4 are devoted to the study of polyhedra and polytopes. A polyhedral wedge in a topological vector space is any wedge generated by a finite number of homogeneous linear inequalities. A polyhedral cone is a polyhedral wedge that is also a cone. In these two sections, among other interesting results, we obtain the following: Every polyhedron has a finite number of extreme points (Lemma 3.27), the compact polyhedra coincide with the polytopes (Theorem 3.32), a wedge in a finite dimensional vector space is polyhedral if and only if it is generated by a finite number of directions (Theorem 3.37). We also provide a proof of the principle of linear programming (Corollary 3.43).

Finally, Section 3.5 collects several alternatives that deal with linear inequalities. It includes the *wedge alternative* (Theorem 3.44), the *operator alternative* (Theorem 3.45), the *Farkas alternative* (Corollary 3.46), and the *concave alternative* (Theorem 3.48).

3.1. Closed cones in finite dimensional vector spaces

In this section we shall present the basic properties of closed cones in finite dimensional vector spaces. As usual, the generic finite dimensional vector space will be $L = \mathbb{R}^\ell$. Keep in mind that the Euclidean topology is the only Hausdorff linear topology that $\mathbb{R}^\ell$ can carry—and hence every finite dimensional vector space admits only one Hausdorff linear topology which is also a norm topology; see Theorem 8.32.

We start with two basic properties of closed cones in finite dimensional vector spaces.

Lemma 3.1. *If $\mathcal{K}$ is a closed cone of a finite dimensional vector space L, then we have*

(1) *the $\mathcal{K}$-order intervals of L are compact and*

(2) *the cone $\mathcal{K}$ is normal.*

Proof. (1) Let $\geq$ denote the vector ordering induced on L by $\mathcal{K}$ and let $[0,x]$ be a $\mathcal{K}$-order interval of L. From $[0,x] = \mathcal{K} \cap (x - \mathcal{K})$, we see that $[0,x]$ is closed; see also Lemma 2.3. To see that $[0,x]$ is also norm bounded, assume that a sequence $\{y_n\} \subseteq [0,x]$ of nonzero vectors satisfies $\|y_n\| \to \infty$. Since the sequence $\left\{\frac{y_n}{\|y_n\|}\right\}$ lies in a compact set, by passing to a subsequence we can assume that $\frac{y_n}{\|y_n\|} \to y$ holds in L. Clearly, $\|y\| = 1$ and so $y \neq 0$. Now from $0 \leq y_n \leq x$, it follows that $0 \leq \frac{y_n}{\|y_n\|} \leq \frac{x}{\|y_n\|}$, and from the closedness of $\mathcal{K}$ and $\frac{x}{\|y_n\|} \to 0$, we see that $0 \leq y \leq 0$ or $y = 0$, which is impossible. Hence, $[0, x]$ is also bounded and thus is a compact set.

(2) It follows immediately from Theorem 2.40. ∎

In finite dimensional spaces the generating cones are precisely the cones with interior points.

Lemma 3.2. *A cone in a finite dimensional vector space is generating if and only if it has an interior point.*

Proof. Let $\mathcal{K}$ be a cone of a finite dimensional vector space L. Assume first that $\mathcal{K}$ has an interior point, say e. By Lemma 2.5, the vector e is a $\mathcal{K}$-order unit, and from this it follows that $\mathcal{K}$ is generating.

For the converse, assume that $\mathcal{K}$ is generating. We claim that $\mathcal{K}$ contains a basis. To see this, let $\{e_1, \ldots, e_k\}$ be a maximal set of linearly independent vectors lying in $\mathcal{K}$. Now let $x \in L$. Since $\mathcal{K}$ is generating, we can choose $y, z \in \mathcal{K}$ such that $x = y - z$. It follows that both y and z lie in the linear span of the set $\{e_1, \ldots, e_k\}$, and from this we infer that x likewise lies in the span of $\{e_1, \ldots, e_k\}$. In other words, $\{e_1, \ldots, e_k\}$ is a basis of L.

Next, put $e = \sum_{i=1}^k e_i$. We claim that e is an interior point of $\mathcal{K}$. To prove this, notice first that the set $V = \{x = \sum_{i=1}^k \lambda_i e_i \colon \ \sum_{i=1}^k |\lambda_i| < 1\}$ is (according to Corollary 8.34) an open neighborhood of zero for the Euclidean topology. Moreover, if $\sum_{i=1}^k |\lambda_i| < 1$, then the vector $x = \sum_{i=1}^k \lambda_i e_i \in V$ satisfies $e + x = \sum_{i=1}^k (1 + \lambda_i) e_i \in \mathcal{K}$. This shows that $e + V \subseteq \mathcal{K}$, so that e is an interior point of $\mathcal{K}$. ∎

Corollary 3.3. *If $\mathcal{K}$ is a closed and generating cone of a finite dimensional vector space L, then the dual cone $\mathcal{K}'$ is closed and generating in L' (and hence $\mathcal{K}'$ has a nonempty interior).*

Proof. By part (2) of Lemma 3.1, we know that $\mathcal{K}$ is a normal cone. So, from Theorem 2.26, the cone $\mathcal{K}'$ is (besides being closed) generating. By Lemma 3.2 the interior of $\mathcal{K}'$ is nonempty. ∎

The closed cones of finite dimensional vector spaces are precisely the Archimedean cones.

Corollary 3.4. *A cone of a finite dimensional vector space is closed if and only if it is Archimedean.*

Proof. Let $\mathcal{K}$ be a cone of a finite dimensional vector space X. If $\mathcal{K}$ is closed, then $\mathcal{K}$ is automatically Archimedean; see Lemma 2.3. For the converse, assume that $\mathcal{K}$ is Archimedean. Since $\mathcal{K}$ is a generating cone in the vector subspace $M = \mathcal{K} - \mathcal{K}$ it generates, it follows from Lemma 3.2 that the interior of $\mathcal{K}$ in M is nonempty. By Lemma 2.4 the cone $\mathcal{K}$ is closed in M. Since M (as a finite dimensional vector subspace of X) is closed, we infer that $\mathcal{K}$ is a closed cone of X. ■

For another characterization of the finite dimensional Archimedean ordered vector spaces see Theorem 3.21.

The strictly positive linear functionals of a closed and generating cone in a finite dimensional space are characterized as follows.

Theorem 3.5. *For a closed and generating cone $\mathcal{K}$ in a finite dimensional vector space L and for some $q \in \mathcal{K}'$ the following statements are equivalent.*

1. *(1) q is $\mathcal{K}$-strictly positive.*
2. *(2) q is an interior point of $\mathcal{K}'$.*
3. *(3) q is an internal point of $\mathcal{K}'$.*
4. *(4) q is an order unit of $\mathcal{K}'$.*

Proof. (1) $\Longrightarrow$ (2) Let q be a $\mathcal{K}$-strictly positive vector and assume by way of contradiction that $q \notin \mathrm{Int}(\mathcal{K}')$, the interior of $\mathcal{K}'$. Since $\mathcal{K}'$ is generating, it follows that $\mathrm{Int}(\mathcal{K}') \neq \emptyset$. By the Separation Theorem 8.16 there exist some nonzero vector $x \in L$ and some $c \in \mathbb{R}$ satisfying $q \cdot x \le c \le p \cdot x$ for all $p \in \mathcal{K}'$. Taking into account that $\mathcal{K}'$ is a cone, we see that $c \le 0$ and $p \cdot x \ge 0$ for all $p \in \mathcal{K}'$. This implies $x \in \mathcal{K}'' = \mathcal{K}$, and so $x >_{\mathcal{K}} 0$. But then, the strict positivity of q implies $q \cdot x > 0$, contrary to $q \cdot x \le c \le 0$. Thus, $q \in \mathrm{Int}(\mathcal{K}')$.

(2) $\Longrightarrow$ (3) Obvious.

(3) $\Longrightarrow$ (4) This follows immediately from Lemma 1.7.

(4) $\Longrightarrow$ (1) Let q be an order unit of $\mathcal{K}'$. Fix $x >_{\mathcal{K}} 0$ and assume by way of contradiction that $q \cdot x = 0$. If $p \in \mathcal{K}'$, then pick some $\lambda \ge 0$ such that $\lambda q \ge \pm p$ and note that $0 = \lambda q \cdot x \ge \pm p \cdot x$ implies $p \cdot x = 0$ for all $p \in \mathcal{K}'$. Since the cone $\mathcal{K}'$ is generating, we see that $p \cdot x = 0$ for all $p \in L'$. Therefore $x = 0$, a contradiction. Thus, $q \cdot x > 0$ and so q is $\mathcal{K}$-strictly positive. ■

Corollary 3.6. *If $\mathcal{K}$ is a closed cone of a finite dimensional vector space L, then the collection of all $\mathcal{K}$-strictly positive vectors is nonempty and dense in $\mathcal{K}'$.*

Proof. Consider the finite dimensional vector subspace $M = \mathcal{K} - \mathcal{K}$. From Corollary 3.3 and Theorem 3.5, it follows that there exists a $\mathcal{K}$-strictly positive functional on M, say q. Now any extension of q to all of L is a $\mathcal{K}$-strictly positive vector of L'. The denseness of the collection of all $\mathcal{K}$-strictly positive vectors in $\mathcal{K}'$ follows immediately from Lemma 2.9. ∎

The norm bounded subsets of a closed cone in a finite dimensional vector space are characterized as follows.

Lemma 3.7. *Let $\mathcal{K}$ be a closed cone of a finite dimensional vector space L. If q is a $\mathcal{K}$-strictly positive vector, then a subset A of $\mathcal{K}$ is bounded if and only if the set of real numbers $q \cdot A = \{q \cdot a\colon\ a \in A\}$ is bounded.*

Proof. Let A be a nonempty subset of $\mathcal{K}$ such that $q \cdot A$ is bounded, where q is a $\mathcal{K}$-strictly positive vector. Replacing L by $\mathcal{K} - \mathcal{K}$, we can assume without loss of generality that the cone $\mathcal{K}$ is also generating. Fix $p \in L'$. Since (by Theorem 3.5) q is an order unit of $\mathcal{K}'$, there exists some $\lambda > 0$ such that $-\lambda q \le p \le \lambda q$. This implies $|p \cdot a| \le \lambda q \cdot a$ for all $a \in A$, and so the set $p \cdot A$ is bounded. Since $p \in L'$ is arbitrary, A must be a bounded subset of L. ∎

An important consequence of the preceding result is that closed cones of finite dimensional vector spaces have compact bases.

Corollary 3.8 (Klee [**61**]). *Every closed cone in a finite dimensional vector space has a compact base.*

Proof. Let $\mathcal{K}$ be a closed cone of a finite dimensional vector space L. By Corollary 3.6 there exists a $\mathcal{K}$-strictly positive linear functional on L, say q. Now a glance at Theorem 1.47 guarantees that $\mathcal{B} = \{x \in \mathcal{K}\colon\ q \cdot x = 1\}$ is a base for $\mathcal{K}$. Since $\mathcal{B}$ is closed, it follows from Lemma 3.7 that $\mathcal{B}$ is a compact subset of L. ∎

Finally, an easy application of the basic duality Theorem 2.13 in connection with the Riesz–Kantorovich Theorem 1.59 yields the following characterization of the Riesz decomposition property in finite dimensional vector spaces. (This is the finite dimensional version of Andô's Theorem 2.47.)

Theorem 3.9. *For a closed and generating cone K in a finite dimensional vector space the following statements are equivalent.*

(1) *$\mathcal{K}$ has the Riesz decomposition property.*

(2) *$\mathcal{K}'$ has the Riesz decomposition property.*

(3) *$\mathcal{K}$ is a lattice cone.*

(4) *$\mathcal{K}'$ is a lattice cone.*

Exercises

1. Give examples of cones in finite dimensional vector spaces whose order intervals
   (a) are bounded but not closed and
   (b) are neither bounded nor closed.
2. If the cone of a finite dimensional ordered vector space is closed, then show that $L^r = L^\sim = L' = L^*$.
3. Can a closed and generating cone in an infinite dimensional Banach space have a compact base?
4. Show that a closed cone in a finite dimensional vector space satisfies the strong Levi property. [*Hint*: Use Theorem 2.45.]
5. Prove that every finite dimensional ordered vector space with a closed cone is Dedekind complete, i.e., every increasing order bounded net has a supremum. [*Hint*: Use the previous exercise in conjunction with Exercise 11 of Section 2.5.]
6. We have seen before that if the cone of an ordered topological vector space (L, τ) is τ-closed, then the order intervals of L are also τ-closed. This exercise discusses the converse of this result.
   (a) Give an example of a Hausdorff ordered topological vector space (L, τ) with τ-closed intervals whose cone is not τ-closed.
   (b) If the order intervals of an ordered topological vector space (L, τ) are τ-closed and L_+ has an interior point, then show that the cone L_+ is τ-closed.
   (c) Establish that a cone $\mathcal{K}$ of a finite dimensional vector space is closed if and only if the $\mathcal{K}$-order intervals of L are closed (in which case the order intervals are also compact).

   [*Hint*: For part (b), let u be an interior point of L_+ and let $x \in \overline{L_+}$. Since $[-u, u]$ is a τ-neighborhood of zero, there exists a net $\{x_\alpha\} \subseteq L_+$ satisfying $x_\alpha \xrightarrow{\tau} x$ and $-u \leq x_\alpha - x \leq u$ for all α. This implies $0 \leq x_\alpha \leq u + x$ for all α.]
7. Use Corollary 3.8 to provide another solution of Exercise 12 of Section 2.1. That is, using Corollary 3.8 show that in a two-dimensional vector space every closed and generating cone is a lattice cone. [*Hint*: Note that a compact base in this case is a line segment and use this in connection with Exercise 7 of Section 1.7.]
8. Prove Theorem 3.9.
9. Show that a closed cone $\mathcal{K}$ of a finite dimensional vector space has the Riesz decomposition property if and only if $\mathcal{K}$ is a lattice cone of $\mathcal{K} - \mathcal{K}$.

3.2. Directional wedges and Yudin cones

In this section X will denote an arbitrary vector space and S a nonempty subset of X. We shall denote by $\mathcal{W}(S)$ the **wedge generated** by S. That is, $\mathcal{W}(S)$ is the smallest (with respect to inclusion) wedge of X that includes S. A moment's thought reveals that $\mathcal{W}(S)$ is the convex subset of X consisting of all vectors $x \in X$ for which there exist vectors $s_1, \ldots, s_n$ in S and nonnegative real numbers $\lambda_1, \ldots, \lambda_n$ such that $x = \sum_{i=1}^{n} \lambda_i s_i$. Also, it should be easy to check that $\mathcal{W}(S) = \mathcal{W}(\operatorname{co} S)$. If $\mathcal{W}$ is an arbitrary wedge of X, then clearly $\mathcal{W}(\mathcal{W}) = \mathcal{W}\big(\mathcal{W}\setminus\{0\}\big) = \mathcal{W}$. This shows that every wedge can be viewed as being generated by a set.

If $\mathcal{W} = \mathcal{W}(S)$, then we can think of the wedge $\mathcal{W}$ as being generated by the "directions" provided by the vectors of S. For this reason the wedge $\mathcal{W}(S)$ is quite often referred to also as the **directional wedge** of the set S (and the vectors of S are called the **directions of the wedge** $\mathcal{W}$).

Lemma 3.10. *For a nonempty convex subset S of a vector space we have the following.*

(1) *The wedge $\mathcal{W}(S)$ is given by $\mathcal{W}(S) = \{\lambda s\colon\ \lambda \geq 0$ and $s \in S\}$.*

(2) *If $0 \notin S$, then $\mathcal{W}(S)$ is a cone.*

Proof. (1) Let $x \in \mathcal{W}(S) \setminus \{0\}$ and write $x = \sum_{i=1}^{n} \lambda_i s_i$ with $s_i \in S$ and $\lambda_i > 0$ for each i. If $\lambda = \sum_{i=1}^{n} \lambda_i > 0$, then the convexity of S implies that the vector $s = \sum_{i=1}^{n} \frac{\lambda_i}{\lambda} s_i$ belongs to S and clearly $x = \lambda s$ holds.

(2) If $\lambda_1 s_1 = -\lambda_2 s_2$ holds with $s_1, s_2 \in S$ and $\lambda_i \geq 0$ for each i and $\lambda_1 + \lambda_2 > 0$, then $0 = \frac{\lambda_1}{\lambda_1+\lambda_2} s_1 + \frac{\lambda_2}{\lambda_1+\lambda_2} s_2 \in S$, which is impossible. This shows that if $0 \notin S$, then $\mathcal{W}(S)$ is a cone. ■

Lemma 3.11. *If a wedge $\mathcal{W}(S)$ is a cone, then every nonzero extremal vector u of $\mathcal{W}(S)$ is a multiple of some direction of the cone, i.e., $\mathcal{R}_u = \mathcal{R}_x$ holds true for some $x \in S$.*

Proof. Let u be a nonzero extremal vector of the cone $\mathcal{W}(S)$. Pick positive scalars $\lambda_1, \ldots, \lambda_k$ and nonzero directions $s_1, \ldots, s_k \in S$ such that $u = \sum_{i=1}^{k} \lambda_i s_i$. From $0 < \lambda_1 s_1 \leq u$ and the fact that u is an extremal vector, it easily follows that there exists some $\mu > 0$ such that $u = \mu s_1$. This implies $\mathcal{R}_u = \mathcal{R}_{s_1}$. ■

We also have the following result concerning wedges generated by weakly compact sets in Banach spaces.

Lemma 3.12. *If S is a nonempty weakly compact convex subset of a Banach space such that $0 \notin S$, then the wedge $\mathcal{W}(S)$ is a closed cone.*

Proof. According to part (2) of Lemma 3.10 the wedge $\mathcal{W}(S)$ is a cone. Now assume that a sequence $\{x_n\}$ of $\mathcal{W}(S)$ satisfies $\|x_n - x\| \to 0$ (and so $x_n \xrightarrow{w} x$). We must show that $x \in \mathcal{W}(S)$. If $x = 0$, then there is nothing to prove. So, we can suppose that $x \neq 0$. For each n pick $\lambda_n > 0$ and

$s_n \in S$ such that $x_n = \lambda_n s_n$. Since S is weakly compact, by passing to a subsequence, we can assume without loss of generality that there exists some $s \in S$ such that $s_n \xrightarrow{w} s$; see Theorem 8.29.

If there exists a subsequence $\{\lambda_{k_n}\}$ of $\{\lambda_n\}$ such that $\lambda_{k_n} \to \infty$, then from the norm boundedness of the sequence $\{x_n\}$ and $s_n = \frac{x_n}{\lambda_n} \xrightarrow{w} 0$, we get $s = 0 \in S$, which is impossible. On the other hand, if some subsequence $\{\lambda_{m_n}\}$ of $\{\lambda_n\}$ satisfies $\lambda_{m_n} \to 0$, then from $x_{m_n} = \lambda_{m_n} s_{m_n} \xrightarrow{w} 0 \cdot s = 0$, we get $x = 0$, which is also a contradiction. These arguments show that we can assume (by passing to a subsequence if necessary) that there exists some $0 < \lambda < \infty$ such that $\lambda_n \to \lambda$.

Now note that $x_n = \lambda_n s_n \xrightarrow{w} \lambda s$, and from this we infer $x = \lambda s \in \mathcal{W}(S)$, as desired. Hence $\mathcal{W}(S)$ is a closed cone. ■

Corollary 3.13. *Every cone of a Banach space with a weakly compact base is closed.*

Recall that a **Hamel basis** of X is a maximal collection of linearly independent vectors $\{e_i\}_{i \in I}$. This implies that for each $x \in X$ there exists a unique family $\{\lambda_i\}_{i \in I}$ of real numbers such that $\lambda_i = 0$ holds for all but a finite number of i and $x = \sum_{i \in I} \lambda_i e_i$. If Y is a vector subspace of X and S is a Hamel basis for Y, then it is not difficult to see that $Y = \mathcal{W}\big(S \cup (-S)\big)$.

Lemma 3.14. *If S is a linearly independent set of a vector space X, then*

(1) *the wedge $\mathcal{W}(S)$ is an Archimedean cone with a base and*

(2) *the extremal rays of $\mathcal{W}(S)$ are precisely the half-rays of the form $\mathcal{R}_x$ with $x \in S$.*

Proof. Assume that S is a linearly independent set.[1]

(1) We shall show first that the wedge $\mathcal{W}(S)$ is a cone. To this end, let a vector $x \in X$ satisfy $x \in \mathcal{W}(S) \cap [-\mathcal{W}(S)]$. Then there exist $s_1, \ldots, s_k \in S$ and nonnegative real numbers $\lambda_1, \ldots, \lambda_k$ and $\mu_1, \ldots, \mu_k$ such that

$$x = \sum_{i=1}^{k} \lambda_i s_i = -\sum_{i=1}^{k} \mu_i s_i \,.$$

The linear independence of S implies $0 \le \lambda_i = -\mu_i \le 0$ for all i, and so $\lambda_i = 0$ for each i. Thus, $x = \sum_{i=1}^{k} \lambda_i s_i = 0$, and so $W(S)$ is a cone.

To see that the cone $\mathcal{K} = \mathcal{W}(S)$ is Archimedean, assume that $nx \le_{\mathcal{K}} y$ holds for each n and some $x \in X$ and $y \in \mathcal{W}(S)$. Also, let $M = \mathcal{K} - \mathcal{K}$, the vector subspace generated by $\mathcal{K}$. From $x \le_{\mathcal{K}} y$ and $x = y - (y - x)$, we see that $x \in M$. Clearly, S is a Hamel basis for M and so we can write

[1] Recall that a subset S of a vector space is said to be **linearly independent** if every finite subset of S is a linearly independent set.

$x = \sum_{s\in S} \lambda_s s$, where $\lambda_s = 0$ for all but a finite number of vectors $s \in S$. Letting $y = \sum_{s\in S} \alpha_s s$ be the representation of y as a member of $\mathcal{W}(S)$, it follows from $nx \leq_{\mathcal{K}} y$ that for each $s \in S$ we have $n\lambda_s \leq \alpha_s$ for all $n \in N$. This implies $\lambda_s \leq 0$ for each $s \in S$ and thus $x = \sum_{s\in S} \lambda_s s \leq_{\mathcal{K}} 0$, proving that $\mathcal{W}(S)$ is an Archimedean cone.

Next, observe that the mapping $f\colon \mathcal{W}(S) \to [0,\infty)$, defined via the formula $f\big(\sum_{s\in S} \mu_s s\big) = \sum_{s\in S} \mu_s$, is strictly positive and additive. Now invoke Theorem 1.47 to infer that $\mathcal{W}(S)$ also has a base.

(2) Let $x \in S$, and fix some $y \in X$ with $0 \leq y \leq x$, i.e., $y \in \mathcal{W}(S)$ and $x - y \in \mathcal{W}(S)$. Choose $s_1,\dots,s_n \in S$ and nonnegative real numbers $\lambda_1,\dots,\lambda_n$ and $\mu_1,\dots,\mu_n$ such that $y = \sum_{i=1}^n \lambda_i s_i$ and $x - y = \sum_{i=1}^n \mu_i s_i$. It follows that $x - \sum_{i=1}^n (\lambda_i + \mu_i) s_i = 0$. Since $x \in S$ and S is a linearly independent set, there exists some i such that $s_i = x$. We can assume that $x = s_1$, and from this we see that $\lambda_1 + \mu_1 = 1$ and $\lambda_i + \mu_i = 0$ for $i = 2,\dots,n$. Clearly, $\lambda_i = 0$ for all $i = 2,\dots,n$, and so $y = \sum_{i=1}^n \lambda_i s_i = \lambda_1 x$. This shows that x is an extremal vector of $\mathcal{W}(S)$, and hence $\mathcal{R}_x$ is an extremal ray of the cone $W(S)$.

To finish the proof, notice that (by Lemma 3.11) every extremal ray of $\mathcal{W}(S)$ is among the half-rays determined by the directions of S. ∎

The following definition is inspired by the work of A. I. Yudin [**92**].

Definition 3.15. *Any cone of a vector space that is generated by a Hamel basis of the space is called a* ***Yudin cone****.*

The Hamel basis that generates a Yudin cone $\mathcal{K}$ is also called a **Yudin basis** (or a **positive basis**) of $\mathcal{K}$. As expected, the Yudin basis of a Yudin cone is essentially uniquely determined.

Lemma 3.16. *A given cone $\mathcal{K}$ of a vector space has essentially at most one Yudin basis in the following sense: If $\{e_i\}_{i\in I}$ and $\{f_j\}_{j\in J}$ are two Yudin bases of $\mathcal{K}$, then each e_i is a scalar multiple of some f_j and each f_j is a scalar multiple of some e_i.*

In other words, if $\{e_i\}_{i\in I}$ is a Yudin basis of a cone $\mathcal{K}$, then every other Yudin basis of $\mathcal{K}$ is of the form $\{\lambda_i e_i\}_{i\in I}$, where $\lambda_i > 0$ for each i.

Proof. According to part (2) of Lemma 3.14 the families $\{e_i\}_{i\in I}$ and $\{f_j\}_{j\in J}$ are (up to scalar multiples) the extremal vectors of $\mathcal{K}$. ∎

Yudin cones are Archimedean lattice cones.

Theorem 3.17. *Every Yudin cone is an Archimedean lattice cone with a base. Specifically, if $\mathcal{K}$ is the Yudin cone of a vector space X generated by* *a Hamel basis* $\{e_i\}_{i\in I}$ *and* $x = \sum_{i\in I} \lambda_i e_i$, $y = \sum_{i\in I} \mu_i e_i$ *are vectors in* X, *then we have the following.*

(1) $x \geq_{\mathcal{K}} y$ *holds in* X *if and only if* $\lambda_i \geq \mu_i$ *for each* $i \in I$.

(2) *The* $\mathcal{K}$*-lattice operations on* x *and* y *are given componentwise, i.e.,*

$$x \vee y = \sum_{i\in I} \max\{\lambda_i, \mu_i\} e_i \quad \text{and} \quad x \wedge y = \sum_{i\in I} \min\{\lambda_i, \mu_i\} e_i \,.$$

(3) *A base for* $\mathcal{K}$ *is the convex set*

$$\mathcal{B} = \Big\{ \sum_{i\in I} \nu_i e_i \in X\colon\ \nu_i \geq 0 \text{ for all } i \text{ and } \sum_{i\in I} \nu_i = 1 \Big\}.$$

(4) *The Riesz space* $(X, \mathcal{K})$ *is Dedekind complete, where if a family of vectors* $\big\{x_\alpha = \sum_{i\in I} \lambda_i^\alpha e_i\big\}_{\alpha\in A}$ *is bounded from above, then its supremum is given by*

$$\sup_{\alpha\in A} x_\alpha = \sum_{i\in I} \big[\sup_{\alpha\in A} \lambda_i^\alpha\big] e_i \,.$$

Proof. The first part is Lemma 3.14(1). The proof of part (1) follows immediately from the fact that a vector $x = \sum_{i\in I} \lambda_i e_i$ satisfies $x \in \mathcal{K}$ if and only if $\lambda_i \geq 0$ holds for each $i \in I$, and the proof of (2) is an easy consequence of (1). The proofs of the other claims are straightforward. ■

Because every vector space has a Hamel basis, every vector space admits a generating Yudin cone. Specifically, we have the following result.

Corollary 3.18. *Every vector space has a Yudin generating cone. In particular, every vector space admits a cone with a base under which it is a Dedekind complete Riesz space.*

Also, if $\mathcal{K}$ is a cone in a vector space generated by a nonempty linearly independent set, then it should be clear that $\mathcal{K}$ is a Yudin cone of $\mathcal{K} - \mathcal{K}$, the vector subspace generated by $\mathcal{K}$.

To continue our discussion on Yudin cones, we need the following result.

Lemma 3.19. *In a Hausdorff topological vector space, the wedge generated by a (nonempty) finite subset is closed.*

Proof. Let $S = \{x_1, \ldots, x_k\}$ be a (nonempty) finite subset of a Hausdorff topological vector space (X, τ). In this case the wedge generated by S is the convex set

$$\mathcal{W} = \Big\{ \sum_{i=1}^{k} \lambda_i x_i\colon\ \lambda_i \geq 0 \text{ for each } i \Big\}.$$

Now fix some nonzero vector $x = \sum_{i=1}^{k} \lambda_i x_i \in \mathcal{W}$, where $\lambda_i \geq 0$ for each i. We claim that there exist a linearly independent subset S_1 of S and nonnegative scalars $\{\alpha_s\colon\ s \in S_1\}$ such that $x = \sum_{s \in S_1} \alpha_s s$.

To see this, notice first that we can assume $\lambda_i > 0$ for each i; otherwise we drop the terms with $\lambda_i = 0$. If the set S is linearly independent, then there is nothing to prove. So assume that S is linearly dependent. This means that there exist scalars $\alpha_1, \ldots, \alpha_k$, not all zero, and $x_1, \ldots, x_k \in S$ such that $\sum_{i=1}^{k} \alpha_i x_i = 0$. We can assume that $\alpha_j > 0$ for some j; otherwise multiply them by -1. Now let $c = \max\{\frac{\alpha_i}{\lambda_i}\colon\ i = 1, \ldots, k\}$, and notice that $c > 0$. In particular, we have $\lambda_i \geq \frac{1}{c}\alpha_i$ (or $\lambda_i - \frac{1}{c}\alpha_i \geq 0$) for each i and $\lambda_i = \frac{1}{c}\alpha_i$ for some i. This implies that

$$x = \sum_{i=1}^{k} \lambda_i x_i = \sum_{i=1}^{k} \lambda_i x_i - \tfrac{1}{c}\sum_{i=1}^{k} \alpha_i x_i = \sum_{i=1}^{k} (\lambda_i - \tfrac{1}{c}\alpha_i) x_i$$

is a linear combination of the x_i with nonnegative coefficients, and one of them is zero. In other words, we have shown that if the set S is not linearly independent, then we can write x as a linear combination with nonnegative coefficients of at most $k-1$ vectors in S. Our claim now can be completed by repeating this process.

Next, assume that a sequence $\{y_n\}$ of $\mathcal{W}$ satisfies $y_n \xrightarrow{\tau} y$ in X. Since the collection of all linearly independent subsets of S is a finite set, by the above discussion, it follows that there exist a linearly independent subset of S, say $\{z_1, \ldots, z_m\}$, and a subsequence of $\{y_n\}$ (which we shall denote by $\{y_n\}$ again) such that $y_n = \sum_{i=1}^{m} \mu_i^n z_i$ with all coefficients μ_i^n nonnegative. If Z denotes the m-dimensional vector subspace generated in X by $\{z_1, \ldots, z_m\}$, then the Hausdorfness of τ implies that Z is a closed subspace of X (see Corollary 8.33) and that τ induces the Euclidean topology on Z. In particular, this implies that for each i we have $\mu_i^n \xrightarrow[n\to\infty]{} \mu_i \geq 0$, and from this we see that $y = \sum_{i=1}^{m} \mu_i z_i \in \mathcal{W}$. Therefore, $\mathcal{W}$ is τ-closed. ∎

We now come to a basic result regarding Yudin cones.

Theorem 3.20. *Assume that (X, τ) is a Hausdorff topological vector space and that S is a finite linearly independent set. If $\mathcal{K} = \mathcal{W}(S)$, then the following hold.*

(1) *The cone $\mathcal{K}$ is τ-closed.*

(2) *The cone $\mathcal{K}$ admits a strictly positive additive function.*

(3) *The cone $\mathcal{K}$ has a compact base.*

(4) *When τ is also locally convex, the topological vector space X admits a continuous $\mathcal{K}$-strictly positive linear functional.*

Proof. (1) It follows immediately from Lemma 3.19.

(2) Let Y be the vector subspace generated by $\mathcal{K}$, i.e., let $Y = \mathcal{K} - \mathcal{K}$. Clearly, Y is finite dimensional and $\mathcal{K}$ generates Y. By Corollary 3.6, there exists a strictly positive linear functional $f\colon Y \to \mathbb{R}$. Its restriction to $\mathcal{K}$ is a strictly positive additive function.

(3) Note that $\mathcal{K}$ is a closed cone of the finite dimensional vector space Y and that τ induces the Euclidean topology on Y. Now a glance at Corollary 3.8 guarantees that $\mathcal{K}$ has a compact base.

(4) Let $f\colon Y \to \mathbb{R}$ be a $\mathcal{K}$-strictly positive linear functional. Since the Hausdorff linear topology τ-induces the Euclidean topology on Y, it follows that f is τ-continuous. Now any τ-continuous linear extension of f to all of X (one of which is always guaranteed by the Hahn–Banach extension theorem) is a continuous $\mathcal{K}$-strictly positive linear functional on X. ■

The Yudin cones in finite dimensional vector spaces are precisely the Archimedean lattice cones. This basic result (that originated the study of Yudin cones) is due to A. I. Yudin [**92**].

Theorem 3.21 (Yudin). *A cone in a finite dimensional vector space is a Yudin cone if and only if it is an Archimedean lattice cone.*

In particular, every Archimedean finite dimensional Riesz space is a Banach lattice.

Proof. Assume that $\mathcal{K}$ is an Archimedean lattice cone of a finite dimensional vector space L. Start by observing that (by Corollary 3.4) the cone $\mathcal{K}$ is closed. Now a glance at Corollary 3.8 shows that $\mathcal{K}$ has a compact base $\mathcal{B}$. If $\mathcal{E}$ denotes the set of extreme points of $\mathcal{B}$, then $\mathcal{K} = \mathcal{W}(\mathcal{E})$.[2] In particular, the set $\mathcal{E}$ (which coincides up to positive scalar multiples with the extremal vectors of $\mathcal{K}$) generates the vector space L. According to Theorem 1.45 the set $\mathcal{E}$ is an independent set, and from this we see that its cardinality must be the same as the dimension of L. Therefore, $\mathcal{K}$ is a Yudin cone.

For the last part consider a finite dimensional vector space L equipped with an Archimedean lattice cone $\mathcal{K}$ and let $\{e_1, \ldots, e_n\}$ be the Yudin basis of the cone. If we define the function $\|\cdot\|\colon L \to \mathbb{R}$ via the formula

$$\Big\|\sum_{i=1}^{n} \lambda_i e_i\Big\| = \sum_{i=1}^{n} |\lambda_i|\,,$$

[2] Here we use the fact that every nonempty compact convex subset of a finite dimensional vector space is the convex hull of its extreme points; see [**12**, Theorem 7.68, p. 297].

then it is easy to see that $\|\cdot\|$ is a norm on L (and so it must be equivalent to the Euclidean norm of L). So, $(L, \|\cdot\|)$ is a Banach space. To complete the proof, it remains to observe that $|x| \leq_{\mathcal{K}} |y|$ in L implies $\|x\| \leq \|y\|$. ■

An important class of vector lattices is the class of vector lattices of type ϕ_I. If I is an arbitrary index set, then ϕ_I is the ideal of $\mathbb{R}^I$ consisting of all "eventually zero" real functions on I. That is,

$$\phi_I = \Big\{\lambda = (\lambda_i)_{i\in I} \in \mathbb{R}^I\colon \ \lambda_i = 0 \ \text{ for all but a finite number of } i\Big\}.$$ [3]

The ordering and the lattice operations in ϕ_I are the pointwise ones. Moreover, ϕ_I equipped with the sup norm, defined by

$$\|\lambda\|_\infty = \sup_{i\in I} |\lambda_i|\,,$$

is a Dedekind complete normed vector lattice. The norm completion of $(\phi_I, \|\cdot\|_\infty)$ is the Banach lattice $c_0(I)$; see Exercise 5 at the end of the section.

Fix a nonempty index set I. We can view every vector $y = (y_i)_{i\in I} \in \mathbb{R}^I$ as a linear functional acting on ϕ_I via the formula

$$\langle x, y\rangle = \sum_{i\in I} x_i y_i\,. \qquad (\star)$$

The duality in $(\star)$ shows that the pair $\langle \phi_I, \mathbb{R}^I\rangle$ is indeed a dual system. As usual, for each $i \in I$ we let $e_i = \chi_{\{i\}}$, i.e., $e_i\colon I \to \mathbb{R}$ is the function that takes the value one on i and zero at every other point. Now if $f \in \phi_I^*$ and if we let $y = (f(e_i))_{i\in I} \in \mathbb{R}^I$, then for each $x = (x_i)_{i\in I} \in \phi_I$ we have

$$f(x) = f\Big(\sum_{i\in I} x_i e_i\Big) = \sum_{i\in I} x_i f(e_i) = \langle x, y\rangle\,.$$

This shows that the algebraic dual of ϕ_I is precisely $\mathbb{R}^I$. In addition, it is not difficult to see that the order dual of ϕ_I is also $\mathbb{R}^I$. Thus, we have the following result.

Lemma 3.22. *If I is a nonempty set, then ϕ_I is an order complete Riesz space and $\phi_I^\sim = \phi_I^* = \mathbb{R}^I$. Moreover, the dual cone of ϕ_I^+ with respect to the dual pair $\langle \phi_I, \mathbb{R}^I\rangle$ is $\mathbb{R}^I_+$.*

It turns out that every vector space X can be identified with a vector lattice of the type ϕ_I. To see this, fix a Hamel basis $\{e_i\}_{i\in I}$ of X and let $\mathcal{K}$ be the Yudin cone generated in X by $\{e_i\}_{i\in I}$. Now notice that the operator $T\colon \phi_I \to X$ defined for each $\lambda = (\lambda_i)_{i\in I} \in \phi_I$ by $T(\lambda) = \sum_{i\in I} \lambda_i e_i$ is a surjective lattice isomorphism. For more about Yudin cones see [**13**].

[3] This definition justifies the letter ϕ used to denote this space. It designates the fact that it consists of "finite" families.

Exercises

1. If $\mathcal{K}$ is a generating cone of a vector space L, then show that there exists a Yudin subcone $\mathcal{K}_1$ of $\mathcal{K}$ such that $\mathcal{K}_1 - \mathcal{K}_1 = L$. [*Hint*: Using Zorn's lemma show that there exists a Hamel basis of L lying in $\mathcal{K}$.]

2. Complete the missing details in the proof of Theorem 3.17.

3. Prove the following property that was used in the last part of the proof of Lemma 3.19. Let $\{u_1, \ldots, u_k\}$ be a finite collection of linearly independent vectors of a Hausdorff topological vector space (X, τ). If $\sum_{i=1}^k \lambda_i^n u_i \xrightarrow[n\to\infty]{\tau} x$ in X, then $\lambda_i^n \xrightarrow[n\to\infty]{} \lambda_i$ holds in $\mathbb{R}$ for each i and $x = \sum_{i=1}^k \lambda_i u_i$.

4. Let I be an infinite index set and consider the dual system $\langle \phi_I, \mathbb{R}^I \rangle$, where for each $x = (x_i)_{i\in I} \in \phi_I$ and $y = (y_i)_{i\in I} \in \mathbb{R}^I$ we let $\langle x, y\rangle = \sum_{i\in I} x_i y_i$. For the standard cone $\mathcal{K} = \phi_I^+$ of ϕ_I establish the following.
   (a) The dual cone of $\mathcal{K}$ coincides with $\mathbb{R}_+^I$, i.e., $\mathcal{K}' = \mathbb{R}_+^I$.
   (b) The cone $\mathcal{K}$ has no $\mathcal{K}$-strictly positive vectors.
   (c) What are the $\mathcal{K}$-strictly positive vectors of $\mathbb{R}^I$?

5. Let I be an infinite set. As usual, let $c_0(I)$ denote the vector subspace of $\mathbb{R}^I$ consisting of all vectors $x = (x_i)_{i\in I} \in \mathbb{R}^I$ such that for each $\epsilon > 0$ there exists a finite subset F of I (depending on x and ϵ) satisfying $|x_i| < \epsilon$ for all $i \notin F$. Alternatively, if we endow I with the discrete topology and if I_∞ is its one-point compactification, then $c_0(I) = \{f \in C(I_\infty)\colon\ f(\infty) = 0\}$. Establish the following.
   (a) $c_0(I)$ with the sup norm is a Banach lattice.
   (b) The norm completion of $(\phi_I, \|\cdot\|_\infty)$ is the Banach lattice $c_0(I)$.

6. Complete the details of the proof of Lemma 3.22.

7. Consider the dual system $\langle \phi_I, \mathbb{R}^I \rangle$, where I is an infinite index set. If τ denotes the Hausdorff locally convex topology of pointwise convergence on ϕ_I, then show that τ is strictly coarser than the weak topology $\sigma(\phi_I, \mathbb{R}^I)$, i.e, $\tau \subseteq \sigma(\phi_I, \mathbb{R}^I)$ and $\tau \neq \sigma(\phi_I, \mathbb{R}^I)$. [*Hint*: If $\{i_1, i_2, \ldots\}$ is a countable subset of I, then for each $i \in I$ we have $\chi_{\{i_n\}}(i) \to 0$ while the constant function $\mathbf{1}$ on I satisfies $\langle \chi_{\{i_n\}}, \mathbf{1}\rangle = 1$ for each n.]

8. Show that the order topology induced on ϕ_I by the cone ϕ_I^+ coincides with the Mackey topology $\tau(\phi_I, \mathbb{R}^I)$, i.e., $\tau_o(\phi_I^+) = \tau(\phi_I, \mathbb{R}^I)$.

9. Establish that the dual pair $\langle \phi_I, \mathbb{R}^I \rangle$ representing a dual system $\langle X, X^*\rangle$ (where X is a vector space and X^* is its algebraic dual) is essentially uniquely determined by proving that for two nonempty sets I_1 and I_2 the following statements are equivalent.
   (a) The sets I_1 and I_2 have the same cardinality.
   (b) The Riesz spaces ϕ_{I_1} and ϕ_{I_2} are lattice isomorphic.
   (c) The Riesz spaces $\mathbb{R}^{I_1}$ and $\mathbb{R}^{I_2}$ are lattice isomoprhic.

3.3. Polyhedral wedges and cones

In this section, X will denote a topological vector space. A **linear inequality** is any inequality of the form

$$x'(x) \leq \alpha$$

(or, equivalently, of the type $x'(x) \geq \alpha$), where $x' \in X'$ and α is a fixed real number. Any linear inequality of the form $x'(x) \leq 0$ (or, equivalently, of the form $x'(x) \geq 0$) is called a **homogeneous linear inequality**. Clearly, the solution set of any linear inequality is a closed half-space of X.

Definition 3.23. *A* ***polyhedron*** Π *of a topological vector space is the solution set of a finite collection of linear inequalities or, equivalently, the intersection of a finite number of closed half-spaces. That is,* Π *is a polyhedron*[4] *if there exist* $x'_1, \ldots, x'_n \in X'$ *and* $\alpha_1, \ldots, \alpha_n \in \mathbb{R}$ *such that*

$$\Pi = \{x \in X\colon\ x'_i(x) \leq \alpha_i \ \ \textit{for } i = 1, \ldots, n\}.$$

A ***polyhedral wedge*** *is a wedge that is the solution set of a finite collection of homogeneous linear inequalities. A polyhedral wedge that is also a cone is called a* ***polyhedral cone****.*

Three polyhedra are shown in the figure below.

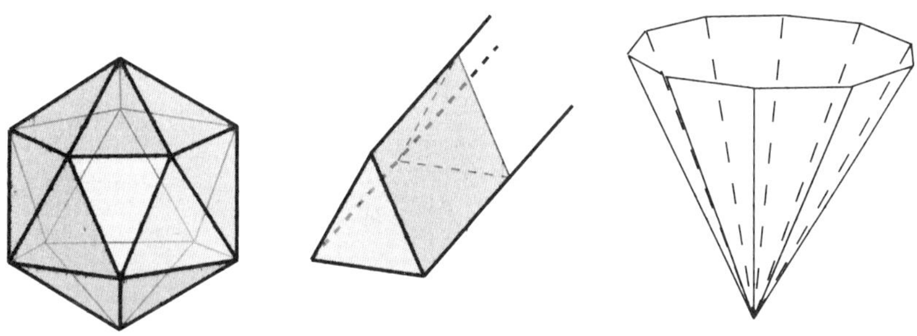

Compact polyhedron Unbounded polyhedron Polyhedral cone

It should be clear that every polyhedron is a (possibly empty) closed convex set and that each polyhedral wedge is a closed wedge. If $x'_1, \ldots, x'_n$ are linear functionals in X', then the polyhedral wedge

$$\{x \in X\colon\ x'_i(x) \geq 0 \ \text{ for } i = 1, \ldots, n\}$$

will be referred to as the **polyhedral wedge generated** by $x'_1, \ldots, x'_n$.

The polyhedral cones are always Archimedean. This follows immediately from Lemma 2.3.

[4] The word "polyhedron" is the Greek word "$\pi o\lambda \acute{\upsilon} \epsilon \delta \rho o \nu$" which means a solid set with many seats or faces.

Lemma 3.24. *Every polyhedral cone is Archimedean—and so if X admits an Archimedean polyhedral cone, then it is necessarily a Hausdorff topological vector space.*

In other words, only Hausdorff topological vector spaces can admit polyhedral cones. Regarding duality properties of polyhedral wedges, we have the following.

Theorem 3.25. *Let $\langle X, X' \rangle$ be a dual system. Then for a wedge $\mathcal{W}$ of X we have the following.*

(1) *The wedge $\mathcal{W}$ is generated by a finite number of directions $x_1, \ldots, x_n$ if and only if its dual wedge $\mathcal{W}'$ is the polyhedral wedge generated by the linear functionals $x_1, \ldots, x_n$. In other words, we have*

$$\mathcal{W} = \Big\{ \sum_{i=1}^{n} \lambda_i x_i \colon \ \lambda_i \geq 0 \ \textit{for each} \ i = 1, \ldots, n \Big\}$$

if and only if

$$\mathcal{W}' = \big\{ x' \in X' \colon \ x_i(x') = x'(x_i) \geq 0 \ \textit{for all} \ i = 1, \ldots, n \big\} .$$

(2) *The wedge $\mathcal{W}$ is a polyhedral wedge generated by a finite number of inequalities $x'_i(x) \geq 0$ $(i = 1, \ldots, n)$ if and only if its dual wedge $\mathcal{W}'$ is generated by the directions $x'_1, \ldots, x'_n$. That is, we have*

$$\mathcal{W} = \Big\{ x \in X \colon \ x'_i(x) \geq 0 \ \textit{for each} \ i = 1, \ldots, n \Big\}$$

if and only if

$$\mathcal{W}' = \Big\{ \sum_{i=1}^{n} \lambda_i x'_i \colon \ \lambda_i \geq 0 \ \textit{for each} \ i = 1, \ldots, n \Big\} .$$

Proof. If $\mathcal{W} = \big\{ \sum_{i=1}^{n} \lambda_i x_i \colon \ \lambda_i \geq 0 \text{ for each } i = 1, \ldots, n \big\}$, then it should be obvious that $\mathcal{W}' = \big\{ x' \in X' \colon \ x_i(x') = x'(x_i) \geq 0 \text{ for all } i = 1, \ldots, n \big\}$.

Now assume that $\mathcal{W} = \big\{ x \in X \colon \ x'_i(x) \geq 0 \text{ for each } i = 1, \ldots, n \big\}$ and let $\mathcal{W}_1 = \big\{ \sum_{i=1}^{n} \lambda_i x'_i \colon \ \lambda_i \geq 0 \text{ for each } i = 1, \ldots, n \big\}$. From Lemma 3.19 we know that $\mathcal{W}_1$ is $\sigma(X', X)$-closed. From the definition of $\mathcal{W}$, it follows that $\mathcal{W}_1 \subseteq \mathcal{W}'$. Now suppose by way of contradiction that there exists some $x'_0 \in \mathcal{W}'$ such that $x'_0 \notin \mathcal{W}_1$. Using the Separation Corollary 8.18, we see that there exist $x_0 \in X$ and $c \in \mathbb{R}$ satisfying $x'_0(x_0) < c \leq x'(x_0)$ for all $x' \in \mathcal{W}_1$. Since $\mathcal{W}_1$ is a wedge, we get $c \leq 0$ and $x'(x_0) \geq 0$ for all $x' \in \mathcal{W}_1$. In particular, we have $x'_i(x_0) \geq 0$ for all $i = 1, \ldots, n$ and so $x_0 \in \mathcal{W}$. However, the latter implies $x'_0(x_0) \geq 0$, contrary to $x'_0(x_0) < c \leq 0$. Hence $\mathcal{W}' = \mathcal{W}_1$ holds true.

To complete the proof, use the above properties in conjunction with the fact (since in both cases the wedge $\mathcal{W}$ is closed) that $\mathcal{W}'' = \mathcal{W}$. ■

Now let $X = \mathbb{R}^m$. If $\mathcal{W}$ is a wedge generated in X by a finite number of directions, say $x_1, \ldots, x_n$, then letting A be the $m \times n$ real matrix A whose columns are the vectors x_i (written in column form), it is not difficult to see that

$$\mathcal{W} = \Big\{ \sum_{i=1}^{n} \lambda_i x_i \colon\ \lambda_i \geq 0 \text{ for each } i \Big\} = \big\{ A\lambda \colon\ \lambda \in \mathbb{R}^n_+ \big\}.$$

Moreover, of course, every subset of the form $\big\{A\lambda \colon\ \lambda \in \mathbb{R}^n_+\big\}$ (where A is an $m \times n$ real matrix) is a wedge of X that is generated by the columns of the matrix A. By Theorem 3.25, we know that the dual wedge of $\mathcal{W}$ is the polyhedral wedge generated by the columns of A (or, equivalently, by the rows of A^t), where A^t denotes the $n \times m$ transpose matrix of A.

In other words, in finite dimensional vector spaces Theorem 3.25 can be stated in "matrix form" as follows.

Corollary 3.26. *For an $m \times n$ real matrix A we have the following.*

1. (1) *If $\mathcal{W}$ is the wedge of $\mathbb{R}^m$ generated by the columns of A (that is, $\mathcal{W} = \{A\lambda \colon\ \lambda \in \mathbb{R}^n_+\}$), then the polyhedral dual wedge of $\mathcal{W}$ is generated by the rows of A^t, i.e., $\mathcal{W}' = \big\{y \in \mathbb{R}^m \colon\ A^t y \geq 0\big\}$.*
2. (2) *If $\mathcal{W}$ is the polyhedral wedge of $\mathbb{R}^m$ generated by the rows of A (that is, $\mathcal{W} = \{x \in \mathbb{R}^n \colon\ Ax \geq 0\}$), then the dual wedge of $\mathcal{W}$ is generated by the columns of A^t, i.e., $\mathcal{W}' = \big\{A^t y \colon\ y \in \mathbb{R}^m_+\big\}$.*

The polyhedra have a finite number of extreme points.

Lemma 3.27. *Every polyhedron has a finite number of extreme points.*

Proof. The proof presented here is due to H. Nikaidô [**72**, p. 41]. Assume that Π is a nonempty polyhedron in a topological vector space X given by

$$\Pi = \big\{x \in X \colon\ x'_i(x) \leq \alpha_i \text{ for } i = 1, \ldots, n\big\},$$

where $x'_i \in X'$ and $\alpha_i \in \mathbb{R}$ for each i. Let $\mathcal{E}$ denote the set of extreme points of Π and assume that $\mathcal{E}$ is nonempty. Define a mapping A from $\mathcal{E}$ into the collection of all subsets of $\{1, \ldots, n\}$ via the formula

$$A(u) = \big\{i \in \{1, \ldots, n\} \colon\ x'_i(u) < \alpha_i\big\}.$$

We claim that A is one-to-one. If this claim is established, then since the set $\{1, \ldots, n\}$ has 2^n subsets, it will follow that $\mathcal{E}$ also has at most 2^n vectors. To establish this claim, let $A(u) = A(v)$. We must show that $u = v$.

Assume at the beginning that $A(u) = A(v) = \emptyset$ holds. This implies $x'_i(u) = x'_i(v) = \alpha_i$ or $x'_i(u - v) = 0$ for all i. Thus, $x'_i\big(v + 2(u - v)\big) = \alpha_i$ for each i and so $v + 2(u - v) \in \Pi$. Now use the identity $u = \frac{1}{2}v + \frac{1}{2}\big[v + 2(u - v)\big]$ and the fact that u is an extreme point of Π to conclude that $u = v$.

Next, suppose that $A(u) = A(v) = B \neq \emptyset$. In this case, let

$$\epsilon = \min\left\{\tfrac{\alpha_i - x_i'(u)}{\alpha_i - x_i'(v)} : \ i \in B\right\},$$

and note that $\epsilon > 0$. Then $\epsilon\bigl[\alpha_i - x_i'(v)\bigr] \leq \alpha_i - x_i'(u)$ holds for each i. (If $i \notin B$, then $\alpha_i - x_i'(v) = \alpha_i - x_i'(u) = 0$.) Assume first that $\epsilon \geq 1$. This implies $\alpha_i - x_i'(v) \leq \alpha_i - x_i'(u)$ or $x_i'(u-v) \leq 0$ for each i. But then we have $x_i'\bigl(v + 2(u-v)\bigr) \leq \alpha_i$ for all i and so $v + 2(u-v) \in \Pi$. As before, from $u = \frac{1}{2}v + \frac{1}{2}\bigl[v + 2(u-v)\bigr]$ it follows that $u = v$. Now suppose that $0 < \epsilon < 1$. Then in this case we have $x_i'(u - \epsilon v) \leq (1-\epsilon)\alpha_i$ or $x_i'\bigl(\frac{1}{1-\epsilon}(u - \epsilon v)\bigr) \leq \alpha_i$ for each i which means that $w = \frac{1}{1-\epsilon}(u - \epsilon v) \in \Pi$. It follows that $u = \epsilon v + (1-\epsilon)w$, so again $u = v$, and the proof is finished. ■

The convex hull of a finite collection of vectors is called a polytope.

Definition 3.28. *A* ***polytope*** *in a vector space is any set that is the convex hull of a finite collection of vectors.*[5]

Compact polyhedra exist only in finite dimensional vector spaces.

Lemma 3.29. *Only finite dimensional Hausdorff locally convex spaces admit nonempty weakly compact polyhedra—in which case they are polytopes.*

Proof. Let Π be a nonempty weakly compact polyhedron in a Hausdorff locally convex space X. Then, according to the Krein–Milman theorem, Π is the closed convex hull of its extreme points. By Lemma 3.27 the set of extreme points of Π is finite, and so Π must coincide with the (compact) set of the convex hull of its extreme points. That is, Π must be a polytope.

To see that X is finite dimensional, let

$$\Pi = \bigl\{x \in X : \ x_i'(x) \leq \alpha_i \ \text{ for } i = 1, \ldots, n\bigr\},$$

where $x_i' \in X'$ and $\alpha_i \in \mathbb{R}$ for each i. Put $Y = \bigcap_{i=1}^n \operatorname{Ker} x_i'$. Clearly, the set $\Pi = \Pi + Y$ is weakly compact. This implies that $Y = \{0\}$ and so for each $x' \in X'$ we have $\bigcap_{i=1}^n \operatorname{Ker} x_i' \subseteq \operatorname{Ker} x'$. This implies that each $x' \in X'$ lies in the linear span of $\{x_1', \ldots, x_n'\}$. Thus, X' is finite dimensional and from this we infer that X'' (and hence X) is likewise finite dimensional. ■

Our final objective is to establish the converse of Lemma 3.29. To do this, we need to introduce the notion of the "one-sided polar" of a set.

Definition 3.30. *For a dual pair $\langle X, X'\rangle$, the* ***one-sided polar*** *of a subset A of X is the subset $A^\bullet$ of X' defined by*

$$A^\bullet = \bigl\{x' \in X' : \ \langle x, x'\rangle \leq 1 \textit{ for all } x \in A\bigr\}.$$

[5] The polytopes play an important role in mathematics since antiquity and have been studied extensively through the ages. For a comprehensive study of polytopes and their history we refer the reader to [**48**] and [**94**].

The one-sided polar of a subset of X' is defined in a similar fashion.[6]

The **one-sided bipolar** *of a subset A of X is the set $(A^\bullet)^\bullet$, written simply as $A^{\bullet\bullet}$. The double one-sided polar of a subset of X' is defined in a similar manner.*

The important result needed for our study here is the following classical "bipolar theorem" for one-sided polars.

Theorem 3.31 (The One-sided Bipolar Theorem). *Let $\langle X, X'\rangle$ be a dual system. If A is a nonempty subset of X that contains zero, then $A^{\bullet\bullet}$ is the convex $\sigma(X, X')$-closed hull of A.*

In particular, if A is a nonempty $\sigma(X, X')$-closed convex subset X containing zero, then $A = A^{\bullet\bullet}$. Similar results hold true for subsets of X'.

Proof. Let A be a nonempty subset of X such that $0 \in A$. It should be clear that $A^\bullet$ is convex and $\sigma(X', X)$-closed. Furthermore, it should also be clear that $A \subseteq A^{\bullet\bullet}$. Let C be the $\sigma(X, X')$-closed convex hull of A and note that $C \subseteq A^{\bullet\bullet}$. Suppose $x \in A^{\bullet\bullet} \setminus C$. Then, by the Separation Corollary 8.18, there exist $x' \in X'$ and $\alpha \neq 0$ satisfying $\langle y, x'\rangle \leq \alpha$ for each $y \in C$ and $\langle x, x'\rangle > \alpha$. Since $0 \in C$, it must be the case that $\alpha > 0$. Replacing x' by $\frac{x'}{\alpha}$, we can assume that $\alpha = 1$. This implies $x' \in A^\bullet$. However, $\langle x, x'\rangle > 1$ yields $x \notin A^{\bullet\bullet}$, a contradiction. Hence, $A^{\bullet\bullet} = C$. ■

We are now ready to state and prove the converse of Lemma 3.29.

Theorem 3.32. *A nonempty convex compact subset of a finite dimensional vector space is a polyhedron if and only if it is a polytope.*

Proof. In view of Lemma 3.29, the only thing that remains to be proven is that a polytope in a finite dimensional vector space is a polyhedron. To this end, let $P = \operatorname{co}\{a_1, \ldots, a_k\}$ be a polytope of $\mathbb{R}^n$. Without loss of generality we can assume that zero is an interior point of P; see Exercise 11 at the end of the section. Clearly, $P^\bullet = \{x \in \mathbb{R}^n\colon\ x \cdot a_i \leq 1 \text{ for } i = 1, \ldots, k\}$, and so $P^\bullet$ is a polyhedron. That is, in general, the one-sided polar of a polytope is a polyhedron.

Next, we claim that $P^\bullet$ is a compact subset of $\mathbb{R}^n$. To see this, notice first that $P^\bullet$ is clearly closed. To establish that $P^\bullet$ is also bounded, fix any vector $u \in \mathbb{R}^n$. Since zero is an interior point of P, there exists some $\alpha > 0$ such that $\pm\alpha u \in P$. This implies $(\pm\alpha u) \cdot x \leq 1$ or $|u \cdot x| \leq \frac{1}{\alpha}$ for all $x \in P^\bullet$, i.e., $P^\bullet$ is pointwise bounded, and thus a bounded subset of $\mathbb{R}^n$. Consequently, $P^\bullet$ is closed and bounded and thus a compact subset of $\mathbb{R}^n$.

[6] The **two-sided polar** (or simply the **polar**) of a nonempty subset A of X is the subset A° of X' defined by $A^\circ = \{x' \in X'\colon\ |\langle x, x'\rangle| \leq 1 \text{ for all } x \in A\}$. Clearly, $A^\circ = A^\bullet \cap (-A)^\bullet$.

Now notice that, according to Lemma 3.29, the compact polyhedron $P^\bullet$ must be a polytope. From this and the One-sided Bipolar Theorem 3.31, we see that $P = P^{\bullet\bullet}$ is a polyhedron. ■

Exercises

1. Let $\mathcal{W}$ be a wedge of a vector space X and let f be a linear functional on X, i.e., $f \in X^*$. Show that $f(x) \geq \alpha$ for all $x \in \mathcal{W}$ if and only if f supports $\mathcal{W}$ at zero, i.e., if and only if $f(x) \geq 0$ holds for each $x \in \mathcal{W}$.
2. Can the set of extreme points of a nonempty polyhedron be empty?
3. Show that in a finite dimensional vector space X the trivial cone $\{0\}$ is a polyhedral cone. Is this conclusion true if X is infinite dimensional? [*Hint*: If X is an n-dimensional vector space and $\{x_1^*, \ldots, x_n^*\}$ is a base of X^*, then note that $\{0\} = \bigcap_{i=1}^n \{x \in X\colon\ x_i^*(x) = 0 \text{ for } i = 1, \ldots, n\}$.]
4. If $T\colon X \to Y$ is a continuous operator between two topological vector spaces and Π is a polyhedron of Y, then show that $T^{-1}(\Pi)$ is a polyhedron of X.
5. Show that a polytope of a vector space X is a compact set for any linear topology on X.
6. Verify that nonzero scalar products and translations of polyhedra (resp. of polytopes) are polyhedra (resp. polytopes).
7. Prove that the intersection of two polyhedra (resp. two polytopes) is a polyhedron (resp. a polytope).
8. Show that the sum of two polytopes is a polytope. [*Hint*: Assume that $A = \operatorname{co}\{x_1, \ldots, x_n\}$ and $B = \operatorname{co}\{y_1, \ldots, y_m\}$. Establish that
$$A + B = \operatorname{co}\{x_i + y_j\colon\ i = 1, \ldots, n,\ j = 1, \ldots, m\}$$
by proving that if $x = \sum_{i=1}^n \lambda_i x_i \in A$ and $y = \sum_{j=1}^m \mu_j y_j \in B$, then $x + y = \sum_{i=1}^n \sum_{j=1}^m \lambda_i \mu_j (x_i + y_j)$.]
9. Show that the one-sided polar of a polytope is a polyhedron.
10. Let $\langle X, X' \rangle$ be a dual system. If A is a nonempty subset of X, then for each wedge $\mathcal{W}$ of X show that $(A + \mathcal{W})^\bullet = A^\bullet \cap \mathcal{W}^\bullet$.
11. Let Y be a vector subspace of a finite dimensional vector space X and let C be a nonempty convex subset of Y. Show that C is a polyhedron of Y if and only if C is a polyhedron of X. [*Hint*: Assume that C is a polyhedron of Y and let
$$C = \{y \in Y\colon\ f_i(y) \leq \alpha_i \text{ for each } i = 1, \ldots, k\},$$
where f_i is a linear functional on Y and α_i is a scalar for each i. Extending each f_i to all of X, we can assume that each f_i is a linear functional on X. Now let $P\colon X \to X$ be a projection on X with range Y and note that
$$C = \{x \in X\colon\ f_i(x) \leq \alpha_i \text{ for } i = 1, \ldots, k \text{ and } Px = x\}.$$
This shows that C is a polyhedron of X.]

12. Let Y be a vector subspace of a finite dimensional vector space X and let $\mathcal{K}$ be a cone of Y. Prove that $\mathcal{K}$ is a polyhedral cone of Y if and only if $\mathcal{K}$ is a polyhedral cone of X. [*Hint*: Argue as in the hint of Exercise 11 above.]

13. Is the sum of two polyhedra a polyhedron? (See also Corollary 3.42.)

14. Show that in $\mathbb{R}^2$ the notions of polytope, nonempty compact polyhedron, and convex "polygon" coincide; see Figure (a) below.

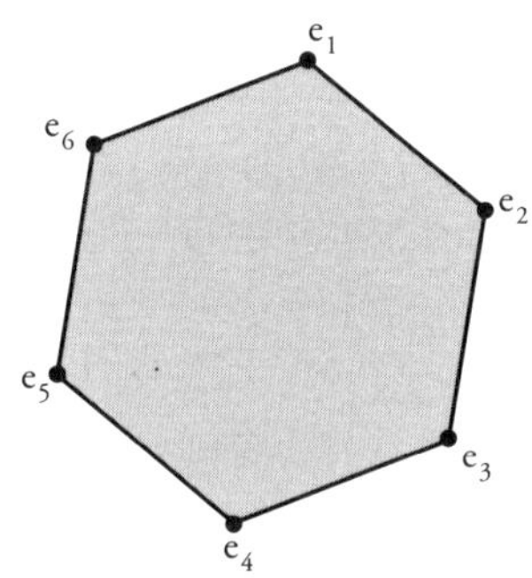

(a) A polygon (hexagon)

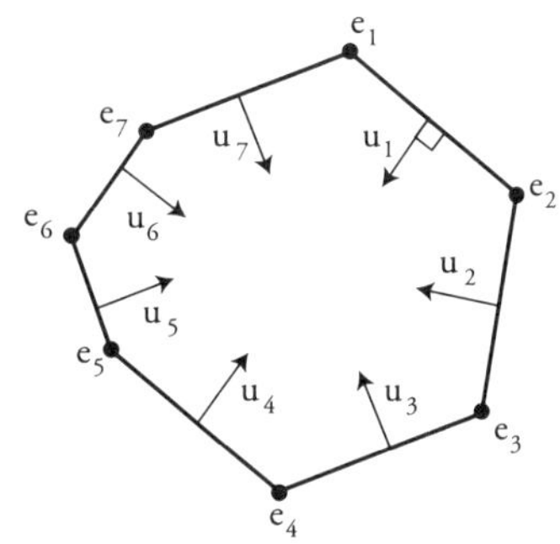

(b) A heptagon

Also, show that for a convex polygon with n vertices the number of minimal linear inequalities that generate it is n and (aside from scalar factors) is uniquely determined. For instance, the minimal collection of linear inequalities that generate the heptagon shown in Figure (b) above is $u_i \cdot x \geq u_i \cdot e_i$, $i = 1, \ldots, 7$, where $e_8 = e_1$ and u_i is perpendicular to the line segment joining e_i and e_{i+1} that is directed "towards" the interior of the heptagon.

3.4. The geometrical structure of polyhedral cones

We now turn our attention to the geometrical structure of polyhedral cones in finite dimensional spaces. Since a cone $\mathcal{K}$ of a finite dimensional vector space X is polyhedral if and only if $\mathcal{K}$ is a polyhedral cone of the vector subspace $Y = \mathcal{K} - \mathcal{K}$ (see Exercises 11 and 12 of Section 3.3), we can assume from the outset that our polyhedral cones are generating. So, for our discussion in this section, $\mathcal{K}$ will denote a fixed generating polyhedral cone of $\mathbb{R}^m$.

Clearly, $\mathcal{K}$ is closed and a glance at Corollary 3.8 informs us that the cone $\mathcal{K}$ has a compact base $\mathcal{B}$. This implies that there exists some vector $a \in \mathbb{R}^m$ that acts as a $\mathcal{K}$-strictly positive linear functional on $\mathbb{R}^m$ satisfying $\mathcal{B} = \{x \in \mathcal{K}\colon\ a \cdot x = 1\}$. But then $\mathcal{B}$ itself is a polyhedron that (according to Lemma 3.27) has a finite number of extreme points, say $e_1, \ldots, e_k$. Now Theorem 1.60 guarantees that $\mathcal{K}$ has exactly k extremal rays—precisely the ones generated by the directions $e_1, \dots, e_k$. Since (by the classical Krein–Milman theorem) the convex hull of $\{e_1, \dots, e_k\}$ coincides with $\mathcal{B}$, it follows that the directions $e_1, \dots, e_k$ also generate the cone $\mathcal{K}$. In other words, we have the following result.

Lemma 3.33. *Every polyhedral cone in a finite dimensional vector space has a finite number of extremal rays whose directions generate the cone.*

How does one determine the extremal rays of a polyhedral cone? The next result of H. Weyl [**90**] provides the answer.

Theorem 3.34 (Weyl). *Assume that a polyhedral cone $\mathcal{K}$ of $\mathbb{R}^m$ with $m \geq 2$ is generated by ℓ homogeneous linear inequalities $a_i \cdot x \geq 0$, $i = 1, \dots, \ell$.*[7] *For each $x \in \mathcal{K}$ consider the possibly empty set $I_x = \{i \in \{1, \dots, \ell\}\colon\ a_i \cdot x = 0\}$ and let A_x denote the matrix with rows $\{a_i\colon\ i \in I_x\}$.*[8] *Then for a nonzero vector $e \in \mathcal{K}$ the following statements are equivalent.*

(1) *The vector e is an extremal vector of the cone $\mathcal{K}$.*

(2) *The set I_e is nonempty and the matrix A_e has rank $m - 1$.*

Proof. (1) $\Longrightarrow$ (2) If the set I_e is empty, then $Ae \gg 0$ and consequently $e \in A^{-1}(\operatorname{Int}(\mathbb{R}^\ell_+)) \subseteq \mathcal{K}$. This implies that e is an interior point of $\mathcal{K}$, and therefore (by Lemma 2.7) it cannot be an extremal vector, a contradiction. Hence $I_e \neq \emptyset$.

Since $A_e e = 0$, we see that the matrix A_e cannot have rank m; otherwise it will follow that $e = 0$, which is impossible. Now assume by way of contradiction that A_e has rank less than $m - 1$. From the classical formula of linear algebra "*rank* + *nullity* = m" (the dimension of the domain space $\mathbb{R}^m$), it follows that the null space of A_e is at least two. This implies that there exists a nonzero vector v such that $A_e v = 0$ and that e and v are linearly independent vectors. From $a_i \cdot e = a_i \cdot v = 0$ for all $i \in I_e$ and $a_i \cdot e > 0$ for all $i \notin I_e$, it is not difficult to see that there exists some $\lambda > 0$ such that $A(e + \lambda v) \geq 0$ and $A(e - \lambda v) \geq 0$, i.e., $e \pm \lambda v \in \mathcal{K}$. Now notice that the vectors $e + \lambda v$ and $e - \lambda v$ are linearly independent and that they satisfy $e = \frac{1}{2}(e + \lambda v) + \frac{1}{2}(e - \lambda v)$, contradicting the fact that e is an extremal vector of $\mathcal{K}$; see Lemma 1.43. Hence the matrix A_e has rank $m - 1$.

(2) $\Longrightarrow$ (1) Assume that $e = u + v$ with the vectors u and v in $\mathcal{K}$. From $A_e u + A_e v = A_e e = 0$, $A_e u \geq 0$ and $A_e v \geq 0$, we see that both u and v lie in the null space of A_e. But if the matrix A_e has rank $m - 1$, it follows from the formula "*rank* + *nullity* = m" that the null space of A_e is

[7] That is, $a_i \in \mathbb{R}^m$ for each i and $\mathcal{K} = \{x \in \mathbb{R}^m\colon\ Ax \geq 0\} = A^{-1}(\mathbb{R}^\ell_+)$, where A is the $\ell \times m$ matrix having rows $a_1, \dots, a_\ell$ (and, of course, here a_i is written as a row m-dimensional vector).

[8] If $I_x = \emptyset$, then the matrix A_x does not exist.

one-dimensional. In particular, u and v must be linearly dependent. Now a glance at Lemma 1.43 shows that e is an extremal vector of the cone $\mathcal{K}$. ■

Next, let $\{x_1, \dots, x_n\}$ be a collection of nonzero vectors that generates the cone $\mathcal{K}$. Fix an extreme point e of $\mathcal{B}$. Since $e \in \mathcal{K}$, there exist nonnegative scalars (not all zero) such that $e = \sum_{i=1}^{n} \lambda_i x_i$. If $\lambda_j > 0$, then it follows that $0 \le_{\mathcal{K}} \lambda_j x_j \le_{\mathcal{K}} e$ and from this (and the fact that e is an extremal vector of $\mathcal{K}$), we see that $\lambda_j x_j = \mu e$ for some $\mu > 0$. This shows that e is a multiple of x_j and so (aside from scalars multiples) the extreme points of $\mathcal{B}$ (i.e., the vectors $e_1, \dots, e_k$) are among the directions $x_1, \dots, x_n$. Thus, we have proven the following.

Lemma 3.35. *The collection of extremal vectors $\{e_1, \dots, e_k\}$ is the "smallest" set of vectors that generates $\mathcal{K}$. This means that if a set $\{x_1, \dots, x_n\}$ generates $\mathcal{K}$, then $n \ge k$ and if another collection of vectors $\{y_1, \dots, y_k\}$ generates the cone $\mathcal{K}$, then (up to scalar multiples) it must be the case that $\{y_1, \dots, y_k\} = \{e_1, \dots, e_k\}$.*

Since the cone $\mathcal{K}$ is polyhedral, there is a minimal set of inequalities, say $\{a_i \cdot x \ge 0\colon\ i = 1, \dots, \ell\}$, that generates the cone $\mathcal{K}$. That is, any other collection of inequalities that generates $\mathcal{K}$ contains at least ℓ inequalities. From Theorem 3.25 we know that the dual cone $\mathcal{K}'$ (which is also generating) is also polyhedral. By the above discussion, $\mathcal{K}'$ is generated by a minimal set of inequalities, say that ℓ' is the number of a minimal set of inequalities, and a minimal collection of directions $f_1, \dots, f_{k'}$ (which are, of course, the directions of the extremal rays of $\mathcal{K}'$). We claim that $\ell' = k$ and $k' = \ell$.

Let us verify that $\ell' = k$; the equality $k' = \ell$ can be proven in a similar manner. Start by observing that (according to Theorem 3.25) the set of inequalities $\{e_i \cdot x \ge 0\colon\ i = 1, \dots, k\}$ generates the cone $\mathcal{K}'$. This shows that $\ell' \le k$. On the other hand, if $\{b_i \cdot x \ge 0\colon\ i = 1, \dots, \ell'\}$ is a minimal set of inequalities that generates $\mathcal{K}'$, then it follows (from Theorem 3.25 again) that the collection of directions $\{b_i\colon\ i = 1, \dots, \ell'\}$ generates $\mathcal{K}$, and so $k \le \ell'$ is also true. Thus, $\ell' = k$.

Since $k' = \ell$ and the set of directions $\{a_i\colon\ i = 1, \dots, \ell\}$ generates (by Theorem 3.25 once more) the cone $\mathcal{K}'$, it follows that each a_i must be a positive scalar of some f_j. This implies that the set of minimal inequalities that generates $\mathcal{K}$ is uniquely determined and (besides scalar multiples) is given by the directions of the extremal rays of the dual cone $\mathcal{K}'$.

Summarizing the above discussion, we have the following result.

Theorem 3.36. *For a polyhedral cone $\mathcal{K}$ of $\mathbb{R}^m$ generated by the minimal set of inequalities $\{a_i \cdot x \ge 0\colon\ i = 1, \dots, \ell\}$ we have the following.*

(1) *The cone $\mathcal{K}$ has a (nonempty) finite number of extremal rays, say k.*

(2) *If* $\mathcal{A}_{e_1}, \ldots, \mathcal{A}_{e_k}$ *are the* k *extremal rays of* $\mathcal{K}$, *then the cone* $\mathcal{K}$ *is generated by the directions* $e_1, \ldots, e_k$.

(3) *The collection of extremal vectors* $\{e_1, \ldots, e_k\}$ *is the unique minimal set that generates* $\mathcal{K}$ *in the following sense:*
   (a) *if a set of vectors* $\{x_1, \ldots, x_r\}$ *generates* $\mathcal{K}$, *then* $r \geq k$, *and*
   (b) *if a collection of* k *vectors* $\{y_1, \ldots, y_k\}$ *generates the cone* $\mathcal{K}$, *then (aside from scalar multiples)* $\{y_1, \ldots, y_k\} = \{e_1, \ldots, e_k\}$.

(4) *The minimal set of linear inequalities that generates the cone* $\mathcal{K}$ *is uniquely determined, and (aside from scalar multiples) its inequalities are precisely the ones given by any collection of the* ℓ *extremal vectors that generate the dual (polyhedral) cone* $\mathcal{K}'$.

The conclusions of Theorem 3.36 can be listed in a table as follows.

Cone	$\mathcal{K}$	$\mathcal{K}'$
Extremal vectors	$e_1, \ldots, e_k$	$a_1, \ldots, a_\ell$
Minimal set of inequalities	$a_1, \ldots, a_\ell$	$e_1, \ldots, e_k$

In $\mathbb{R}^3$ the number of extremal rays of a polyhedral cone and the cardinality of a minimal set of linear inequalities needed to generate it coincide, i.e., $\ell = k$; see Exercise 9 at the end of the section. However, in general $\ell \neq k$; see Exercise 10 at the end of the section.

It turns out that the polyhedral wedges and the wedges generated by a finite number of directions are the same objects.

Theorem 3.37. *In a finite dimensional vector space, a wedge is polyhedral if and only if it is generated by a finite number of directions.*

Proof. Let X be a finite dimensional vector space; we can assume without loss of generality that $X = \mathbb{R}^n$. Also, let $\mathcal{W}$ be a wedge of X.

Assume first that $\mathcal{W}$ is a polyhedral wedge. In this case, it is not difficult to see that there exists an $m \times n$ real matrix A such that

$$\mathcal{W} = \{x \in \mathbb{R}^n: \ Ax \geq 0\}.$$

Suppose at the beginning that the matrix A has rank n. In this case, it follows that the closed wedge $\mathcal{W}$ is also a cone of X. Now a glance at Theorem 3.36 shows that $\mathcal{W}$ is generated by the (finite) set of its extremal directions.

Next, consider the case where the null space of A, i.e., the vector space $N = \{x \in \mathbb{R}^n: \ Ax = 0\}$, is nontrivial. Let Q be any projection on $\mathbb{R}^n$ with range N and view Q as an $n \times n$ matrix and define the $(m+n) \times n$ matrix

$$A_1 = \begin{bmatrix} A \\ Q \end{bmatrix}.$$

We claim that A_1 has rank n. To see this, assume that $A_1 x = 0$ or, equivalently, $Ax = 0$ and $Qx = 0$. Clearly, $Ax = 0$ means $x \in N$ and so $x = Qx = 0$. This shows that A_1 has rank n. Applying the preceding case to the matrix A_1, we see that the cone

$$\{x \in \mathbb{R}^n\colon\ A_1 x \geq 0\}$$

is generated by a finite number of vectors, say $u_1, \ldots, u_\ell$. Next, choose a base $\{v_1, \ldots, v_r\}$ for N, and we claim that the vectors

$$v_1, \ldots, v_r, -v_1, \ldots, -v_r, u_1, \ldots, u_\ell$$

generate the wedge $\mathcal{W}$. To see this, let $\mathcal{W}_1$ be the wedge generated by the above vectors. Clearly, $\mathcal{W}_1 \subseteq \mathcal{W}$ is trivially true. Now let $x \in \mathcal{W}$, i.e., $Ax \geq 0$. Then we have $A(x - Qx) = Ax \geq 0$ and $Q(x - Qx) = 0$. So, $A_1(x - Qx) \geq 0$ or $x - Qx \in \mathcal{W}_1$. Since $Qx \in N \subseteq \mathcal{W}_1$, it easily follows that $x = Qx + (x - Qx) \in \mathcal{W}_1 + \mathcal{W}_1 = \mathcal{W}_1$. Thus, $\mathcal{W} = \mathcal{W}_1$.

For the converse, assume that $\mathcal{W}$ is generated by a finite collection of vectors. From Theorem 3.25 it follows that the dual wedge $\mathcal{W}'$ is a polyhedral wedge. But then, by the previous case, the wedge $\mathcal{W}'$ is generated by a finite number of vectors. Repeating the preceding argument once more, we see that $\mathcal{W} = \mathcal{W}'' = (\mathcal{W}')'$ is indeed a polyhedral wedge, and the proof is finished. ■

Corollary 3.38. *The sum of two polyhedral wedges of a finite dimensional vector space is a polyhedral wedge.*

The next result deals with an important decomposition property of polyhedra in finite dimensional vector spaces.

Theorem 3.39. *Let* $\Pi = \{x \in \mathbb{R}^n\colon\ x \cdot a_i \leq \alpha_i \text{ for } i = 1, \ldots, k\}$ *be a nonempty polyhedron of* $\mathbb{R}^n$. *If we consider the associated polyhedral wedge*

$$\mathcal{W} = \{x \in \mathbb{R}^n\colon\ x \cdot a_i \leq 0 \text{ for all } i = 1, \ldots, k\},$$

then there exists a polytope P *of* $\mathbb{R}^n$ *such that* $\Pi = P + \mathcal{W}$.

Moreover, in this case, if the polyhedron Π *does not contain any straight lines, then* P *is precisely the polytope generated by the extreme points of* Π.

Proof. Without loss of generality, we can assume that $0 \in \Pi$ (why?). Consider the polyhedral wedge W of $\mathbb{R}^{n+1}$ defined by

$$W = \{(x, r) \in \mathbb{R}^n \times \mathbb{R}\colon\ x \cdot a_i - \alpha_i r \leq 0 \text{ for } i = 1, \ldots, k \text{ and } r \geq 0\}.$$

It should be clear that

$$x \in \Pi \iff (x,1) \in W\,. \qquad (\star)$$

Assume first that Π is a polytope. Since $\Pi + \mathcal{W} \subseteq \Pi$ (why?), it follows that $\mathcal{W} = \{0\}$. Thus, in this case we have $\Pi = \Pi + \{0\}$, and we are done. So, we can suppose that Π is not a polytope.

By Theorem 3.37 there is a finite collection $\{(x_i, r_i)\colon\ i = 1, \ldots, m\}$ of directions (that we assume to be minimal) that generates W. Since $\Pi \neq \emptyset$, it follows from $(\star)$ that $r_i > 0$ is true for some i. Without loss of generality, we can assume that $r_i = 1$ for all r_i for which $r_i > 0$. If $r_i = 1$ were true for each i, then it follows from $(\star)$ that Π must coincide with the polytope generated by $x_1, \ldots, x_m$, which is a contradiction. Therefore, there exists a minimal collection of directions

$$(x_1, 1), \ldots, (x_\ell, 1), (x_{\ell+1}, 0), \ldots, (x_m, 0)$$

with $1 \leq \ell < m$ that generates W. If we let P be the polytope generated by $\{x_1, \ldots, x_\ell\}$ and $\mathcal{W}_1$ the wedge generated by the directions $\{x_{\ell+1}, \ldots, x_m\}$, then it is easy to see that $\Pi = P + \mathcal{W}_1$.

We claim that $\mathcal{W} = \mathcal{W}_1$. To see this, note that

$$\begin{aligned} x \in \mathcal{W} \quad &\iff \quad x \cdot a_i \leq 0 \ \text{ for } i = 1, \ldots, k \iff (x, 0) \in W \\ &\iff \quad \exists\, \lambda_i \geq 0 \text{ such that } (x,0) = \sum_{i=1}^{\ell} \lambda_i (x_i, 1) + \sum_{i=\ell+1}^{m} \lambda_i (x_i, 0) \\ &\iff \quad \exists\, \lambda_i \geq 0,\ i = \ell+1, \ldots, m, \text{ such that } x = \sum_{i=\ell+1}^{m} \lambda_i x_i \\ &\iff \quad x \in \mathcal{W}_1\,. \end{aligned}$$

Therefore, we have proven that $\Pi = P + \mathcal{W}$.

Now assume that Π does not contain any straight lines. This implies that $\mathcal{W}$ does not contain any straight lines or, equivalently, that $\mathcal{W} \cap (-\mathcal{W}) = \{0\}$; see Exercise 8 of Section 2.1. Since the collection of generators of the wedge $\mathcal{W}$ is minimal, it follows that for each $i = 1, \ldots, \ell$ the point x_i is an extreme point of P.

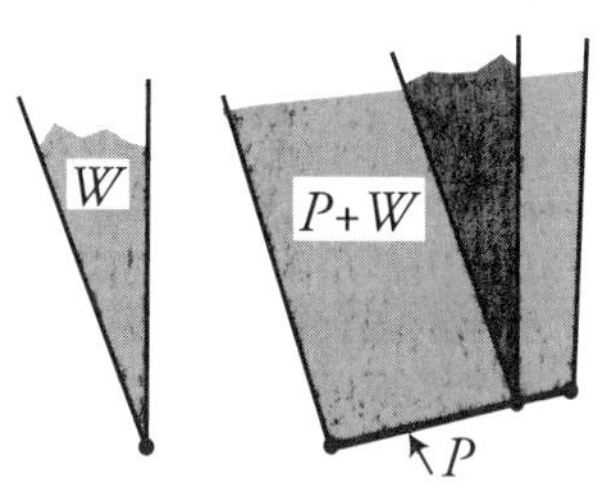

We shall establish that the vectors $x_1, \ldots, x_m$ are the extreme points of Π. Start by observing that each extreme point of Π is an extreme point of P. To see this, let e be an extreme point of $\Pi = P + \mathcal{W}$ and write $e = p + w$ with $p \in P$ and $w \in \mathcal{W}$. From $e = \frac{1}{2}(p + 0) + \frac{1}{2}(p + 2w)$, we obtain $w = 0$ and consequently $e \in P$. Now notice that e is an extreme point of P. (The accompanying figure clarifies the geometry of the situation.)

Next, fix any x_i with $1 \le i \le \ell$ and let $x_i = \alpha(p_1+w_1)+(1-\alpha)(p_2+w_2)$, where $p_j \in P$ and $w_j \in \mathcal{W}$ for all $j = 1,2$ and $0 < \alpha < 1$. Now writing $p_j = \sum_{s=1}^{\ell} \mu_s^j x_s$ and $w_j = \sum_{s=\ell+1}^{m} \mu_s^j x_s$ with $\mu_j^s \ge 0$ and $\sum_{s=1}^{\ell} \mu_s^j = 1$, we get

$$x_i = \sum_{s=1}^{\ell} \big[\alpha\mu_s^1 + (1-\alpha)\mu_s^2\big] x_s + \alpha w_1 + (1-\alpha) w_2 \,. \qquad (\star\star)$$

If $\alpha\mu_i^1 + (1-\alpha)\mu_i^2 < 1$, then it is easy to see from $(\star\star)$ that the vector $(x_i, 1)$ can be deleted from the collection of generators of W, which is impossible. So, $\alpha\mu_i^1 + (1-\alpha)\mu_i^2 = 1$. This implies $\mu_i^1 = \mu_i^2 = 1$, from which it follows that $x_i = p_1 = p_2$ and $\alpha w_1 + (1-\alpha) w_2 = 0$. Now from

$$\alpha w_1 = (1-\alpha)(-w_2) \in \mathcal{W} \cap (-\mathcal{W}) = \{0\} \,,$$

we infer that $w_1 = w_2 = 0$ and so $p_1 + w_1 = p_2 + w_2 = x_i$, proving that x_i is indeed an extreme point of Π. ■

Since every wedge is the sum of a vector space and a cone (see Lemma 1.4), the following result should be immediate.

Corollary 3.40. *If Π is a nonempty polyhedron in a finite dimensional vector space, then there exist a vector subspace V, a polytope P and a cone $\mathcal{K}$ such that*

$$\Pi = V + P + \mathcal{K} \,.$$

In finite dimensional vector spaces the convex sets that are polyhedra are characterized as follows.

Theorem 3.41. *A nonempty convex subset of a finite dimensional vector space is a polyhedron if and only if it is the sum of a polytope and a polyhedral wedge.*

Proof. Let A be a nonempty subset of a finite dimensional vector space X. Without loss of generality we can assume that zero belongs to the set A. If A is a polyhedron, then it follows immediately from Theorem 3.39 that A is the sum of the polytope and a polyhedral wedge.

For the converse, assume that $A = P + \mathcal{W}$, where P is a polytope and $\mathcal{W}$ is a polyhedral wedge of X. Since A is a sum of a compact set and a closed set, it is closed. Now notice that the one-sided polars $P^\bullet$ and $\mathcal{W}^\bullet$ of P and $\mathcal{W}$, respectively, are polyhedra. This implies that the one-sided polar $A^\bullet = P^\bullet \cap \mathcal{W}^\bullet$ is likewise a polyhedron, which according to Theorem 3.39 can be written as a sum of a polytope and a polyhedral wedge. But then, by what we have just proven, its one-sided polar $A = (A^\bullet)^\bullet$ is indeed a polyhedron. ■

Corollary 3.42. *In a finite dimensional vector space the sum of two polyhedra is a polyhedron.*

Proof. Let Π_1 and Π_2 be two nonempty polyhedra. According to Theorem 3.41 there exist polytopes P_1 and P_2 and polyhedral wedges $\mathcal{W}_1$ and $\mathcal{W}_2$ such that $\Pi_i = P_1 + \mathcal{W}_i$, $i = 1, 2$. Hence $\Pi_1 + \Pi_2 = (P_1 + P_2) + (\mathcal{W}_1 + \mathcal{W}_2)$. Now notice that $P_1 + P_2$ is a polytope (see Exercise 8 of Section 3.3) and that $\mathcal{W}_1 + \mathcal{W}_2$ is a polyhedral wedge (see Corollary 3.38), and then invoke Theorem 3.41 once more to conclude that $\Pi_1 + \Pi_2$ is a polyhedron. ■

We shall close the section by presenting one more application of Theorem 3.39 that is known as *the Principle of Linear Programming.*

Corollary 3.43 (The Principle of Linear Programming). *If Π is a polyhedron of some $\mathbb{R}^n$ that is bounded below, i.e., there exists some $a \in \mathbb{R}^n$ satisfying $\pi \geq a$ for each $\pi \in \Pi$, and p is a nonzero given vector, then the minimization problem*

$$\begin{aligned} &\min p \cdot x \\ &\text{s.t.} : x \in \Pi \end{aligned}$$

has a solution if and only if p (as a linear functional on $\mathbb{R}^n$) is bounded below on Π.

Moreover, in case the above minimization problem has a solution, then one of its solutions is always an extreme point of Π.

Proof. If the minimization problem has a solution, then clearly p is bounded below on Π. For the converse, assume that p is bounded below on Π. If Π is a polytope, then p attains a minimum value on Π at an extreme point of Π. So, we can assume that Π is not a polytope. According to Theorem 3.39, we can write $\Pi = P + \mathcal{W}$, where P is a polytope and $\mathcal{W}$ is a wedge that does not contain any straight lines since Π does not contain one.

Using Theorem 3.39 once more, we see that the extreme points of Π are exactly the extreme points of P. Moreover, since p is bounded below on $\mathcal{W}$, it follows that $p \cdot w \geq 0$ holds for all $w \in \mathcal{W}$. But then the solutions of the minimization problem

$$\begin{aligned} &\min p \cdot x \\ &\text{s.t.} : x \in \Pi = P + \mathcal{W} \end{aligned}$$

include the solutions of the minimization problem

$$\begin{aligned} &\min p \cdot x \\ &\text{s.t.} : x \in P\,. \end{aligned}$$

But the latter minimization problem has a solution at an extreme point of P which is also an extreme point of Π. ∎

Since every polyhedron is simply the solution set of a matrix inequality of the form $Ax \geq a$, where A is an $m \times n$ real matrix and a is an m-dimensional vector, the Linear Programming Optimization Problem is usually written in the form

$$\begin{aligned} &\min p \cdot x \\ &\text{s.t.}: Ax \geq a\ . \end{aligned}$$

Its geometrical interpretation is shown in the figure below. There is a vast literature on linear programming and related topics. For more on these areas see [**41, 42, 76, 82**].

Minimize $p{\cdot}x$ subject to $Ax \geq a$

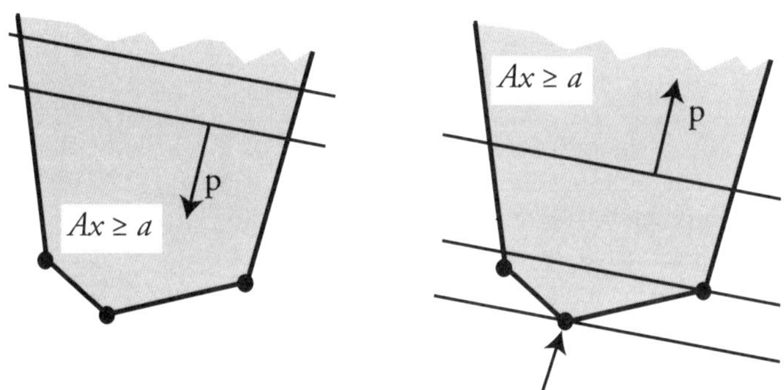

Case with no solution　　　Case with a solution

Exercises

1. Consider the matrix

$$A = \begin{bmatrix} 5 & 2 & 3 & 1 & 0 \\ 1 & 2 & 1 & 5 & 6 \\ 4 & 3 & 8 & 6 & 2 \\ 2 & 0 & 1 & 5 & 9 \end{bmatrix}.$$

Find the extremal rays of the $\mathbb{R}^4$-polyhedral cone $\mathcal{K} = \{A\lambda\colon\ \lambda \in \mathbb{R}^5_+\}$.

2. Let $\mathcal{K}$ be a generating polyhedral cone of $\mathbb{R}^m$. If k is the minimal number of directions generating $\mathcal{K}$ and ℓ is the minimal number of inequalities that generate $\mathcal{K}$, then show that $k \geq m$ and $\ell \geq m$.

3. Prove Corollary 3.38.

4. Give an alternate proof of Lemma 3.19 by using Theorem 3.37.

5. Consider the polyhedron Π of $\mathbb{R}^2$ defined by the linear inequalities
(a) $x \geq 0$,
(b) $x + y \geq 2$,
(c) $x - 2y \geq -1$,
(d) $x - y \geq 1$.

According to Theorem 3.39 the polyhedron Π has a decomposition of the form $\Pi = P + \mathcal{W}$, where P is a polytope and $\mathcal{W}$ is a polyhedral wedge. Find such a decomposition for Π.

6. Find a decomposition of the form $\Pi = P + \mathcal{W}$, where P is a polytope and $\mathcal{W}$ is a polyhedral wedge, for the polyhedron of $\mathbb{R}^3$ defined by

$$\Pi = \big\{(x, y, z) \in \mathbb{R}^3 \colon\ 1 \leq x + y + z \leq 3\big\}.$$

7. Show that in a finite dimensional vector space any two nonempty disjoint polyhedra can be strongly separated. [*Hint*: Let Π_1 and Π_2 be two nonempty disjoint polyhedra of a finite dimensional vector space X. According to Corollary 3.42 the nonempty convex set $\Pi_1 - \Pi_2$ is a polyhedron and so it is a closed convex set. Now notice that $0 \notin \Pi_1 - \Pi_2$ and then use Corollary 8.18.]

8. Prove Corollary 3.40.

9. Let $\mathcal{K}$ be a generating polyhedral cone of $\mathbb{R}^3$. Show that the smallest number of homogeneous linear inequalities that generate $\mathcal{K}$ coincides with the number of extremal rays of $\mathcal{K}$. [*Hint*: Let $\mathcal{B}$ be a base of $\mathcal{K}$. It follows that $\mathcal{B}$ is a polygon; see the accompanying figure and Exercise 14 of Section 3.3. Let $e_1, e_2, \ldots, e_n$ be an enumeration of the extreme points of $\mathcal{B}$ ordered "counterclockwise." The vectors e_i are all extremal and they generate the extremal rays of $\mathcal{K}$. Now notice that the n cross product vectors $u_i = e_i \times e_{i+1}$, where $i = 1, \ldots, n$ and $e_{n+1} = e_1$, give rise to the minimal collection (aside from scalar multiples) of linear homogeneous inequalities that generates the cone $\mathcal{K}$. That is,

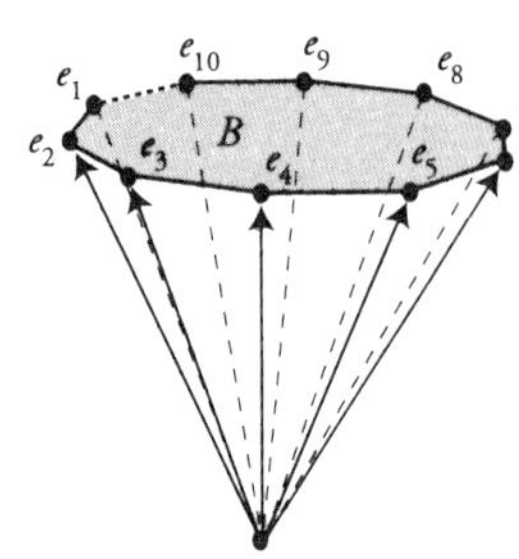

$$\mathcal{K} = \{x \in \mathbb{R}^3 \colon\ u_i \cdot x \geq 0 \text{ for each } i = 1, \ldots, n\}.$$

Next, use Theorem 3.34 to establish that $\{u_1, \ldots, u_n\}$ is the unique minimal set of linear homogeneous inequalities that generates the cone $\mathcal{K}$.]

10. (K. C. Border [**32**]) In the previous exercise we saw that the number of minimal inequalities generating a polyhedral cone in $\mathbb{R}^3$ and the number of extremal rays of the cone are the same. In this exercise, we present an example of a polyhedral cone in $\mathbb{R}^4$ with five extremal rays whose generating minimal set consists of six linear inequalities.

The desired example is the polyhedral cone $\mathcal{K}$ that is generated in $\mathbb{R}^4$ by the five columns of the 4×5 matrix

$$A = \begin{bmatrix} 1 & 1 & 1 & 1 & 1 \\ 1 & 2 & 3 & 4 & 5 \\ 1 & 4 & 9 & 16 & 25 \\ 1 & 8 & 27 & 64 & 125 \end{bmatrix}.$$

We shall denote the columns of the matrix A by a_1, a_2, a_3, a_4, a_5. To establish that $\mathcal{K}$ satisfies the above claims, verify the following steps.

(a) The cone $\mathcal{K}$, i.e., the cone $\mathcal{K} = \{Ax\colon\ x \in \mathbb{R}^5_+\}$, is a closed and generating cone of $\mathbb{R}^4$ whose extremal rays are precisely the ones generated by the five column vectors a_1, a_2, a_3, a_4, a_5 (and so $\mathcal{K}$ has precisely five extremal rays).

(b) The dual cone $\mathcal{K}'$ is generated by the five linear inequalities $a_i \cdot x \geq 0$ for each $i = 1, \ldots, 5$. That is,

$$\mathcal{K}' = \{x \in \mathbb{R}^4\colon\ a_i \cdot x \geq 0 \text{ for each } i = 1, \ldots, 5\}.$$

(c) Consider the 4×10 matrix

$$C = \begin{bmatrix} 60 & 30 & 10 & -6 & -12 & -20 & 40 & 24 & 15 & 8 \\ -47 & -31 & -17 & 11 & 19 & 29 & -38 & -26 & -23 & -14 \\ 12 & 10 & 8 & -6 & -8 & -10 & 11 & 9 & 9 & 7 \\ -1 & -1 & -1 & 1 & 1 & 1 & -1 & -1 & -1 & -1 \end{bmatrix}.$$

We designate the ten columns of C by c_i, $i = 1, \ldots, 10$. This is a computer generated matrix whose columns satisfy the following orthogonality properties:

$$\begin{gathered} c_1 \perp \{a_3, a_4, a_5\},\ c_2 \perp \{a_2, a_3, a_5\},\ c_3 \perp \{a_1, a_2, a_5\}, \\ c_4 \perp \{a_1, a_2, a_3\},\ c_5 \perp \{a_1, a_3, a_4\},\ c_6 \perp \{a_1, a_4, a_5\}, \\ c_7 \perp \{a_2, a_4, a_5\},\ c_8 \perp \{a_2, a_3, a_4\},\ c_9 \perp \{a_1, a_3, a_5\}, \\ c_{10} \perp \{a_1, a_2, a_4\}. \end{gathered}$$

Since any four column vectors of the matrix A are linearly independent, every three columns of A are linearly independent. Notice that there are $\binom{5}{3} = 10$ possible collections of three vectors from the columns of A (the ones shown above). Use Theorem 3.34 to conclude that the possible extreme vectors (up to scalar multiples) of the cone $\mathcal{K}'$ are the vectors c_i, $i = 1, \ldots, 10$.

(d) The vectors $c_1, c_2, c_3, c_4, c_5, c_6$ belong to the dual cone $\mathcal{K}'$ and the vectors c_7, c_8, c_9, c_{10} do not belong to $\mathcal{K}'$ (and so again, using Theorem 3.34, infer that $\mathcal{K}'$ has six extremal rays).

(e) Now use Theorem 3.36 to conclude that the minimal set of linear inequalities that generates $\mathcal{K}$ is six.

3.5. Linear inequalities and alternatives

In this section, we shall present a number of classical characterizations concerning solutions of linear inequalities that are associated with linear operators. Because of their usefulness the characterizations will be presented in the form of alternatives.

We start our discussion with the wedge alternative.

Theorem 3.44 (The Wedge Alternative). *Assume that $\langle L, L' \rangle$ is a dual system. If $\mathcal{W}$ is a wedge of L, then for each $x \in L$ one of the following two mutually exclusive alternatives holds true.*

(1) *$x \in \overline{\mathcal{W}}$, i.e., x belongs to the $\sigma(L, L')$-closure of $\mathcal{W}$.*

(2) *There exists some $x' \in L'$ satisfying*
   (a) *$\langle x, x' \rangle > 0$ and*
   (b) *$\langle w, x' \rangle \leq 0$ for all $w \in \mathcal{W}$.*

Proof. It is easy to check that statements (1) and (2) are mutually exclusive. Assume $x \notin \overline{\mathcal{W}}$. Then, by the Strong Separating Hyperplane Theorem 8.17, there exist a nonzero $x' \in L'$ and some constant α satisfying $\langle x, x' \rangle > \alpha$ and $\langle w, x' \rangle \leq \alpha$ for all $w \in \overline{\mathcal{W}}$. Since $\mathcal{W}$ is a wedge, it follows that $\alpha \geq 0$ and $\langle w, x' \rangle \leq 0$ for all $w \in \overline{\mathcal{W}}$. Thus, $\langle x, x' \rangle > 0$ and $\langle w, x' \rangle \leq 0$ for all $w \in \mathcal{W}$. In other words, we have shown that if (1) is false, then (2) is true. ■

The next alternative result deals with continuous linear operators. Keep in mind that if $T\colon X \to Y$ is a continuous operator between two Hausdorff locally convex spaces, then their **adjoint operator** $T'\colon Y' \to X'$ is the operator defined via the bilinear identity $[T'y'](x) = y'(Tx)$ or

$$\langle x, T'y' \rangle = \langle Tx, y' \rangle$$

for all $y' \in Y'$ and all $x \in X$.

Theorem 3.45 (The Operator Alternative). *Let $T\colon X \to Y$ be a continuous operator between two Hausdorff locally convex spaces and let $\mathcal{W}$ be a wedge of X. Then for any given vector $y \in Y$ one of the following two mutually exclusive alternatives holds true.*

(1) *$y \in \overline{T(\mathcal{W})}$, i.e., y belongs to the closure of $T(\mathcal{W})$ in Y.*

(2) *There exists some $y' \in Y'$ satisfying*

$$y'(y) < 0 \quad \text{and} \quad T'y' \in \mathcal{W}'.$$

Proof. Fix $y \in Y$ and note that $T(\mathcal{W})$ is a wedge of Y. Clearly, the alternatives (1) and (2) are mutually exclusive. According to the Wedge Alternative (Theorem 3.44) we have either $y \in \overline{T(\mathcal{W})}$ or there exists some

$z' \in Y'$ satisfying $z'(y) > 0$ and $z'(Tw) = T'z'(w) \leq 0$ for all $w \in \mathcal{W}$. Letting $y' = -z'$, we see that $y'(y) < 0$ and $T'y' \in \mathcal{W}'$. ∎

The next result is a matrix alternative and is a classical result due to J. Farkas [**40**].

Corollary 3.46 (The Farkas Alternative). *If A is an $m \times n$ real matrix, then for each vector $b \in \mathbb{R}^m$ one of the following two mutually exclusive alternatives holds true.*

(1) *There exists some $x \in \mathbb{R}^n_+$ such that $b = Ax$.*

(2) *There exists some vector $a \in \mathbb{R}^m$ satisfying*

$$a \cdot b < 0 \quad \textit{and} \quad A^t a \geq 0\,.$$

Proof. Notice first that statements (1) and (2) are indeed mutually exclusive. Consider the polyhedral wedge $\mathcal{W} = \mathbb{R}^n_+$ of $\mathbb{R}^n$ and let $T\colon \mathbb{R}^n \to \mathbb{R}^m$ be the operator determined by the matrix A. Clearly, $\mathcal{W}' = \mathbb{R}^n_+$ and $T(\mathcal{W})$ is a wedge of $\mathbb{R}^m$ that is generated by a finite number of directions. By Lemma 3.19 the wedge $T(\mathcal{W})$ is also closed. Moreover, note that T' is represented by the transpose A^t of A. Now our conclusion follows immediately from the Operator Alternative (Corollary 3.45). ∎

The next alternative is also a classical one and is due to E. Stiemke [**87**].

Theorem 3.47 (The Stiemke Alternative). *If A is an $m \times n$ real matrix, then one of the following two mutually exclusive alternatives holds true.*

(1) *There exists a strictly positive vector $x \in \mathbb{R}^n$ such that $Ax = 0$.*

(2) *There exists some $y \in \mathbb{R}^m$ such that $A^t y > 0$.*

Proof. The proof will be based on an identity regarding orthogonal complements of sets. If D is a nonempty subset of some $\mathbb{R}^\ell$, then we define its **orthogonal complement** $D^\perp$ by

$$D^\perp = \big\{x \in \mathbb{R}^\ell\colon\ a \cdot x = 0 \text{ for all } a \in D\big\}\,.$$

The **double orthogonal complement** $D^{\perp\perp}$ is the orthogonal complement of $D^\perp$, i.e., $D^{\perp\perp} = (D^\perp)^\perp$. It is not difficult to see that $D^\perp$ is a vector subspace of $\mathbb{R}^\ell$ and that $D^{\perp\perp}$ coincides with the vector subspace generated by D in $\mathbb{R}^\ell$.

Now recall that if $T\colon X \to Y$ is an operator, then its **null space** $N(T)$ and **range** $R(T)$ are the vector subspaces defined by

$$N(T) = \{x \in X\colon\ T(x) = 0\} \quad \text{and} \quad R(T) = \big\{T(x)\colon\ x \in X\big\}\,.$$

The identity we need for our proof here is the following: If the matrix A is viewed as an operator from $\mathbb{R}^n$ to $\mathbb{R}^m$, then

$$[N(A)]^{\perp} = R(A^t)\,, \tag{$\dagger$}$$

where A^t is the transpose of A. (See Exercise 2 at the end of the section.)

Now we return to our proof. Notice first that statements (1) and (2) are indeed mutually exclusive. We must show that if (1) is not true, then (2) is true. So, assume that (1) is false. That is, we assume that the equation $Ax = 0$ has no strictly positive solutions. This is equivalent to $N(A) \cap \mathrm{Int}(\mathbb{R}^n_+) = \emptyset$. So, by the Separation Theorem 8.16, there exist some nonzero vector $p \in \mathbb{R}^n$ and some $c \in \mathbb{R}$ satisfying $p \cdot x \le c \le p \cdot z$ for all $x \in N(A)$ and all $z \in \mathbb{R}^n_+$. Since $N(A)$ is a vector space, it follows that $p \cdot x = 0$ for all $x \in N(A)$ and $p \cdot z \ge 0$ for all $z \in \mathbb{R}^n_+$. That is, $p \in [N(A)]^{\perp}$ and $p > 0$. Now a glance at (†) shows that there exists some vector $y \in \mathbb{R}^m$ satisfying $A^t y = p > 0$. ■

Our final alternative deals with concave functions and is due to K. Fan, I. Glicksberg, and A. J. Hoffman [**39**].

Theorem 3.48 (The Concave Alternative). *Let $f_1, \ldots, f_k \colon C \to \mathbb{R}$ be concave functions defined on a nonempty convex subset of some vector space. Then exactly one of the following two alternatives is true.*

(1) *There exists some $x \in C$ such that $f_i(x) > 0$ for each $i = 1, \ldots, k$.*

(2) *There exist nonnegative scalars $\lambda_1, \ldots, \lambda_k$, not all zero, such that*

$$\sum_{i=1}^{k} \lambda_i f_i(x) \le 0$$

for each $x \in C$.

Proof. It is easy to see that both statements cannot be true simultaneously. Now consider the subset of $\mathbb{R}^k$:

$$A = \big\{y \in \mathbb{R}^k \colon \exists x \in C \text{ such that } y_i \le f_i(x) \text{ for each } i\big\}.$$

Clearly A is nonempty. To see that A is convex, let $y, z \in A$, and pick $x_1, x_2 \in C$ satisfying $y_i \le f_i(x_1)$ and $z_i \le f_i(x_2)$ for each i. Now if $0 \le \alpha \le 1$, then the concavity of the functions f_i implies

$$\alpha y_i + (1-\alpha) z_i \le \alpha f_i(x_1) + (1-\alpha) f_i(x_2) \le f_i\big(\alpha x_1 + (1-\alpha) x_2\big)$$

for each i. Since $\alpha x_1 + (1-\alpha)x_2 \in C$, the latter inequalities show that $\alpha y + (1-\alpha) z \in A$. That is, A is a convex subset of $\mathbb{R}^k$. Now notice that if

(1) is not true, then $A \cap \mathrm{Int}(\mathbb{R}^k_+) = \emptyset$. But then, according to Theorem 8.16, there exists a nonzero vector $\lambda = (\lambda_1, \ldots, \lambda_k)$ such that

$$\lambda \cdot y = \sum_{i=1}^{k} \lambda_i y_i \geq \sum_{i=1}^{k} \lambda_i f_i(x)$$

holds for all $y \in \mathbb{R}^k_+$ and all $x \in C$. Clearly, $\sum_{i=1}^{k} \lambda_i f_i(x) \leq 0$ for all $x \in C$ and $\lambda \cdot y \geq 0$ for all $y \in \mathbb{R}^k_+$. The latter yields $\lambda_i \geq 0$ for each i and the proof of the theorem is complete. ■

An immediate consequence of the preceding result is the following alternative.

Corollary 3.49. *If A is a real $m \times n$ matrix, then one of the following mutually exclusive alternatives is true.*

(a) *There exists some $0 < x \in \mathbb{R}^m$ such that $A^t x \leq 0$.*

(b) *There exists some $y \in \mathbb{R}^n_+$ such that $Ay \gg 0$.*

Proof. Let a_i, $i = 1, \ldots, m$, denote the row vectors of the real matrix A and consider the concave functions $f_i \colon \mathbb{R}^n_+ \to \mathbb{R}$ defined by $f_i(x) = a_i \cdot x$. According to the Concave Alternative, either there exists some $y \in \mathbb{R}^n_+$ satisfying $f_i(y) > 0$ for each i (which is equivalent to $Ay \gg 0$) or else there exists a nonzero vector $x = (x_1, \ldots, x_m) \in \mathbb{R}^m_+$ such that

$$\sum_{i=1}^{m} x_i f_i(y) = x \cdot (Ay) = (A^t x) \cdot y \leq 0$$

holds for all $y \in \mathbb{R}^n_+$. This implies $A^t x \leq 0$ and the proof is finished. ■

Exercises

1. Prove the following alternative of Farkas's lemma: For a real $m \times n$ matrix A and an arbitrary vector $b \in \mathbb{R}^n$ exactly one of the following statements is true.
   (a) There exists some $x \in \mathbb{R}^m_+$ such that $A^t x \leq b$.
   (b) There exists some $y \in \mathbb{R}^n_+$ satisfying $Ay \geq 0$ and $b \cdot y < 0$.
   [*Hint*: Assume that (a) is false. This implies that there do not exist vectors $x, z \in \mathbb{R}^m_+$ such that $A^t x + z = b$ or $\begin{bmatrix} A^t \\ I \end{bmatrix} \begin{bmatrix} x \\ z \end{bmatrix} = b$, where I denotes the $n \times n$ identity matrix. Now consider the $(m+n) \times n$ real matrix $B = \begin{bmatrix} A^t \\ I \end{bmatrix}$ and then use the Farkas Alternative (see Corollary 3.46) to conclude that there exists some vector $y \in \mathbb{R}^n$ satisfying $b \cdot y < 0$ and $B = \begin{bmatrix} Ay \\ y \end{bmatrix} \geq 0$.]

2. For a nonempty subset D of $\mathbb{R}^\ell$ establish the following properties that were used in the proof of Theorem 3.47.

(a) $D^\perp$ is a vector subspace of $\mathbb{R}^\ell$.

(b) $D^{\perp\perp}$ is the vector subspace generated by D in $\mathbb{R}^\ell$.

(c) If A is an $m \times n$ real matrix, then $[N(A)]^\perp = R(A^t)$.

[*Hint*: For part (c) note first that for $y \in \mathbb{R}^m$ and $x \in N(A)$ we have $(A^t y) \cdot x = y \cdot (Ax) = y \cdot 0 = 0$. This implies $R(A^t) \subseteq [N(A)]^\perp$. If equality does not hold, then pick some $x \in [N(A)]^\perp \setminus R(A^t)$ and then use the separation theorem to select some nonzero vector $u \in \mathbb{R}^n$ satisfying $u \cdot x > 0$ and $(A^t y) \cdot u = y \cdot Au = 0$ for all $y \in \mathbb{R}^m$. It follows that $Au = 0$ or $u \in N(A)$. But then $x \in [N(A)]^\perp$ yields $u \cdot x = 0$, which is impossible.]

3.6. Pull-back cones of operators

We shall discuss here a class of cones that are associated with operators. Assume that $T\colon X \to Y$ is an operator between two vector spaces and let $\mathcal{W}$ be a wedge of Y. Then it is easy to see that the inverse image of $\mathcal{W}$ under T is a wedge of X. That is, the set

$$T^{-1}(\mathcal{W}) = \big\{x \in X\colon\ T(x) \in \mathcal{W}\big\}$$

is a wedge of X. If T is also one-to-one and $\mathcal{W}$ is a cone, then the wedge $T^{-1}(\mathcal{W})$ is likewise a cone of X. We state this simple result below as a lemma for future use.

Lemma 3.50. *Let $T\colon X \to Y$ be a one-to-one operator between two vector spaces. If $\mathcal{K}$ is a cone of Y, then $T^{-1}(\mathcal{K})$ is a cone of X and the operator $T\colon \big(X, T^{-1}(\mathcal{K})\big) \to (Y, \mathcal{K})$ is a positive operator.*

Moreover, if $\mathcal{K}$ is an Archimedean cone, then so is the cone $T^{-1}(\mathcal{K})$.

A cone $\mathcal{K}$ of a vector space X is called the **pull-back cone** of the cone of an ordered vector space L if there exists a one-to-one operator $T\colon X \to L$ such that $\mathcal{K} = T^{-1}(L_+)$. Alternatively, $\mathcal{K}$ is the pull-back of the cone of an ordered vector space L if and only if the ordered vector space $(X, \mathcal{K})$ is order-embeddable in L.

Likewise, a cone $\mathcal{K}$ of a topological vector space X is called the **continuous pull-back cone** of the cone of an ordered topological vector space L if there exists a continuous one-to-one operator $T\colon X \to L$ such that $\mathcal{K} = T^{-1}(L_+)$.[9]

[9] There is another name for the pull-back cones that originates from the theory of finance. Let $T\colon X \to Y$ be a one-to-one operator between two vector spaces and let $\mathcal{K}$ be a cone of Y. Then the pull-back cone $T^{-1}(\mathcal{K})$ is also called the T-**dominance cone** of $\mathcal{K}$. The vector ordering induced on X by the cone $T^{-1}(\mathcal{K})$ is referred to as the T-**dominance** ordering on X induced by the cone K (or simply the ordering of T-**dominance** on X if there is no confusion about the cone $\mathcal{K}$) and it is denoted by $\geq_T$, i.e., $x \geq_T y$ holds in X if and only if $T(x) \geq_{\mathcal{K}} T(y)$ holds in Y.

The rest of the discussion in this section is devoted to proving the following basic result. (As usual, **1** will denote the constant function one on $[0,1]$, i.e., $\mathbf{1}(t)=1$ for all $t\in[0,1]$.)

Theorem 3.51. *Every closed cone of a finite dimensional vector space is the pull-back cone of the (standard) cone of $C[0,1]$.*

Moreover, if $\mathcal{K}$ is a closed and generating cone of a finite dimensional vector space X, then $\mathcal{K}$ can be taken to be the pull-back cone of a one-to-one operator $T\colon X\to C[0,1]$ such that $Tu=\mathbf{1}$ for some vector $u\in \operatorname{Int}(\mathcal{K})$.

To prove this theorem, we need several results that will be presented in the form of lemmas. As usual, the sup norm on any $C(\Omega)$-space will be denoted by $\|\cdot\|_\infty$, i.e., $\|x\|_\infty=\sup_{\omega\in\Omega}|x(\omega)|$ for each $x\in C(\Omega)$.

Lemma 3.52. *If $\phi\colon\Omega_1\to\Omega_2$ is a continuous surjective function between two compact topological spaces, then the mapping $x\mapsto x\circ\phi$ is a norm-preserving order-embedding of $C(\Omega_2)$ into $C(\Omega_1)$ satisfying $\mathbf{1}_{\Omega_2}\circ\phi=\mathbf{1}_{\Omega_1}$.*

Moreover, $x\mapsto x\circ\phi$ is a lattice isomorphism.

Proof. The proof of the first part is straightforward. For the second part note that for each pair $x,y\in C(\Omega_2)$ and each $\omega\in\Omega_1$ we have

$$\begin{aligned}\big[(x\circ\phi)\vee(y\circ\phi)\big](\omega) &= \max\{(x\circ\phi)(\omega),(y\circ\phi)(\omega)\}\\ &= \max\{x(\phi(\omega)),y(\phi(\omega))\}=(x\vee y)(\phi(\omega))\\ &= \big[(x\vee y)\circ\phi\big](\omega)\,.\end{aligned}$$

Thus, $(x\circ\phi)\vee(y\circ\phi)=(x\vee y)\circ\phi$ and consequently $x\mapsto x\circ\phi$ is a lattice isomorphism. ■

For the rest of the discussion $\mathcal{C}$ will denote the classical Cantor set. For details about the Cantor set see [**12**, Section 3.13, p. 98].

Lemma 3.53. *If Ω is a compact metrizable topological space, then there exists a norm-preserving order-embedding of $C(\Omega)$ into $C(\mathcal{C})$ that carries $\mathbf{1}_\Omega$ to $\mathbf{1}_{\mathcal{C}}$.*

Proof. By a well-known theorem of P. Alexandroff and P. Urysohn [**11**] (see also [**12**, Theorem 3.60, p. 100]) there exists a surjective continuous function $\phi\colon\mathcal{C}\to\Omega$, and the desired conclusion is an immediate consequence of Lemma 3.52. ■

Lemma 3.54. *There is a norm-preserving order-embedding of $C(\mathcal{C})$ into $C[0,1]$ that maps $\mathbf{1}_{\mathcal{C}}$ to $\mathbf{1}$.*

In particular, if Ω is any compact metrizable topological space, then there exists a norm-preserving order-embedding of $C(\Omega)$ into $C[0,1]$ in such a way that $\mathbf{1}_\Omega$ is mapped to $\mathbf{1}$.

Proof. Recall that the complement of the Cantor set $\mathcal{C}$ is a countable union of pairwise disjoint open intervals. That is, we can write $\mathcal{C}^c = \bigcup_{n=1}^{\infty}(a_n, b_n)$, where $(a_n, b_n) \cap (a_m, b_m) = \emptyset$ for $n \neq m$. Now each $x \in C(\mathcal{C})$ can be extended to a function $\widehat{x} \in C[0,1]$ by extending the graph of x on each open interval (a_n, b_n) to coincide with the graph of the line segment joining the points $(a_n, x(a_n))$ and $(b_n, x(b_n))$. That is, for each $a_n < t < b_n$ we let

$$\widehat{x}(t) = \tfrac{x(b_n)-x(a_n)}{b_n-a_n}(t-a_n) + x(a_n)\,.$$

Some easy verifications show the following.

(a) $\widehat{x}$ is a continuous function.

(b) If $x = \mathbf{c}$, the constant function c, then $\widehat{x} = \mathbf{c}$. In particular, $\widehat{\mathbf{1}}_{\mathcal{C}} = \mathbf{1}$.

(c) $\widehat{x} \geq 0$ holds in $C[0,1]$ if and only if $x \geq 0$ holds in $C(\mathcal{C})$.

(d) If $x, y \in C(\mathcal{C})$ and $\lambda \in \mathbb{R}$, then $\widehat{x+y} = \widehat{x} + \widehat{y}$ and $\widehat{\lambda x} = \lambda\widehat{x}$.

(e) $\max_{t\in\mathcal{C}} |x(t)| = \max_{t\in[0,1]} |\widehat{x}(t)|$.

The above properties show that $x \mapsto \widehat{x}$ is a norm-preserving order-embedding of $C(\mathcal{C})$ into $C[0,1]$ satisfying $\hat{\mathbf{1}}_{\mathcal{C}} = \mathbf{1}$.

The last part follows easily from the above conclusion and Lemmas 3.52 and 3.53. ∎

Lemma 3.55. *If $\Omega_1, \ldots, \Omega_m$ are compact metrizable topological spaces, then there exists a metrizable compact topological space Ω and a surjective lattice isometry $T\colon C(\Omega_1)\times\cdots\times C(\Omega_m) \to C(\Omega)$ satisfying $T(\mathbf{1}_{\Omega_1}, \ldots, \mathbf{1}_{\Omega_m}) = \mathbf{1}_{\Omega}$.*

In particular, in this case, there exists a norm-preserving order-embedding of $C(\Omega_1) \times \cdots \times C(\Omega_m)$ into $C[0,1]$ that carries $(\mathbf{1}_{\Omega_1}, \ldots, \mathbf{1}_{\Omega_m})$ to $\mathbf{1}$.

Proof. The Banach lattice $C(\Omega_1) \times \cdots \times C(\Omega_m)$ (with the maximum norm $\|(x_1, \ldots, x_m)\| = \max\{\|x_1\|_\infty, \ldots, \|x_m\|_\infty\}$) is an AM-space with order unit $(\mathbf{1}_{\Omega_1}, \ldots, \mathbf{1}_{\Omega_m})$. By a well-known theorem (see [**12**, Theorem 9.32, p. 359]), we know that there exist a compact topological space Ω and a lattice isometry $T\colon C(\Omega_1) \times \cdots \times C(\Omega_m) \to C(\Omega)$ satisfying $T(\mathbf{1}_{\Omega_1}, \ldots, \mathbf{1}_{\Omega_m}) = \mathbf{1}_{\Omega}$. Now since each Ω_i is a compact metrizable topological space, the AM-space $C(\Omega_i)$ is a separable Banach space for each i; see Theorem 8.25. This implies that $C(\Omega_1) \times \cdots \times C(\Omega_m)$ is a separable Banach space, and so $C(\Omega)$ is likewise a separable Banach space which in turn guarantees that Ω is a compact metrizable topological space; see again Theorem 8.25.

The proof now can be completed by invoking Lemma 3.54. (For a direct proof, see Exercise 6 at the end of the section.) ∎

Lemma 3.56. *For each k there exists a linearly independent set of functions $\{x_1, \ldots, x_k\}$ of $C[0,1]$ such that* $\operatorname{Span}\{x_1, \ldots, x_k\} \cap C^+[0,1] = \{0\}$.

Proof. Consider a partition of $[0,1]$ into k closed subintervals, that is, let $0 = a_0 < a_1 < \cdots < a_k = 1$. For each $j = 1,\ldots,k$ fix σ_j and τ_j such that $a_{j-1} < \sigma_j < \tau_j < a_j$, and then pick a function $x_j \in C[0,1]$ satisfying $x_j(\sigma_j) = 1$, $x_j(\tau_j) = -1$, and $x_j(t) = 0$ for all $t \notin [a_{j-1}, a_j]$.

Note that if $\sum_{j=1}^k c_j x_j \geq 0$, then for each $i = 1,\ldots,k$ we have

$$c_i = c_i x_i(\sigma_i) = \Big[\sum_{j=1}^k c_j x_j\Big](\sigma_i) \geq 0 \ \text{ and } \ -c_i = c_i x_i(\tau_i) = \Big[\sum_{j=1}^k c_j x_j\Big](\tau_i) \geq 0\,.$$

Therefore, $c_i = 0$ for each $i = 1,\ldots,k$. This shows that the set of continuous functions $\{x_1,\ldots,x_k\}$ of $C[0,1]$ is linearly independent and that $\operatorname{Span}\{x_1,\ldots,x_k\} \cap C^+[0,1] = \{0\}$. ■

We are now ready to prove Theorem 3.51. We shall break the proof into two parts.

(I) *Assume first that $\mathcal{K}$ is a closed and generating cone of a finite dimensional vector space X.*

We can assume that $X = \mathbb{R}^J$ for some J. Pick a compact convex base $\mathcal{B}$ of the dual cone $\mathcal{K}'$ of $\mathcal{K}$. (By Corollary 3.8 such a base $\mathcal{B}$ always exists.) Also, pick the interior point u of $\mathcal{K}$ satisfying $\mathcal{B} = \{x \in \mathcal{K}' \colon\ x \cdot u = 1\}$. Now define the operator $T\colon \mathbb{R}^J \to C(\mathcal{B})$ by

$$[Tx](b) = b \cdot x \ \text{ for all } \ b \in \mathcal{B}\,.$$

Clearly, T is a linear operator and $Tx = 0$ implies $b \cdot x = 0$ for all $b \in \mathcal{B}$. Since $\mathcal{B}$ is a base of $\mathcal{K}'$, it follows that $b \cdot x = 0$ for all $b \in \mathcal{K}'$. Now using that $\mathcal{K}'$ is generating, we get $b \cdot x = 0$ for all $b \in \mathbb{R}^J$ and from this we obtain $x = 0$. Thus, T is one-to-one. Moreover, notice that for $x \in \mathbb{R}^J$ we have

$$\begin{aligned} Tx \geq 0 \quad &\Longleftrightarrow \quad [Tx](b) = b \cdot x \geq 0 \ \text{ for all } \ b \in \mathcal{B} \\ &\Longleftrightarrow \quad y \cdot x \geq 0 \ \text{ for all } \ y \in \mathcal{K}' \\ &\Longleftrightarrow \quad x \in \mathcal{K}'' = \mathcal{K}\,. \end{aligned}$$

Thus, $T\colon \mathbb{R}^J \to C(\mathcal{B})$ is an order-embedding satisfying $[Tu](b) = b \cdot u = 1$ for all $b \in \mathcal{B}$, i.e., $Tu = \mathbf{1}_{\mathcal{B}}$. The rest of the proof follows from Lemma 3.54 by observing that $\mathcal{B}$ is a compact metrizable topological space.

(II) *Assume $\mathcal{K}$ is a nongenerating closed cone of a Euclidean space $\mathbb{R}^J$.*

In this case, let $Y = \mathcal{K} - \mathcal{K}$ and let Z be a complement of Y, i.e., $\mathbb{R}^J = Y \oplus Z$. By part (I), there exists an order-embedding $T_1\colon Y \to C[0,1]$ such that $\mathcal{K} = T_1^{-1}(C^+[0,1])$.

Now let $k = \dim Z \geq 1$ and then use Lemma 3.56 to pick a k-dimensional vector subspace V of $C[0,1]$ such that $V \cap C^+[0,1] = \{0\}$ and let $T_2\colon Z \to V$ be a linear isomorphism. Define $T_3\colon \mathbb{R}^J = Y \oplus Z \to C[0,1] \times C[0,1]$ by $T_3(y \oplus z) = (T_1 y, T_2 z)$. Clearly, T_3 is a one-to-one linear operator satisfying $T_3^{-1}(C^+[0,1] \times C^+[0,1]) = \mathcal{K}$. Now if $T_4\colon C[0,1] \times C[0,1] \to C[0,1]$ is any order-embedding (one of which is always guaranteed by Lemma 3.55), then the operator $T = T_4 \circ T_3\colon \mathbb{R}^J \to C[0,1]$ is an order-embedding satisfying $T^{-1}(C^+[0,1]) = \mathcal{K}$.[10]

Let us say that an ordered topological vector space (L, τ) is **topologically order-embeddable** into another ordered topological vector space (M, τ_1) if there exists a one-to-one operator $T\colon (L, \tau) \to (M, \tau_1)$ that is an order-embedding and at the same time $T\colon (L, \tau) \to (T(L), \tau_1)$ is a homeomorphism.

Theorem 3.51 is also a simple consequence of the following general result.

Theorem 3.57. *Every separable ordered normed space whose cone is closed and normal is topologically order-embeddable in $C[0,1]$.*

Proof. Let L be a separable ordered normed space whose cone L_+ is closed and (norm) normal. Put $\Omega = U' \cap L'_+ = \{x' \in L'_+ \colon \|x'\| \leq 1\}$. From the separability of L and the Alaoglu–Bourbaki Theorem 8.8, it follows that Ω equipped with its w^*-topology is a compact metrizable topological space (see also [**12**, Theorem 6.30, p. 239]).

Now define the mapping $T\colon L \to C(\Omega)$ by letting $(Tx)(\omega) = \omega(x)$ for all $x \in L$ and all $\omega \in \Omega$. It should be clear that T is a linear operator. The normality of the cone L_+ implies that the wedge L'_+ is generating in L'; see Theorem 2.26. This guarantees that a linear functional on L' is the zero functional if and only if it vanishes on Ω. Consequently, from

$$Tx = 0 \iff \omega(x) = 0 \text{ for all } \omega \in \Omega \iff x = 0\,,$$

it follows that T is one-to-one. Moreover, using the closedness of L_+, we see that

$$\begin{aligned} Tx \geq 0 \quad &\iff \quad \omega(x) \geq 0 \text{ for all } \omega \in \Omega \\ &\iff \quad x'(x) \geq 0 \text{ for all } x' \in L'_+ \\ &\iff \quad x \in L''_+ = L_+\,, \end{aligned}$$

where L''_+ is the dual cone in L of the wedge L'_+ with respect to the dual system $\langle L, L' \rangle$; see part (3) of Theorem 2.13. This implies that $T\colon L \to C(\Omega)$ is an order-embedding.

---

[10] It should be immediate that every order-embedding $T\colon \mathbb{R}^J \to L$, where L is a Hausdorff ordered topological vector space, is automatically a topological order-embedding.

To see that $T\colon L \to C(\Omega)$ is also a topological-embedding, it suffices to show that a sequence $\{x_n\}$ of L satisfies $\|x_n\| \to 0$ if and only if the sequence $\{x_n\}$ converges uniformly to zero on Ω. To establish this, let $\Omega' = \Omega - \Omega = U' \cap L'_+ - U' \cap L'_+$ and note that Ω' is a convex, circled, and $\sigma(L', L)$-compact subset of L'. From $L' = L'_+ - L'_+$ it follows that Ω' is also an absorbing subset of L' and that $L' = \bigcup_{n=1}^{\infty} n\Omega'$.

Since L' is a Banach space, the classical Baire category theorem guarantees that there exist $x' \in L'$, $\lambda > 0$ and some $n \in \mathbb{N}$ such that $x' + \lambda U' \subseteq n\Omega'$. This implies the existence of some $k \in \mathbb{N}$ such that

$$U' \subseteq k\Omega' = k[U' \cap L'_+ - U' \cap L'_+]\,.$$

From this, it easily follows that a sequence $\{x_n\}$ of L satisfies $\|x_n\| \to 0$ (i.e., $\{x_n\}$ converges uniformly to zero on U') if and only if it converges uniformly to zero on $U' \cap L'_+ = \Omega$, as desired.

Now to complete the proof apply Lemma 3.54. ■

Corollary 3.58 (Polyrakis [**78**]). *Every separable Banach lattice is topologically order-embeddable in $C[0, 1]$.*

Finally, we close the section with a general representation of locally convex spaces with normal cones (due to H. H. Schaefer) that also implies Theorem 3.51. We state this important result below (which also can be viewed as a characterization of normal cones in Hausdorff locally convex spaces) and for a proof we refer the reader to [**85**, Theorem 4.4, p. 224].

Theorem 3.59 (Schaefer). *For an ordered Hausdorff locally convex space (L, τ) whose cone L_+ is τ-closed the following statements are equivalent.*

(1) *The cone L_+ is τ-normal.*

(2) *There exists a Hausdorff locally compact topological space Ω such that (L, τ) is topologically order-embeddable in $C(\Omega)$, where $C(\Omega)$ is equipped with the Hausdorff locally convex topology of uniform convergence on the compact subsets of Ω.*

If (L, τ) is a separable normed space, then Ω can be taken to be $[0, 1]$.

Exercises

1. For an operator $T\colon X \to Y$ between two vector spaces establish the following. (See also Exercise 10 of Section 1.1.)

(a) If a wedge $\mathcal{W}$ of X is generated by a finite set of directions, then the wedge $T(\mathcal{W})$ of Y is also generated by a finite set of directions.

(b) If X and Y are finite dimensional and $\mathcal{W}$ is a polyhedral wedge of Y, then $T^{-1}(\mathcal{W})$ is likewise a polyhedral wedge of X.

2. Prove Lemma 3.50.

3. This exercise characterizes the pull-back cones that are lattice cones. Let $T\colon X \to L$ be a one-to-one operator from a vector space X to an ordered vector space L. Show that the pull-back cone of T is a lattice cone if and only if the range of T is a lattice-subspace of the ordered vector space L.

4. Complete the details of the proof of Lemma 3.52.

5. Prove the properties of the function $\widehat{x}$ listed in the proof of Lemma 3.54.

6. This exercise presents an elementary way of proving Lemma 3.55. Notice first that it follows from Lemma 3.54 that there exists a norm-preserving order-embedding from $C(\Omega_1) \times \cdots \times C(\Omega_m)$ into $C[0,1] \times \cdots \times C[0,1]$ (a product of $C[0,1]$ with m-factors) in such a way that $(\mathbf{1}_{\Omega_1}, \ldots, \mathbf{1}_{\Omega_m})$ is mapped to $(\mathbf{1}, \ldots, \mathbf{1})$.

Show that there exists a norm-preserving order-embedding from $C[0,1] \times C[0,1]$ into $C[0,1]$ in such a way that $(\mathbf{1}, \mathbf{1})$ is mapped to $\mathbf{1}$ by establishing the following claims.

(a) If $[a,b]$ is a closed interval, then the mapping $T_{a,b}\colon C[a,b] \to C[0,1]$, defined by $[T_{a,b}x](t) = x((b-a)t+a)$ for all $x \in C[a,b]$ and $t \in [0,1]$, is a surjective lattice isometry satisfying $T_{a,b}(\mathbf{1}_{[a,b]}) = \mathbf{1}$.

(b) Define the mapping $R\colon C[0,1] \times C[2,3] \to C[0,3]$ for each (x,y) in $C[0,1] \times C[2,3]$ via the formula

$$[R(x,y)](t) = \begin{cases} x(t) & \text{if } \ t \in [0,1], \\ [y(2) - x(1)](t-1) + x(1) & \text{if } \ t \in (1,2), \\ y(t) & \text{if } \ t \in [2,3]. \end{cases}$$

Then the mapping R is a norm-preserving order-embedding satisfying $R(\mathbf{1}_{[0,1]}, \mathbf{1}_{[2,3]}) = \mathbf{1}_{[0,3]}$.

(c) If we consider the scheme of order-embeddings

$$C[0,1] \times C[0,1] \xrightarrow{I \times (T_{2,3})^{-1}} C[0,1] \times C[2,3] \xrightarrow{R} C[0,3] \xrightarrow{T_{0,3}} C[0,1],$$

then their composition gives a norm-preserving order-embedding of $C[0,1] \times C[0,1]$ into $C[0,1]$ that maps $(\mathbf{1}, \mathbf{1})$ to $\mathbf{1}$.

7. Let $x_1, \ldots, x_J$ be J linearly independent vectors in an ordered vector space L and consider the one-to-one operator $T\colon \mathbb{R}^J \to L$ defined by $T(\theta_1, \ldots, \theta_J) = \sum_{j=1}^J \theta_j x_j$. We shall call the cone $T^{-1}(L_+)$ the **pull-back cone** of the collection of vectors $\{x_1, \ldots, x_J\}$.

Find the pull-back cone for the following collections of linearly independent functions in $C[0,1]$:

(a) $\{\mathbf{1}, x\}$, where $x(t) = t$ for all $t \in [0,1]$,

(b) $\{\mathbf{1}, x, x^2\}$, where $x(t) = t$ and $x^2(t) = t^2$ for all $t \in [0,1]$.

8. Consider the three linearly independent functions $\mathbf{1}, u, v \in C[0,1]$ defined for each $t \in [0,1]$ by $\mathbf{1}(t) = 1$, $u(t) = \cos t$, and $v(t) = \sin t$.

Show that their pull-back cone is the standard ice cream cone of $\mathbb{R}^3$, i.e., the cone $\{(x,y,z) \in \mathbb{R}^3\colon\ z \geq \sqrt{x^2+y^2}\}$.

9. Prove Theorem 3.51 by employing Theorem 3.57.

10. If L is a separable Banach lattice, then show that there exists a closed lattice-subspace of $C[0,1]$ that is lattice isomorphic to L.

Chapter 4

Krein operators

Recall that a *Krein space* is an ordered Banach space having order units and a closed cone. A *Krein operator* is a positive operator T on a Krein space such that for every $x > 0$ some vector of the form $T^n x$ is an order unit. The purpose of this short chapter is to present some remarkable results regarding fixed points and eigenvalues, on one hand, of Krein operators and, on the other hand, of families of positive operators on Krein spaces. In this chapter, L will denote an arbitrary Krein space and $\mathcal{U}$ the convex subset of L_+ consisting of all order units of L. Here is a list of some representative results that are obtained in this chapter.

(1) Krein operators are continuous and carry order units to order units, i.e., they leave $\mathcal{U}$ invariant (Lemma 4.9).

(2) The adjoint of a positive operator on a Krein space has a positive eigenvector corresponding to a nonnegative eigenvalue (Theorem 4.12).

(3) If every member of a family of pairwise commuting positive operators on a Krein space fixes a common order unit, then their corresponding family of adjoint operators fixes a common positive linear functional (Theorem 4.19).

(4) If a positive contraction on a Krein space has 1 as an eigenvalue, then its adjoint fixes a positive linear functional (Corollary 4.23).

Most of the results in this chapter are due to the Russian mathematician Mark G. Krein. However, the presentation of the material in this chapter is taken from [**5**]. For generalizations of some of the results in this chapter see [**73**]. For terminology and notation concerning operator theory we follow [**3**] and [**4**].

4.1. The concept of a Krein operator

Recall that a **Krein space** is an ordered Banach space having a closed cone and order units. In particular, in a Krein space L,

(a) the cone L_+ is closed and generating,

(b) the order units of L_+ are precisely the interior vectors of L_+ (Theorem 2.8), and

(c) its dual cone L'_+ is closed and normal (Theorem 2.42).

However, the dual cone L'_+ need not be generating; see Example 2.72. We now turn to the definition of a Krein operator.

Definition 4.1. *A positive operator $T\colon L \to L$ on a Krein space is said to be a **Krein operator** whenever for each $x > 0$ there exists some positive integer n (depending on x) such that $T^n x$ is an order unit.*

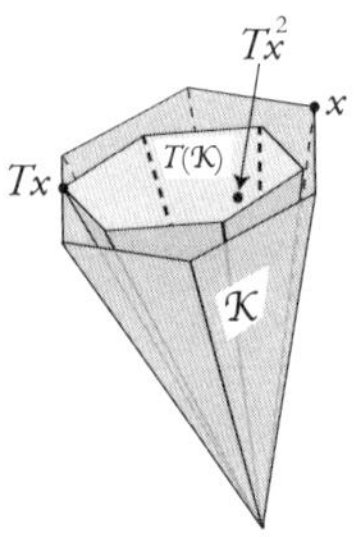

In finite dimensions this is an operator which after a fixed number of iterations takes the positive cone into its interior. In the infinite dimensional setting it is an operator that takes each postive vector into an order unit after some finite number of iterations. This is a geometrical property and is illustrated in the companion figure.

For finite dimensional Krein spaces the natural number n in Definition 4.1 can be chosen independently of x.

Theorem 4.2. *A positive operator $T\colon L \to L$ on a finite dimensional Krein space is a Krein operator if and only if there exists some (fixed) $k \in \mathbb{N}$ such that for each $x > 0$ the vector $T^k x$ is an order unit.*

Proof. The "only if" part needs proof. So, let $T\colon L \to L$ be a Krein operator with L finite dimensional. Fix a compact base $\mathcal{B}$ for L_+. (By Corollary 3.8 a compact base always exists for L_+.)

Now assume by way of contradiction that there is no such $k \in \mathbb{N}$. It easily follows that there exists a sequence $\{x_n\}$ of $\mathcal{B}$ and a sequence $\{k_n\}$ of natural numbers such that $k_n \uparrow \infty$, $T^{k_n} x_n$ is an order unit (i.e., an interior point of L_+), and $T^k x_n$ is not an order unit for all $1 \le k < k_n$. Let $x \in \mathcal{B}$ be an accumulation point of the sequence $\{x_n\}$ and then choose some $\ell \in \mathbb{N}$ such that $T^\ell x$ is an interior point of L_+. This implies that $T^\ell x_n$ is also an interior point of L_+ for infinitely many n and from this we easily obtain a contradiction (how?). ■

Here are a few examples of Krein operators.

Example 4.3. Let $A = [a_{ij}]$ be a real $n \times n$ matrix such that $a_{ij} > 0$ holds for all (i, j). Then A as an operator from $\mathbb{R}^n$ to $\mathbb{R}^n$ is a Krein operator.

The reader should notice that there are positive matrices that define Krein operators without having all their entries positive. For instance, the matrix $\begin{bmatrix} 0 & 1 \\ 1 & 1 \end{bmatrix}$ defines a Krein operator on $\mathbb{R}^2$.

It is not difficult to see that a nonnegative $n \times n$ matrix A defines a Krein operator if and only if for some $k \geq 1$ the matrix A^k has strictly positive entries.[1] (See also Theorem 4.2.) ■

Example 4.4. Let L be a Krein space, let u be an order unit, and let $f \in L'$ be a strictly positive linear functional on L (i.e., $x > 0$ implies $f(x) > 0$). Then the **rank-one operator** $T = f \otimes u$ [defined by $(f \otimes u)(x) = f(x)u$] is a Krein operator.

More generally, any **finite-rank operator** $T = \sum_{i=1}^n f_i \otimes u_i$, where each f_i is a strictly positive linear functional on L, $u_i > 0$ holds for each i, and $\sum_{i=1}^n u_i$ is an order unit, is a Krein operator. ■

Example 4.5. Let μ be a nonzero Borel measure on a Hausdorff compact topological space Ω and let $T\colon C(\Omega) \to C(\Omega)$ be a positive integral operator generated by a continuous strictly positive kernel. That is, there exists a continuous real-valued function $T\colon \Omega \times \Omega \to \mathbb{R}$ satisfying $T(t, s) > 0$ for all $(t, s) \in \Omega \times \Omega$ and

$$(Tf)(t) = \int_\Omega T(t, s) f(s)\, d\mu(s)$$

for each $f \in C(\Omega)$.[2] Then T is a compact[3] Krein operator. ■

Example 4.6. Let $T\colon [a, b] \times [a, b] \to [0, \infty)$ be a continuous function such that $T(t, s) > 0$ holds for all $|s - t| < \epsilon$, where $\epsilon > 0$ is fixed. Then the integral operator generated by $T(t, s)$, i.e., the operator $T\colon C[a, b] \to C[a, b]$ defined by

$$(Tf)(t) = \int_a^b T(t, s) f(s)\, ds,$$

is a compact Krein operator.

To see this, note first that T^2 is also an integral operator whose kernel is given by the formula

$$T^2(t, s) = \int_a^b T(t, u) T(u, s)\, du.$$

---

[1] Any nonnegative matrix with this property is known as a **primitive matrix**.

[2] As usual, we use the same letter for the kernel and the integral operator it generates.

[3] Recall that an operator on a Banach space X is said to be a **compact operator** if it carries the closed unit ball of X to a norm totally bounded subset of X.

Now we claim that $T^2(t,s) > 0$ holds for all $|s-t| < 2\epsilon$. To see this, let $s < t$ and $t - s = |s-t| < 2\epsilon$. Let $u_0 = \frac{s+t}{2}$, the midpoint of the interval $[s,t]$, and put $\delta = \epsilon - \frac{|s-t|}{2}$. Clearly, $0 < \delta < \epsilon$, and if $|u_0 - u| < \delta$, then

$$|s-u| \leq |s-u_0| + |u-u_0| < \tfrac{|s-t|}{2} + \delta = \epsilon\,,$$

and similarly $|t-u| < \epsilon$. Consequently, $T(t,u) > 0$ and $T(u,s) > 0$ for every $u \in (u_0 - \delta, u_0 + \delta)$, and so

$$T^2(t,s) = \int_a^b T(t,u)T(u,s)\,du \geq \int_{u_0-\delta}^{u_0+\delta} T(t,u)T(u,s)\,du > 0\,.$$

Repeating the above process, we see that there exists some positive integer n such that $T^n(t,s) > 0$ holds for all $(t,s) \in [a,b] \times [a,b]$. It follows that T is a compact Krein operator. ■

Example 4.7. This example is an important special case of the preceding example: If $T\colon [a,b] \times [a,b] \to [0,\infty)$ is a continuous function such that $T(t,t) > 0$ for all $a \leq t \leq b$, then the integral operator generated by T on $C[a,b]$ is a compact Krein operator. ■

Example 4.8. Consider a continuous function $T\colon [a,b] \times [a,b] \to [0,\infty)$ such that the partial derivatives $\frac{\partial^n T}{\partial t^n}$ and $\frac{\partial^n T}{\partial s^n}$ exist and are continuous for all $n = 1, \ldots, k$. Also, assume that $T(t,t) > 0$ holds for all t. Then the operator $T\colon C^k[a,b] \to C^k[a,b]$, defined by

$$(Tf)(t) = \int_a^b T(t,s)f(s)\,ds,$$

is a (compact) Krein operator. ■

Three basic properties of Krein operators are included in the next lemma.

Lemma 4.9. *If $T\colon L \to L$ is a Krein operator, then*

(1) *T is a continuous operator,*

(2) *T carries order units of L to order units of L, and*

(3) *T^n is a Krein operator for each $n \geq 1$. That is, for each $n \geq 1$ the operator T^n leaves the interior of L_+ invariant.*

Proof. (1) This follows immediately from Theorem 2.32.

(2) Let u be an order unit. Since T is a Krein operator, there exists some positive integer $k \geq 1$ such that $T^k u$ is also an order unit. Next, choose some $\alpha > 0$ such that $T^{k-1}u \leq \alpha u$ and note that $T^k u \leq \alpha T u$ holds. This implies that Tu is an order unit.

(3) This is an easy consequence of part (2). ■

Exercises

1. Establish the claim of Example 4.3. That is, show that a nonnegative $n \times n$ matrix defines a Krein operator on $\mathbb{R}^n$ if and only if for some $k \geq 1$ the matrix A^k has strictly positive entries.
2. Complete the missing details in Examples 4.4–4.8.
3. If $T\colon L \to L$ is a Krein operator and $S\colon L \to L$ is an arbitrary positive operator, then show that $S + T$ is a Krein operator. [*Hint*: Show by induction that for each $n \geq 1$ there is some positive operator $R_n\colon L \to L$ such that $(S+T)^n = S^n + R_n$.]
4. Prove part (3) of Lemma 4.9. That is, show that if $T\colon L \to L$ is a Krein operator, then T carries interior points of L_+ to interior points of L_+.
5. Let $T\colon L \to L$ be a Krein operator. Establish that for each $n \geq 1$ we have $T^n \neq I$, where I denotes the identity operator on L. [*Hint*: Use part (3) of Lemma 4.9.]
6. Show that the composition of two commuting Krein operators is also a Krein operator. However, the product of two noncommuting Krein operators need not be a Krein operator. [*Hint*: For the last part, consider the positive matrices $A = \begin{bmatrix} 0 & 1 \\ 1 & 1 \end{bmatrix}$ and $B = \begin{bmatrix} 1 & 1 \\ 1 & 0 \end{bmatrix}$ and note that both define Krein operators on $\mathbb{R}^2$. Their product matrix $AB = \begin{bmatrix} 1 & 0 \\ 2 & 1 \end{bmatrix}$ does not define a Krein operator.]
7. Complete the last part of the proof of Theorem 4.2.

4.2. Eigenvalues of Krein operators

In this section, we shall study some basic properties of the eigenvalues of Krein operators. Recall that if $T\colon V \to V$ is an operator on a vector space, then a scalar λ is said to be an **eigenvalue** of T if there exists a nonzero vector v such that $Tv = \lambda v$. The nonzero vector v is called an **eigenvector** of T corresponding to the eigenvalue λ.

We start our discussion in this section with a lemma.

Lemma 4.10. *If $T\colon L \to L$ is a Krein operator and $x_0 > 0$ is a positive eigenvector corresponding to an eigenvalue λ_0, then*

(1) $\lambda_0 > 0$,

(2) *x_0 is an order unit, and*

(3) *the eigenspace corresponding to λ_0, i.e., the vector subspace*

$$N_T(\lambda_0) = \{x \in L\colon\ Tx = \lambda_0 x\},$$

is one-dimensional.

Proof. Assume $Tx_0 = \lambda_0 x_0$ with $x_0 > 0$. From this and the positivity of T, we infer immediately that $\lambda_0 > 0$. Since T is a Krein operator, there exists some positive integer n such that $T^n x_0$ is an order unit. From $T^n x_0 = (\lambda_0)^n x_0$, it follows that x_0 is also an order unit and that $\lambda_0 > 0$. So (1) and (2) are true.

Now suppose that $Tx = \lambda_0 x$ holds for some $x \neq 0$. Pick some $\alpha \in \mathbb{R}$ such that $x_0 + \alpha x \in \partial L_+$ (see Exercise 9 of Section 2.1) and note that $\alpha \neq 0$. Indeed, if $\alpha = 0$, then Theorem 2.8 implies $x_0 + \alpha x = x_0 \in \operatorname{Int}(L_+)$, which is absurd.

If $x_0 + \alpha x > 0$, then from $T(x_0 + \alpha x) = \lambda_0(x_0 + \alpha x)$ and (2) above, we infer that $x_0 + \alpha x$ is an order unit, i.e., $x_0 + \alpha x \in \operatorname{Int}(L_+)$, which is impossible. Thus, $x_0 + \alpha x = 0$ or $x = \left(-\frac{1}{\alpha}\right)x_0$ proving that $N_T(\lambda_0)$ is one-dimensional. ■

Now we come to a fundamental property of Krein operators.

Theorem 4.11 (M. G. Krein). *A Krein operator $T\colon L \to L$ has (aside from scalar multiples) at most one positive eigenvector. Moreover, if $x_0 > 0$ is the positive eigenvector of T with corresponding eigenvalue λ_0, then $\lambda_0 > 0$ and every other real eigenvalue λ of T satisfies $|\lambda| \leq \lambda_0$.*

Proof. Assume that $x_0 > 0$ is a positive eigenvector for T corresponding to the eigenvalue λ_0. By Lemma 4.10, we know that $\lambda_0 > 0$ and that x_0 is an order unit and (essentially) the only eigenvector corresponding to λ_0.

Now suppose that λ is a real eigenvalue of T satisfying $|\lambda| > \lambda_0$. Pick a nonzero vector x such that $Tx = \lambda x$. Since x_0 is an order unit, there exists some $\alpha > 0$ such that $\pm x \leq \alpha x_0$. Hence, $T^n(\pm x) \leq \alpha T^n x_0$ for each n, or $\lambda^n(\pm x) \leq \alpha(\lambda_0)^n x_0$ for all n. It follows that $\left|\frac{\lambda}{\lambda_0}\right|^n(\pm x) \leq \alpha x_0$ for all n. Since $\left|\frac{\lambda}{\lambda_0}\right| > 1$ and L is Archimedean, we infer that $\pm x \leq 0$ or $x = 0$, which is a contradiction. Hence $|\lambda| \leq \lambda_0$ holds true, and thus T does not have any real eigenvalue with absolute value greater than λ_0.

If now $x_1 > 0$ is an eigenvector corresponding to an eigenvalue λ_1, then by Lemma 4.10 the eigenvector x_1 is an order unit and $\lambda_1 > 0$. By the above, $\lambda_1 \leq \lambda_0$ and $\lambda_0 \leq \lambda_1$ both hold, and so $\lambda_1 = \lambda_0$. Consequently, by Lemma 4.10(3), x_1 is a scalar multiple of x_0, and the proof is finished. ■

The next theorem of M. G. Krein [**63**] reveals an important eigenvalue property of the adjoint of a Krein operator.

Theorem 4.12 (M. G. Krein). *The adjoint of a positive operator on a Krein space has a positive eigenvector corresponding to a nonnegative eigenvalue.*

Proof. Let $T\colon L \to L$ be a positive operator on a Krein space. By Theorem 2.32 the operator T is automatically continuous. Fix an order unit u

and then pick some $r > 0$ such that $\|x\| \leq r$ implies $u \pm x \in L_+$. Thus, if $f \in L'_+$ and $x \in L$ satisfies $\|x\| \leq 1$, then $f(u \pm rx) \geq 0$, or $|f(x)| \leq \frac{f(u)}{r}$. Consequently, $\|f\| \leq \frac{f(u)}{r}$ holds for each $f \in L'_+$.

Next, consider the set

$$C = \left\{ f \in L'_+ \colon\ f(u) = 1 \right\}.$$

Clearly, C is nonempty, convex, and w^*-closed. By the above, we also have $\|f\| \leq \frac{1}{r}$ for each $f \in C$, and so C is a norm bounded subset of L'. It follows that C is a nonempty, convex, and w^*-compact subset of L'. (See the Alaoglu–Bourbaki Theorem 8.8.)

Next, define the mapping $F\colon C \to C$ by

$$F(f) = \tfrac{f+T'f}{[f+T'f](u)} = \tfrac{f+T'f}{1+T'f(u)}.$$

A straightforward verification shows that F indeed maps C into C and that $F\colon (C, w^*) \to (C, w^*)$ is a continuous function. Therefore, by Tychonoff's classical Fixed Point Theorem[4] there exists some $\phi \in C$ such that $F(\phi) = \phi$. That is, $\phi + T'\phi = [1 + T'\phi(u)]\phi$, or $T'\phi = [T'\phi(u)]\phi$, which shows that the linear functional $0 < \phi \in L'_+$ is an eigenvector of T' corresponding to the nonnegative eigenvalue $T'\phi(u)$. (A proof that does not use fixed point theorems can be found in [**60**].) ∎

Now let $T\colon X \to X$ be a bounded operator on a Banach space. Recall that a vector subspace V of X is said to be T-**invariant** or **invariant under** T if $T(V) \subseteq V$. A vector subspace V of X is T-**hyperinvariant** or **hyperinvariant under** T if it is invariant under every continuous operator on X that commutes with T. A vector subspace of X is **nontrivial** if it is different from X and $\{0\}$.

Theorem 4.13. *On a Krein space, every positive operator which is not a multiple of the identity has a nontrivial hyperinvariant closed subspace.*

Proof. Let L be a Krein space and let $T\colon L \to L$ be a positive operator which is not a multiple of the identity. By Theorem 4.12 there exist some $\lambda \geq 0$ and some $0 < \phi \in L'$ such that $T'\phi = \lambda\phi$. Put $R = T - \lambda I$ and note that $R'\phi = 0$. It follows that $Y = \overline{R(L)}$ is a nontrivial closed vector subspace of L. To complete the proof, notice that if a continuous operator $S\colon L \to L$ satisfies $ST = TS$, then $S(Y) \subseteq Y$ holds. ∎

Since a Krein operator cannot be a multiple of the identity operator, we have the following consequence.

Corollary 4.14. *Every Krein operator has a nontrivial hyperinvariant closed subspace.*

---

[4] See for instance [**12**, Corollary 17.56, p. 583].

The next result describes a large class of nonquasinilpotent operators. (Recall that an operator $T: X \to X$ on a normed space is **quasinilpotent** if its spectral radius is zero, i.e., $r(T) = 0$.[5])

Theorem 4.15. *Every Krein operator has positive spectral radius.*

Proof. Let $T: L \to L$ be a Krein operator. By Theorem 4.12 there exist a positive eigenvector $0 < f \in L'$ and some nonnegative scalar λ such that $T'f = \lambda f$. We claim that $\lambda > 0$ (and hence $r(T) = r(T') \geq \lambda > 0$ must hold).

Fix $0 < x \in L_+$ and then pick some integer n such that $T^n x$ is an order unit. If $\lambda = 0$, then

$$0 < \langle f, T^n x \rangle = \langle T'f, T^{n-1}x \rangle = \lambda \langle f, T^{n-1}x \rangle = 0,$$

which is impossible. Hence, $\lambda > 0$ must hold, as desired. ∎

A positive eigenvalue λ_0 of a continuous operator on a Banach space is said to be the **leading eigenvalue** whenever $\lambda_0 \geq |\lambda|$ holds for every other eigenvalue λ of the operator.

Corollary 4.16. *If $T: L \to L$ is a compact Krein operator, then the spectral radius $r(T)$ of T is the one and only leading eigenvalue of T.*

Proof. By Theorem 4.15, we have $r(T) > 0$. So, by the classical Krein–Rutman Theorem [**63**] (see also [**55**, Theorem 5.3.7, p. 182]), we know that the spectral radius $r(T)$ is an eigenvalue, and we are done. ∎

Corollary 4.16 applied to matrices yields the classical result of O. Perron [**77**] and G. Frobenius [**43**]; see also [**3**, Chapter 8] and [**93**, Chapter 19].

Corollary 4.17 (Perron–Frobenius). *If A is a nonnegative $n \times n$ matrix such that for some $k \geq 1$ the matrix A^k has strictly positive entries, then the spectral radius of A is a strictly positive leading eigenvalue of multiplicity one having a strictly positive eigenvector.*

Exercises

1. Verify that the mapping $F: (C, w^*) \to (C, w^*)$ defined in Theorem 4.12 is indeed continuous.
2. Generalize the result used in the proof of Theorem 4.13 as follows: *If the adjoint of bounded operator $T: X \to X$ on a Banach space has an eigenvalue λ, then the closed vector subspace $\overline{(T - \lambda I)(X)}$ of X is nontrivial and T-hyperinvariant.*

[5] The **spectral radius** of T is defined by $r(T) = \lim_{n\to\infty} \|T^n\|^{\frac{1}{n}}$.

3. (**Oikhberg–Troitsky** [**73**]) Assume that L is an ordered normed space such that its closed unit ball is dominated by a unit vector, i.e., there exists some $u \in L$ with $\|u\| = 1$ satisfying $u \geq U$. Show that the adjoint of every positive operator on L has a nonzero positive eigenvector.
4. (**Krein–Rutman** [**63**]) Let L be an ordered normed space whose cone has a nonempty interior. Show that the adjoint of any positive operator has a nonzero positive eigenvector.

4.3. Fixed points and eigenvectors

In this section, we shall discuss several interrelationships between fixed points and eigenvectors of families of commuting positive operators on Krein spaces. Before stating the results, we shall clarify a crucial connection between fixed points and eigenvectors.

Let $T\colon L \to L$ be a positive operator on a Krein space, let $u \in L$ be an order unit and consider the nonempty, convex, and w^*-compact set

$$C = \left\{f \in L'_+ \colon\ f(u) = 1\right\}.$$

Also, as in the proof of Theorem 4.12, we consider the continuous mapping $F\colon (C, w^*) \to (C, w^*)$ defined by

$$F(f) = \tfrac{f+T'f}{[f+T'f](u)} = \tfrac{f+T'f}{1+T'f(u)}\,.$$

The important thing to emphasize here is that a linear functional $\phi \in C$ is a fixed point for F if and only if ϕ is an eigenvector for the adjoint operator T' corresponding to the eigenvalue $T'\phi(u)$. Indeed, if $F(\phi) = \phi$, then it should be clear that $T'\phi = [T'\phi(u)]\phi$ holds. Conversely, if $\phi \in C$ satisfies $T'\phi = \lambda\phi$, then

$$F(\phi) = \tfrac{\phi+T'\phi}{[\phi+T'\phi](u)} = \tfrac{\phi+\lambda\phi}{1+\lambda} = \phi\,.$$

The next result generalizes Theorem 4.12.

Theorem 4.18 (M. G. Krein). *Let $\{T_\alpha\}_{\alpha\in A}$ be a family of pairwise commuting positive operators on a Krein space L. Then the family of adjoint operators $\{T'_\alpha\}_{\alpha\in A}$ has a common nonzero positive eigenvector, i.e., there exist some $0 < \phi \in L'$ and a family of scalars $\{\lambda_\alpha\}_{\alpha\in A}$ (which is necessarily a family of nonnegative scalars) such that $T'_\alpha\phi = \lambda_\alpha\phi$ holds for each α.*

Proof. Fix an order unit u of L and consider the nonempty, convex and w^*-compact set

$$C = \left\{f \in L'_+ \colon\ f(u) = 1\right\}.$$

For each index α consider the continuous mapping $F_\alpha\colon (C, w^*) \to (C, w^*)$ defined by

$$F_\alpha(f) = \tfrac{f+T'_\alpha f}{[f+T'_\alpha f](u)} = \tfrac{f+T'_\alpha f}{1+T'_\alpha f(u)}\,.$$

As noticed before, every fixed point of F_α is an eigenvector of T'_α corresponding to a nonnegative eigenvalue. Let D_α denote the set of all fixed points of F_α, i.e., let

$$D_\alpha = \{f \in C\colon\ F_\alpha(f) = f\}.$$

By Theorem 4.12 each D_α is a nonempty and w^*-compact subset of C. To complete the proof, we must show that $\bigcap_{\alpha\in A} D_\alpha \neq \emptyset$, or equivalently that the family of w^*-compact sets $\{D_\alpha\}_{\alpha\in A}$ has the finite intersection property.

The proof goes by induction. As mentioned above, each D_α is nonempty. So, for the induction step, assume that every intersection of any n members of the family $\{D_\alpha\}_{\alpha\in A}$ has a nonempty intersection and let $\alpha_1,\dots,\alpha_n,\alpha_{n+1}$ be $n+1$ indices. We must show that $\bigcap_{i=1}^{n+1} D_{\alpha_i} \neq \emptyset$.

To this end, start by noting that (by our induction hypothesis) there exists some $0 < \phi \in D_{\alpha_1} \cap \cdots \cap D_{\alpha_n}$. So, if $\lambda_i = T'_{\alpha_i}\phi(u) \geq 0$, then it follows from $F_{\alpha_i}(\phi) = \phi$ that $T'_{\alpha_i}\phi = \lambda_i\phi$ for each $i = 1,\dots,n$. Now for each $1 \leq i \leq n$ define the set

$$C_i = \{f \in C\colon\ T'_{\alpha_i} f = \lambda_i f\}.$$

Clearly, each C_i is convex, w^*-compact, and (since $\phi \in C_i$) nonempty. Next, we claim that $F_{\alpha_{n+1}}(C_i) \subseteq C_i$ holds for each $1 \leq i \leq n$. Indeed, if $f \in C_i$, then

$$\begin{aligned}
T'_{\alpha_i}(F_{\alpha_{n+1}}f) &= T'_{\alpha_i}\Big(\tfrac{f+T'_{\alpha_{n+1}}f}{1+T'_{\alpha_{n+1}}f(u)}\Big) = \tfrac{T'_{\alpha_i}f+T'_{\alpha_i}T'_{\alpha_{n+1}}f}{1+T'_{\alpha_{n+1}}f(u)}\\
&= \tfrac{T'_{\alpha_i}f+T'_{\alpha_{n+1}}T'_{\alpha_i}f}{1+T'_{\alpha_{n+1}}f(u)} = \tfrac{\lambda_i f+\lambda_i T'_{\alpha_{n+1}}f}{1+T'_{\alpha_{n+1}}f(u)}\\
&= \lambda_i\Big(\tfrac{f+T'_{\alpha_{n+1}}f}{1+T'_{\alpha_{n+1}}f(u)}\Big) = \lambda_i\big(F_{\alpha_{n+1}}f\big),
\end{aligned}$$

which (in view of $F_{\alpha_{n+1}} \in C$) shows that $F_{\alpha_{n+1}}f \in C_i$.

Now if we let $G = \bigcap_{i=1}^{n} C_i$, then G is a convex, w^*-compact, and (in view of $\phi \in G$) nonempty subset of C such that $F_{\alpha_{n+1}}(G) \subseteq G$. By Tychonoff's Fixed Point Theorem there exists some $0 < \psi \in G$ with $F_{\alpha_{n+1}}(\psi) = \psi$. Since $\psi \in C_i$, we have $T'_{\alpha_i}\psi = \lambda_i\psi$, and hence $F_{\alpha_i}(\psi) = \psi$ for each $1 \leq i \leq n$. So, $\psi \in \bigcap_{i=1}^{n+1} D_{\alpha_i}$ and hence $\bigcap_{i=1}^{n+1} D_{\alpha_i} \neq \emptyset$. The proof of the theorem is now complete. ■

The fixed point analogue of Theorem 4.18 is as follows.

Theorem 4.19 (M. G. Krein). *Let $\{T_\alpha\}_{\alpha\in A}$ be a family of pairwise commuting positive operators on a Krein space L. If there exists an order unit $u \in L$ such that $T_\alpha(u) = u$ for all α, then there exists some $0 < f \in L'$ such that $T'_\alpha f = f$ for all α.*

Proof. Assume that for some order unit u we have $T_\alpha(u) = u$ for each $\alpha \in A$. Since $-tu \le x \le tu$ implies $-tu = -tT_\alpha(u) \le T_\alpha(x) \le tT_\alpha(u) = tu$, it follows that $\|T_\alpha(x)\|_u \le \|x\|_u$ holds for each $x \in L$ and each $\alpha \in A$. In particular, if $k_1, \dots, k_n$ are arbitrary nonnegative integers and $\alpha_1, \dots, \alpha_n$ are arbitrary indices, then

$$\big\|T_{\alpha_1}^{k_1} T_{\alpha_2}^{k_1} \cdots T_{\alpha_n}^{k_n} x\big\|_u \le \|x\|_u \qquad (\star)$$

holds for each $x \in X$.

Next, we consider the vector subspace Y of X defined by

$$Y = \Big\{x \in L\colon\ \exists\, \alpha_1, \dots, \alpha_n \in A \ \&\ x_1, \dots, x_n \in L \text{ with } x = \sum_{i=1}^n \big(T_{\alpha_i} x_i - x_i\big)\Big\}.$$

We claim that $Y \cap \mathcal{U} = \emptyset$, where $\mathcal{U} = \operatorname{Int}(L_+)$ is the convex set of all order units. To see this, assume by way of contradiction that $Y \cap \mathcal{U} \ne \emptyset$. So, there exist indices $\alpha_1, \dots, \alpha_n$ and vectors $x_1, \dots, x_n$ such that the vector $e = \sum_{i=1}^n \big(T_{\alpha_i} x_i - x_i\big)$ is an order unit. Now fix $m > n$ and define the operator

$$T_m = \tfrac{1}{m^n} \sum_{k_1=1}^m \sum_{k_2=1}^m \cdots \sum_{k_n=1}^m T_{\alpha_1}^{k_1} T_{\alpha_2}^{k_2} \cdots T_{\alpha_n}^{k_n},$$

where $\alpha_1, \dots, \alpha_n$ are the fixed indices appearing in the representation of e as an element of Y. Clearly, T_m is a positive operator and $T_m(u) = u$ holds. Since the family $\{T_\alpha\}_{\alpha \in A}$ consists of pairwise commuting operators, we see that

$$\begin{aligned}
T_m(e) &= \tfrac{1}{m^n} \sum_{i=1}^n \sum_{k_1=1}^m \sum_{k_2=1}^m \cdots \sum_{k_n=1}^m T_{\alpha_1}^{k_1} T_{\alpha_2}^{k_2} \cdots T_{\alpha_n}^{k_n} \big(T_{\alpha_i} x_i - x_i\big) \\
&= \tfrac{1}{m^n} \sum_{i=1}^n \sum_{k_1=1}^m \cdots \sum_{k_{i-1}=1}^m \sum_{k_{i+1}=1}^m \cdots \sum_{k_n=1}^m \prod_{j \ne i} T_{\alpha_j}^{k_j} \sum_{k_i=1}^m \big(T_{\alpha_i}^{k_i+1} - T_{\alpha_i}^{k_i}\big) x_i \\
&= \tfrac{1}{m^n} \sum_{i=1}^n \sum_{k_1=1}^m \cdots \sum_{k_{i-1}=1}^m \sum_{k_{i+1}=1}^m \cdots \sum_{k_n=1}^m \prod_{j \ne i} T_{\alpha_j}^{k_j} \big(T_{\alpha_i}^{m+1} - T_{\alpha_i}\big) x_i.
\end{aligned}$$

If we let $c = \max_{1 \le i \le n} \|x_i\|_u$, then from $(\star)$ it follows that

$$\begin{aligned}
\Big\|\prod_{j \ne i} T_{\alpha_j}^{k_j} \big(T_{\alpha_i}^{m+1} - T_{\alpha_i}\big) x_i\Big\|_u &= \Big\|T_{\alpha_1}^{k_1} \cdots T_{\alpha_{i-1}}^{k_{i-1}} T_{\alpha_{i+1}}^{k_{i+1}} \cdots T_{\alpha_n}^{k_n} \big(T_{\alpha_i}^{m+1} - T_{\alpha_i}\big) x_i\Big\|_u \\
&\le 2c\,.
\end{aligned}$$

Consequently, from the triangle inequality, we infer that

$$\big\|T_m(e)\big\|_u \le \tfrac{1}{m^n} n m^{n-1} 2c = \tfrac{2cn}{m}\,.$$

Now, note that if $\eta > 0$ satisfies $u \le \eta e$, then

$$1 = \|u\|_u = \|T_m(u)\|_u \le \eta \|T_m(e)\|_u \le \tfrac{2\eta cn}{m}$$

holds for each $m > n$, which is impossible. Thus, indeed $Y \cap \mathcal{U} = \emptyset$.

Finally, according to the Interior Separating Hyperplane Theorem 8.16, there exists some nonzero $f \in L'$ such that $f(y) = 0$ for all $y \in Y$ and $f(x) \geq 0$ for all $x \in L_+$. So $0 < f \in L'_+$ and from

$$\langle x, T'_\alpha f - f \rangle = \langle T_\alpha x - x, f \rangle = 0 ,$$

we see that $T'_\alpha f = f$ for each $\alpha \in A$, i.e., the linear functional $f > 0$ is a fixed point for the family of operators $\{T'_\alpha\}_{\alpha \in A}$, as desired. ■

Corollary 4.20. *Let $T\colon L \to L$ be a positive operator on a Krein space. If a nonzero real number λ is an eigenvalue of T having an eigenvector which is an order unit, then $\lambda > 0$ and the adjoint operator T' also has λ as an eigenvalue.*

The next result is a famous theorem of A. A. Markov [**67**] and is a surprisingly simple consequence of Theorem 4.19. In order to formulate it, we need some preliminary discussion.

Recall that if S is an arbitrary nonempty set, then $\ell_\infty(S)$ denotes the Banach lattice of all bounded real-valued functions on S. In fact, with the sup norm $\ell_\infty(S)$ is an AM-space with unit the constant function $\mathbf{1}$ on S. In particular, $\ell_\infty(S)$ is a Krein space. Now if $\phi\colon S \to S$ is a mapping, then ϕ defines a positive continuous operator $T_\phi\colon \ell_\infty(S) \to \ell_\infty(S)$ via the formula

$$T_\phi(x)(s) = x(\phi(s)) = (x \circ \phi)(s), \ x \in \ell_\infty(S), \ \ s \in S.^6$$

That is, $(T_\phi)(x) = x \circ \phi$ and so $(T_\psi T_\phi)(x) = x \circ \phi \circ \psi$. Also, recall that the dual $\ell'_\infty(S)$ can be identified with the space of all finitely additive measures on S; see for instance [**12**, p. 499].

Theorem 4.21 (Markov). *Let $\{\phi_\alpha\}_{\alpha \in A}$ be a family of pairwise commuting mappings on an arbitrary nonempty set S into itself, and let $T_\alpha = T_{\phi_\alpha}$ for each α. Then there exists a positive functional $0 < f \in \ell'_\infty(S)$ such that*

$$f(T_\alpha x) = f(x)$$

for all α and all $x \in \ell_\infty(S)$.

Proof. Notice again that $\ell_\infty(S)$ is a Krein space. Also, the family $\{T_\alpha\}_{\alpha \in A}$ consists of pairwise commuting positive operators on $\ell_\infty(S)$. In addition, if $\mathbf{1}$ denotes the constant function one on S, then $\mathbf{1}$ is an order unit of $\ell_\infty(S)$ and $T_\alpha(\mathbf{1}) = \mathbf{1}$ holds for each α. The conclusion now is a direct consequence of Theorem 4.19. ■

A companion theorem to Theorem 4.19 is the following result.

[6] Operators of this type are called **composition operators**.

Theorem 4.22. *Let $\{T_\alpha\}$ be a family of pairwise commuting positive contractions*[7] *on a Krein space L. If there exists a nonzero vector $v \in L$ such that $T_\alpha(v) = v$ holds for each α, then there exists a nonzero positive linear functional $f \in L'$ satisfying $T'_\alpha f = f$ for each α.*

Proof. Start by fixing some nonzero vector $v \in L$ such that $T_\alpha(v) = v$ holds for each α. From Exercise 16 of Section 2.7, it follows that there exists an order unit u such that $\|x\| \leq 1$ implies $-u \leq x \leq u$ (and hence $\|x\|_u \leq 1$). In particular, since $\|x\| \leq 1$ implies $\|T_\alpha(x)\| \leq \|T_\alpha\| \cdot \|x\| \leq \|x\| \leq 1$, we see that $\|T_\alpha(x)\|_u \leq 1$ holds for each $\|x\| \leq 1$.

Next, consider the positive operators $T_\alpha\colon (L, \|\cdot\|) \to (L, \|\cdot\|_u)$ and notice that $\|T_\alpha\|_u = \sup_{\|x\|\leq 1} \|T_\alpha(x)\|_u \leq 1$. Consequently,

$$\|T_\alpha(x)\|_u \leq \|T_\alpha\|_u \cdot \|x\| \leq \|x\|$$

holds for each $x \in X$. Moreover, note that if $k_1, k_2, \ldots, k_n$ are arbitrary positive integers and $\alpha_1, \alpha_2, \ldots, \alpha_n$ are arbitrary indices, then

$$\begin{aligned} \big\|T^{k_1}_{\alpha_1} T^{k_2}_{\alpha_2} \cdots T^{k_n}_{\alpha_n} x\big\|_u &= \big\|T_{\alpha_1}\big(T^{k_1-1}_{\alpha_1} T^{k_2}_{\alpha_2} \cdots T^{k_n}_{\alpha_n} x\big)\big\|_u \\ &\leq \big\|T^{k_1-1}_{\alpha_1} T^{k_2}_{\alpha_2} \cdots T^{k_n}_{\alpha_n} x\big\| \\ &\leq \|x\|. \end{aligned}$$

The rest of the arguments are as in the proof of Theorem 4.19. We consider the same vector subspace Y. Assume that $Y \cap \mathcal{U} \neq \emptyset$, and then for each $m > n$ construct the operator T_m. If we let $c = \max_{1\leq i\leq n} \|x_i\|$, then

$$\big\|T_m(e)\big\|_u \leq \tfrac{2cn}{m}.$$

Now if $\eta > 0$ satisfies $\pm v \leq \eta e$, then

$$0 < \|v\|_u = \big\|T_\alpha(v)\big\|_u \leq \eta\big\|T_\alpha(e)\big\|_u \leq \tfrac{2\eta cn}{m}$$

holds for each $m > n$, which is impossible. The remaining proof is exactly as in the last part of Theorem 4.19. ■

Corollary 4.23. *If a positive contraction $T\colon L \to L$ on a Krein space has 1 as an eigenvalue, then there exists some $0 < f \in L'$ such that $T'f = f$.*

Finally, the following example shows that the commutativity of the family of operators in Theorems 4.19 and 4.22 cannot be dropped.

Example 4.24. Let $\mathcal{P}$ denote the family of all (Borel) probability measures on $[0,1]$. For each $\mu \in \mathcal{P}$ consider the positive operator $T_\mu\colon C[0,1] \to C[0,1]$ defined by $T_\mu f(t) = \int_0^1 f(s)\,d\mu(s)$, i.e., $T_\mu = \mu \otimes \mathbf{1}$, where $\mathbf{1}$ denotes the constant function one on $[0,1]$. Clearly, $T_\mu T_\nu = T_\nu$ holds for all $\mu, \nu \in \mathcal{P}$.

[7] Recall that an operator $T\colon X \to X$ on a normed space is called a **contraction** if $\|T\| \leq 1$.

In particular, it follows that $\{T_\mu\}_{\mu\in\mathcal{P}}$ is a family of noncommuting positive projections on $C[0,1]$. In addition, notice that $T_\mu\mathbf{1} = \mathbf{1}$ for each $\mu \in \mathcal{P}$.

Now assume by way of contradiction that there is some $0 < \theta \in \mathrm{ca}[0,1]$ such that $T'_\mu\theta = \theta$ holds for each μ; we can suppose that $\theta([0,1]) = 1$. Hence,

$$\theta = (T'_\mu)\theta = (\mathbf{1}\otimes\mu)\theta = \langle \mathbf{1}, \theta\rangle\mu = \mu$$

holds for each $\mu \in \mathcal{P}$, which is impossible. So, the family $\{T'_\mu\}_{\mu\in\mathcal{P}}$ does not have any common fixed point. ■

Finally, we mention that some of the results presented in this chapter have been generalized by I. A. Bakhtin in [**24, 25**].

Exercises

1. Theorem 4.18 was proved by M. G. Krein under the extra hypothesis that $T_\alpha(\mathcal{U}) \subseteq \mathcal{U}$ holds for each α. Theorem 4.18 implies, of course, this version of M. G. Krein. Show that M. G. Krein's version also implies Theorem 4.18. [*Hint*: For each $0 < \epsilon < 1$ let $T_{\alpha,\epsilon} = T_\alpha + \epsilon I$, where I is the identity operator. Clearly, $T_{\alpha,\epsilon}(\mathcal{U}) \subset \mathcal{U}$ holds, and so for each $\epsilon > 0$ there exist some $0 < \phi_\epsilon \in C$ and a family of nonnegative scalars $\{\lambda_{\alpha,\epsilon}\}_{\alpha\in A}$ such that

$$T'_\alpha\phi_\epsilon + \epsilon\phi_\epsilon = \lambda_{\alpha,\epsilon}\phi_\epsilon\,. \tag{†}$$

The net $\{\phi_\epsilon\}_{0<\epsilon<1}$ (where $\epsilon_1 \succeq \epsilon_2$ means $\epsilon_1 \le \epsilon_2$) of C has a w^*-convergent subnet. So, by passing to a subnet if necessary, we can assume that $\phi_\epsilon \xrightarrow{w^*} \phi$ holds in C. Now fix $\alpha \in A$ and then pick some $t > 0$ with $T_\alpha u \le tu$. From (†) it follows that

$$0 \le \lambda_{\alpha,\epsilon} = \lambda_{\alpha,\epsilon}\phi_\epsilon(u) = T'_\alpha\phi_\epsilon(u) + \epsilon\phi_\epsilon(u) \le t + 1\,.$$

This implies that the net $\{\lambda_{\alpha,\epsilon}\}_{0<\epsilon<1}$ has an accumulation point $\lambda_\alpha \ge 0$. Finally, a glance at (†) guarantees that $T'_\alpha\phi = \lambda_\alpha\phi$.

2. Complete the details missing from the proof of Theorem 4.22.

3. Show that Theorem 4.22 does not follow from Theorem 4.19. [*Hint*: For instance, if $L = \mathbb{R}^2$ and $T\colon L \to L$ is the positive projection represented by the matrix $\begin{bmatrix}1 & 0\\ 0 & 0\end{bmatrix}$, then T has 1 as an eigenvalue with eigenvector $\begin{bmatrix}1\\ 0\end{bmatrix}$. However, T does not have any order unit as an eigenvector.]

Chapter 5

$\mathcal{K}$-lattices

This chapter introduces a class of ordered vector spaces, called $\mathcal{K}$-lattices, that can be viewed as a generalization of Riesz spaces. These ordered vector spaces retain the properties of vector lattices with the exception of the associativity of the lattice operations. In the finite dimensional setting, an ordered vector space L with a generating closed cone is a $\mathcal{K}$-lattice if and only if there is a super cone $\mathcal{K}$ of L_+, i.e., $L_+ \subseteq \mathcal{K}$, such that for any non-empty finite subset A of L and any $\mathcal{K}$-strictly positive linear functional f the minimization problem

$$\min\{f(a)\colon\ a \geq A\}$$

has a unique solution that is independent of the $\mathcal{K}$-strictly positive linear functional f.

The $\mathcal{K}$-lattices are interesting in their own right. However, in this chapter we develop the theory of $\mathcal{K}$-lattices and use it to study some fundamental duality properties of spaces that are not vector lattices. In particular, we are interested in the following problem that remains unsolved:

($\bullet$) *suppose that L is an ordered vector space and N is a Dedekind complete vector lattice. When is it the case that for any $T \in \mathcal{L}_{\rm b}(L,N)$, i.e., for every order bounded operator T from L to N, we have*

$$\inf\{S(x) \in \mathcal{L}_{\rm b}(L,N)\colon\ S \geq \{0,T\}\} = \sup\{T(y)\colon\ y \in [0,x]\}$$

for all $x \in L_+$?

This is a problem that is related directly to the lattice structure of the vector space $\mathcal{L}_{\rm b}(L,N)$, of all order bounded operators from L to N, and has its roots in the classical Riesz–Kantorovich Theorem 1.59. Specifically, the above problem can be specialized as follows.

($\bigstar$) *If $\mathcal{L}_b(L, N)$ is a vector lattice, where L is an ordered vector space and N is a Dedekind complete vector lattice, is it then the case that for each $T \in \mathcal{L}_b(L, N)$ and each $x \in L_+$ we have*

$$T^+(x) = \sup\{T(y)\colon\ y \in [0, x]\}\,?$$

This problem has been studied extensively in the literature of vector lattices and ordered vector spaces; see for instance [**1, 2, 7, 8, 9, 10, 83**]. Using the theory of $\mathcal{K}$-lattices, we provide a solution to this problem in very special cases.

The lay out of this chapter is as follows. We first introduce the notion of a $\mathcal{K}$-lattice and study its properties. Then we introduce the notion of the Riesz–Kantorovich transform, establish several of its important properties, and prove interesting extentions of Andô's Theorem 2.47. Subsequently, we define an important $\mathcal{K}$-lattice (that we call the *order extension* of $\mathcal{L}_b(L, N)$) and use it to study duality properties of $\mathcal{L}_b(L, N)$.

5.1. The notion and properties of $\mathcal{K}$-lattices

Recall that if A is a nonempty subset of a partially ordered set, then the symbol $x \geq A$ means that $x \geq a$ holds for all $a \in A$. Similarly, $x \leq A$ designates that $x \leq a$ for all $a \in A$.

The purpose of this section is to introduce the notion of a $\mathcal{K}$-lattice and to study the basic properties of $\mathcal{K}$-lattices. The framework for the $\mathcal{K}$-lattices is as follows. We consider an ordered vector space with cone L_+ (whose induced vector ordering is denoted, as usual, by $\geq$ or $\leq$) and we assume that $\mathcal{K}$ is another cone of L satisfying $L_+ \subseteq \mathcal{K}$, i.e., $\mathcal{K}$ is a "super cone" of L_+. We shall also call any such cone $\mathcal{K}$ a **super cone** of the ordered vector space L.

The vector ordering induced by the super cone $\mathcal{K}$ on L will be indicated by the symbol $\succcurlyeq$ (or $\preccurlyeq$) instead of $\geq_{\mathcal{K}}$ (or $\leq_{\mathcal{K}}$). That is, $x \succcurlyeq y$ (or, equivalently, $y \preccurlyeq x$) means $x - y \in \mathcal{K}$. Clearly, $x \geq y$ implies $x \succcurlyeq y$ but $x \succcurlyeq y$ need not imply $x \geq y$.

For the rest of our discussion in this section, unless otherwise specified, L will be a fixed ordered vector space and $\mathcal{K}$ a super cone of L_+.

Definition 5.1. *A nonempty subset A of L is said to have a* ***$\mathcal{K}$-supremum*** *if there exists a vector $u \in L$ satisfying*

(a) *$u \geq A$ and*

(b) *whenever $v \geq A$, then $v \succcurlyeq u$.*

That is, the $\mathcal{K}$-supremum is the $\mathcal{K}$-minimum of the (nonempty) collection of all L_+-upper bounds of A.

*Similarly, a nonempty subset A of L is said to have a $\mathcal{K}$-**infimum** if there exists a vector $w \in L$ satisfying*

(a) $w \leq A$ *and*

(b) *whenever* $v \leq A$, *then* $v \preccurlyeq w$.

As expected, the $\mathcal{K}$-suprema and $\mathcal{K}$-infima are uniquely determined.

Lemma 5.2. *A nonempty subset of L has at most one $\mathcal{K}$-supremum and at most one $\mathcal{K}$-infimum.*

Proof. Assume that u and w are two $\mathcal{K}$-suprema of a nonempty subset A of L. Then it is immediate from the definition of the $\mathcal{K}$-supremum that $u \succcurlyeq w$ and $w \succcurlyeq u$ must be true, in which case we get $u = w$. ■

The $\mathcal{K}$-supremum of a nonempty subset A of L (if it exists) will be denoted by $\veebar A$. Similarly, the $\mathcal{K}$-infimum of a nonempty subset A of L (if it exists) will be denoted by $\barwedge A$. If there is a need to indicate the cone $\mathcal{K}$, then we shall write $\veebar_{\mathcal{K}} A$ instead of $\veebar A$ and $\barwedge_{\mathcal{K}} A$ instead of $\barwedge A$. Also, we shall write $x \veebar y$ and $x \barwedge y$ instead of $\veebar\{x, y\}$ and $\barwedge\{x, y\}$, respectively. In particular, we have $x \veebar y = y \veebar x$ and $x \barwedge y = y \barwedge x$ for all x and y.

We mention once more that the $\mathcal{K}$-supremum of a set A (if it exists) is precisely the smallest vector with respect to the vector ordering induced by the cone $\mathcal{K}$ of the set of all L_+-upper bounds of A. That is, if we let $B = \{x \in L\colon\ x \geq A\}$, then $\veebar A = \mathcal{K}\text{-}\min B$. The $\mathcal{K}$-infimum of A has an analogous interpretation: If $C = \{x \in L\colon\ x \leq A\}$, then $\barwedge A = \mathcal{K}\text{-}\max C$.

In the next result we gather a few basic properties of the $\mathcal{K}$-suprema and $\mathcal{K}$-infima. These properties are reminiscent of the corresponding properties for vector lattices; see Theorem 1.17.

Lemma 5.3. *Let $\mathcal{K}$ be a super cone of an ordered vector space L. Then for any nonempty subset A of L, vectors $x, y \in L$ and scalars $\alpha \geq 0$ we have the following properties.*

(1) $\barwedge(-A) = -(\veebar A)$ *and* $\veebar(-A) = -(\barwedge A)$.

(2) $\veebar(x + A) = x + \veebar A$ *and* $\barwedge(x + A) = x + \barwedge A$.

(3) $\veebar(\alpha A) = \alpha(\veebar A)$ *and* $\barwedge(\alpha A) = \alpha(\barwedge A)$.

(4) *The $\mathcal{K}$-supremum $x \veebar y$ exists if and only if the $\mathcal{K}$-infimum $x \barwedge y$ exists, in which case we have*
$$x + y = x \veebar y + x \barwedge y\,.$$
In particular, if $x \veebar 0$ exists, then $x \barwedge 0$ also exists and we have
$$x = x \veebar 0 + x \barwedge 0\,.$$

(5) $x \veebar y = \frac{1}{2}[x+y+(x-y)\veebar(y-x)]$ *and* $x \barwedge y = \frac{1}{2}[x+y+(x-y)\barwedge(y-x)]$.

Proof. (1) The first identity (as well as the other one) should be interpreted as follows: *if for a nonempty subset A of L the $\mathcal{K}$-supremum $\veebar A$ exists, then the nonempty set $-A$ has a $\mathcal{K}$-infimum and $\barwedge(-A) = -(\veebar A)$ holds true.* Equivalently, *if for a nonempty subset A of L the $\mathcal{K}$-infimum $\barwedge(-A)$ exists, then the nonempty set A has a $\mathcal{K}$-supremum and $\barwedge(-A) = -(\veebar A)$.*

We verify the first part of (1) only. Let $u = \veebar A$. From $u \geq A$ we get $-u \leq -A$. Now if $x \in L$ satisfies $x \leq -A$, then $-x \geq A$ and so $-x \succcurlyeq u$. This implies $x \preccurlyeq -u$ and from this we see that $-u = \barwedge A$.

(2) Again, these formulas should be given the obvious appropriate interpretation. For instance, the first part of (2) should be read as follows: *If the $\mathcal{K}$-supremum $\veebar A$ of the set A exists, then for each $x \in L$ the set $x + A$ has a $\mathcal{K}$-supremum and $\veebar(x + A) = x + \veebar A$.*

We prove only this property and leave the proof of the other for the reader. So, let $u = \veebar A$ exist in L. Clearly, $x + u \geq x + A$ holds true. Now assume that some $v \in L$ satisfies $v \geq x + A$. Then, $v - x \geq A$ and so $v - x \succcurlyeq u$. This implies $v \succcurlyeq x + u$ and consequently the $\mathcal{K}$-supremum of the set $x + A$ exists and $\veebar(x + A) = x + \veebar A$.

(3) This is straightforward and is left for the reader.

(4) Using (2) and (1) consecutively, we get

$$\begin{aligned} x + y - x \veebar y &= x + y + (-x) \barwedge (-y) \\ &= [x + y + (-x)] \barwedge [x + y + (-y)] \\ &= y \barwedge x\,. \end{aligned}$$

The special case follows from the above property by letting $y = 0$.

(5) Using (2) and (3) in succession, we get

$$\tfrac{1}{2}[x + y + (x - y) \veebar (y - x)] = \tfrac{1}{2}[(2x) \veebar (2y)] = \tfrac{1}{2}[2(x \veebar y)] = x \veebar y\,.$$

The correct formulation of this property is the following: *If the $\mathcal{K}$-supremum $x \veebar y$ exists, then $(x - y) \veebar (y - x)$ also exists and*

$$\tfrac{1}{2}[x + y + (x - y) \veebar (y - x)] = x \veebar y\,.$$

Equivalently, *if $(x - y) \veebar (y - x)$ exists, then $x \veebar y$ also exists and we have $\frac{1}{2}[x + y + (x - y) \veebar (y - x)] = x \veebar y$.* ■

The $\mathcal{K}$-suprema and $\mathcal{K}$-infima enjoy the following "invariance" property.

Lemma 5.4. *Suppose that L is an ordered vector space and that $\mathcal{K}$ and $\mathcal{C}$ are two super cones satisfying $L_+ \subseteq \mathcal{C} \subseteq \mathcal{K}$. If for some nonempty subset A of L the $\mathcal{C}$-supremum $\veebar_{\mathcal{C}} A$ of A exists, then the $\mathcal{K}$-supremum $\veebar_{\mathcal{K}} A$ of A also exists and they coincide, i.e., $\veebar_{\mathcal{C}} A = \veebar_{\mathcal{K}} A$.*

A similar result is true for infima.

Proof. Let $x = \veebar_{\mathcal{C}} A$. This means that $x \geq A$ and if $y \geq A$, then $y \geq_{\mathcal{C}} x$. Since $\mathcal{C} \subseteq \mathcal{K}$, it follows that $y \geq_{\mathcal{K}} x$. The latter shows that $x = \veebar_{\mathcal{K}} A$. ■

An immediate consequence is that the ordinary suprema and infima are $\mathcal{K}$-suprema and $\mathcal{K}$-infima.

Corollary 5.5. *Assume that $\mathcal{K}$ is a super cone of an ordered vector space L and that A is a nonempty subset of L. If the supremum $\vee A$ of A exists in L, then the $\mathcal{K}$-supremum of A also exists and the two suprema coincide, i.e., $\veebar A = \vee A$.*

Similarly, if the infimum $\wedge A$ of A exists in L, then the $\mathcal{K}$-infimum of A also exists and the two infima coincide, i.e., $\barwedge A = \wedge A$.

Now we come to the definition of a $\mathcal{K}$-lattice.[1]

Definition 5.6. *An ordered vector space L is a **$\mathcal{K}$-lattice** (where $\mathcal{K}$ is a super cone of L_+) if every nonempty finite subset of L has a $\mathcal{K}$-supremum.*

*In case L is a $\mathcal{K}$-lattice, we shall also say that L_+ is a **$\mathcal{K}$-lattice cone**.*

The geometric interpretation of $\mathcal{K}$-lattices is described in the next result whose straightforward proof is left for the reader.

Lemma 5.7. *An ordered vector space L is a $\mathcal{K}$-lattice for a super cone $\mathcal{K}$ if and only if for any finite collection $\{x_1, \ldots, x_n\}$ of vectors of L there exists some $z \in L$ such that*

$$z \in (x_1 + L_+) \cap (x_2 + L_+) \cap \cdots \cap (x_n + L_+) \subseteq z + \mathcal{K}\,.$$

Three observations are in order here.

(1) An ordered vector space L is (in view of part (1) of Lemma 5.3) a **$\mathcal{K}$-lattice** if and only if every nonempty finite subset of L has a $\mathcal{K}$-infimum.

(2) An ordered vector space L is (in view of Lemma 5.2) a Riesz space L if and only if it is an L_+-lattice.

(3) Since in a $\mathcal{K}$-lattice every finite set of L must have at least one upper bound, it follows immediately that the cone L_+ is generating.

We continue with two consequences of Corollary 5.5.

Corollary 5.8. *If L is a Riesz space, then L is a $\mathcal{K}$-lattice for every super cone $\mathcal{K}$ of L.*

---

[1] We use the name $\mathcal{K}$-lattice to honor L. V. Kantorovich whose works and ideas inspired us to introduce these type of cones. Kantorovich used similar techniques in optimization problems and he shared (with T. C. Koopmans) the 1975 Nobel Prize in Economics "for his contributions to the theory of optimum allocation of resources."

Since in a Riesz space the operations $\vee$ and $\wedge$ satisfy the associative law, the following is an easy consequence of Corollary 5.5.

Corollary 5.9. *If a Riesz space L is a $\mathcal{K}$-lattice, then the operations $\underline{\vee}$ and $\overline{\wedge}$ satisfy the associative laws. That is, for all $x, y, z \in L$ we have*

$$(x \underline{\vee} y) \underline{\vee} z = x \underline{\vee} (y \underline{\vee} z) \quad \text{and} \quad (x \overline{\wedge} y) \overline{\wedge} z = x \overline{\wedge} (y \overline{\wedge} z).$$

It turns out that the Riesz spaces are precisely the $\mathcal{K}$-lattices whose join operation $\underline{\vee}$ satisfies the associative law.

Theorem 5.10. *An ordered vector space L is a Riesz space if and only if there exists a super cone $\mathcal{K}$ such that*

(1) *L is a $\mathcal{K}$-lattice and*
(2) *the operation $\underline{\vee}$ satisfies the associative law on L_+, i.e., for all $x, y, z \in L_+$ we have $(x \underline{\vee} y) \underline{\vee} z = x \underline{\vee} (y \underline{\vee} z)$.*

Proof. If L is a Riesz space, then the super cone $\mathcal{K} = L_+$ satisfies the desired properties. For the converse, assume that a super cone $\mathcal{K}$ satisfies properties (1) and (2). Notice first that (since L is a $\mathcal{K}$-lattice) the cone L_+ is generating.

Now suppose by way of contradiction that L_+ is not a lattice cone. Since L_+ is generating, this means that there exist two vectors $x, y \in L_+$ for which $x \vee y$ does not exist in L. Using that L_+ is generating once more, we see that the set $\{z \in L_+ \colon\ z \geq \{x, y\}\}$ is nonempty. In particular, since $x \vee y$ does not exist, there exists some $z \in L_+$ satisfying $z \geq \{x, y\}$ and $z \not\geq x \underline{\vee} y$. Otherwise, if for each $z \geq \{x, y\}$ we have $z \geq x \underline{\vee} y$, then $x \vee y$ exists and $x \vee y = x \underline{\vee} y$, which is impossible. By (2) we have

$$(x \underline{\vee} y) \underline{\vee} z = x \underline{\vee} (y \underline{\vee} z). \tag{$\star$}$$

Now from $z \geq \{x, y\}$ we get $z = y \vee z$. So, by Corollary 5.5, we see that $z = y \underline{\vee} z$. Similarly, $z = x \underline{\vee} z$. Hence $x \underline{\vee} (y \underline{\vee} z) = x \underline{\vee} z = z$. But then from $(\star)$ we obtain $(x \underline{\vee} y) \underline{\vee} z = z$, which implies $z \geq x \underline{\vee} y$, a contradiction. ■

The Dedekind complete $\mathcal{K}$-lattices are defined as follows.

Definition 5.11. *A $\mathcal{K}$-lattice L is called* ***Dedekind complete*** *if every nonempty subset of L that is bounded from above in L has a $\mathcal{K}$-supremum.*

Using property (1) of Lemma 5.3 once more, we see that an ordered vector space L with a generating cone is a Dedekind complete $\mathcal{K}$-lattice if and only if every nonempty subset of L that is bounded from below with respect to $\geq$ has a $\mathcal{K}$-infimum.

Not every $\mathcal{K}$-lattice is Dedekind complete. For instance, every vector lattice L that fails to be Dedekind complete is not a Dedekind complete $\mathcal{K}$-lattice for any super cone of L_+.

The next result presents a useful class of Dedekind complete $\mathcal{K}$-lattices.

Lemma 5.12. *Let M be a Dedekind complete vector lattice and let L be a vector subspace of M. Assume that there exists a cone L_+ of L satisfying the following properties.*

(a) *The cone L_+ generates L, i.e., $L = L_+ - L_+$.*

(b) *The induced cone $\mathcal{K} = L \cap M_+$ is a super cone of L_+, i.e., $L_+ \subseteq \mathcal{K}$.*

(c) *For any nonempty subset A of L_+ the infimum $\inf_M A$ of the set A in M belongs to L_+, i.e., $\inf_M A \in L_+$.*

Then the ordered vector space L is a Dedekind complete $\mathcal{K}$-lattice. Moreover, for any nonempty subset A of L_+ that is bounded from above in L the infimum of the set $\{x \in L\colon\ x \geq A\}$ in M coincides with $\veebar A$, i.e.,

$$\veebar A = \inf_M \{x \in L\colon\ x \geq A\}.$$

Proof. We denote the vector ordering on L induced by the cone L_+ by $\geq$ and the one induced by the cone M_+ on M by $\succeq$, and so $\succeq$ also denotes the vector ordering induced by the cone $\mathcal{K}$ on L.

Now let A be a nonempty subset of L that is bounded from above in L. Clearly, the set $B = \{x \in L\colon\ x \geq A\}$ is nonempty. In particular, the set B is bounded from below in L by any $a \in A$. But then (since $L_+ \subseteq \mathcal{K}$) B is also bounded from below in M by any $a \in A$. Because M is a Dedekind complete vector lattice, the vector $u = \inf_M B$ exists in M and clearly $x \succeq u$ holds in M for each $x \geq A$.

Next notice that for each $a \in A$ we have

$$\begin{aligned} u - a &= \inf_M \{x \in L\colon\ x \geq A\} - a \\ &= \inf_M [\{x \in L\colon\ x \geq A\} - a] \\ &= \inf_M \{x - a\colon\ x \geq A\}. \end{aligned}$$

From this and property (c), we see that $u - a \in L_+$ for each $a \in A$. This implies that $u \in L$ and that u is an upper bound of A in L. Now observe that for each $x \in L$ satisfying $x \geq A$ we have $x \succeq u$. This shows that $\veebar A$ exists and $\veebar A = u$. ■

This result allows us to construct a large class of examples of Dedekind complete $\mathcal{K}$-lattices as follows. Let Ω be a nonempty set and let N be a vector lattice. Also, let $M = N^\Omega$ be the vector space of all functions from Ω to N. (Alternatively, M consists of families $(x_\omega)_{\omega\in\Omega}$, where $x_\omega \in N$ for all $\omega \in \Omega$.) With the usual definitions of addition and scalar multiplication of functions, it is easy to see that M is a vector space. Moreover, with the pointwise lattice operations $\curlywedge$ and $\curlyvee$, defined for each $f, g \in M$ by

$$[f \curlyvee g](\omega) = [f(\omega)] \vee [g(\omega)] \quad \text{and} \quad [f \curlywedge g](\omega) = [f(\omega)] \wedge [g(\omega)],$$

the vector space M is a Riesz space whose positive cone M_+ consists of all nonnegative functions from Ω to N. It is also a routine matter to verify that if N is Dedekind complete, then M is likewise a Dedekind complete Riesz space, where the supremum and the infimum of an arbitrary nonempty subset A of M (if they exist) are the pointwise ones, i.e., for each $\omega \in \Omega$ we have

$$\Big[\bigcurlyvee A\Big](\omega) = \bigvee_{f \in A} f(\omega) \quad \text{and} \quad \Big[\bigcurlywedge A\Big](\omega) = \bigwedge_{f \in A} f(\omega).$$

We summarize the above discussion in the following result.

Lemma 5.13. *If Ω is a nonempty set and L is a Riesz space, then the vector space $M = N^\Omega$ under the pointwise algebraic and lattice operations, i.e., under the operations*

$$[f + g](\omega) = f(\omega) + g(\omega) \quad \textit{and} \quad (\lambda f)(\omega) = \lambda f(\omega),$$

$$[f \curlyvee g](\omega) = [f(\omega)] \vee [g(\omega)] \quad \textit{and} \quad [f \curlywedge g](\omega) = [f(\omega)] \wedge [g(\omega)],$$

is also a Riesz space. Its positive cone M_+ consists of all nonnegative functions from Ω to N.

Moreover, the Riesz space M is Dedekind complete if and only if N is Dedekind complete. In addition, the supremum and infimum of a nonempty subset A of M (if they exist) are given for each $\omega \in \Omega$ by the formulas

$$\Big[\bigcurlyvee A\Big](\omega) = \bigvee_{f \in A} f(\omega) \quad \textit{and} \quad \Big[\bigcurlywedge A\Big](\omega) = \bigwedge_{f \in A} f(\omega).$$

From this and Lemma 5.12 we have the following.

Corollary 5.14. *Assume that*

(a) *$M = N^\Omega$, where N is a Dedekind complete Riesz space and Ω is a nonempty set,*

(b) *L_+ is a subcone of M_+ that contains the pointwise infima of its nonempty subsets.*

If $L = L_+ - L_+$ and $\mathcal{K} = L \cap M_+$, then the vector space L is a Dedekind complete $\mathcal{K}$-lattice. Moreover, for any nonempty subset A of L_+ that is bounded from above in L we have

$$\underline{\curlyvee} A = \curlywedge\{x \in L\colon\ x \geq A\}.$$

We shall close the section with the introduction and the properties of $\mathcal{K}$-sublattices.

Definition 5.15. *A given vector subspace V of a $\mathcal{K}$-lattice L is called a* ***$\mathcal{K}$-sublattice*** *if for any nonempty finite subset A of V we have $\underline{\vee} A \in V$ and $\overline{\wedge} A \in V$.*

Two basic properties of $\mathcal{K}$-sublattices are included in the next result.

Lemma 5.16. *For a $\mathcal{K}$-lattice L we have the following.*

(1) *Every $\mathcal{K}$-sublattice V of L is a $\mathcal{K}$-lattice—where V is considered ordered by the induced cone $V_+ = V \cap L_+$ and the role of $\mathcal{K}$ is played by $V \cap \mathcal{K}$.*

(2) *If $\{V_i\}_{i \in I}$ is a family of $\mathcal{K}$-sublattices of L, then their intersection $V = \bigcap_{i \in I} V_i$ is likewise a $\mathcal{K}$-sublattice of L.*

Proof. (1) It is straightforward.

(2) If A is a nonempty finite subset of $V = \bigcap_{i \in I} V_i$, then $\underline{\vee} A$ exists and belongs to V_i for each $i \in I$, i.e., $\underline{\vee} A \in V$. This shows that V is a $\mathcal{K}$-sublattice. ■

The above result shows that every nonempty subset A of a $\mathcal{K}$-lattice is included in a smallest (with respect to inclusion) $\mathcal{K}$-sublattice. This is called the **$\mathcal{K}$-sublattice generated by** A in L and is denoted $\mathcal{K}(A)$. That is, $\mathcal{K}(A)$ is the intersection of the family of all $\mathcal{K}$-sublattices that include A.

Exercises

1. Complete the missing details in the proof of Lemma 5.3.

2. If an ordered vector space L is a $\mathcal{K}$-lattice, then show that the cone L_+ (and hence the cone $\mathcal{K}$) is generating.

3. Prove Lemma 5.7 and Corollary 5.8.

4. Show that every Riesz space L satisfies the associative laws for its lattice operations $\vee$ and $\wedge$ by proving that for all $x, y, z \in L$ we have

$$(x \vee y) \vee z = x \vee (y \vee z) = \vee\{x, y, z\} \text{ and}$$
$$(x \wedge y) \wedge z = x \wedge (y \wedge z) = \wedge\{x, y, z\}.$$

5. Let $\mathcal{K}$ be a super cone of an ordered vector space and let $x, y, z \in L$. Show with appropriate interpretations that

$$(x \,\underline{\vee}\, y) \,\underline{\vee}\, z = x \,\underline{\vee}\, (y \,\underline{\vee}\, z) \iff (x \,\overline{\wedge}\, y) \,\overline{\wedge}\, z = x \,\overline{\wedge}\, (y \,\overline{\wedge}\, z).$$

6. In $L = \mathbb{R}^3$ let $L_+ = \{(x_1, x_2, x_3) \in \mathbb{R}^3_+ \colon \ x_1 \geq x_2 \text{ and } x_1 \geq x_3\}$ and let $\mathcal{K} = R^3_+$.

(a) Show that L_+ is a generating cone.
(b) Show that L_+ is not a lattice cone.
(c) Show that L_+ is a $\mathcal{K}$-lattice cone.

(d) Give an example of three vectors $x, y, z \in L_+$ such that

$$(x \veebar y) \veebar z \neq \veebar\{x, y, z\}.$$

(e) Is every polyhedral subcone in $\mathbb{R}^3_+$ a $\mathcal{K}$-lattice cone?
[*Hint*: For (a) show that $(2, 1, 1)$ is an order unit. For (b) verify that L_+ has four extremal rays. For (c) use Lemma 5.12.]

7. Show that every Dedekind complete $\mathcal{K}$-lattice L is a $\mathcal{K}$-lattice but the converse is not true. [*Hint*: Note that (since the cone L_+ is generating) every nonempty finite subset of L is bounded from above in L.]

8. Assume that (L, τ) is an arbitrary topological vector space and let $\mathcal{K}$ be a lattice cone whose τ-interior $\mathrm{Int}(\mathcal{K})$ is nonempty. If L is ordered by the cone $L_+ = \mathrm{Int}(\mathcal{K}) \cup \{0\}$, then show that L is not a $\mathcal{K}$-lattice.

9. Define the absolute value $|x|$ of the vector x in a $\mathcal{K}$-lattice as $x \veebar (-x)$. Which of the corresponding statements in Theorem 1.17 can you prove?

10. Recall that a function $f: C \to \mathbb{R}$, where C is a convex subset of a vector space, is said to be **concave** if

$$f(\alpha x + (1-\alpha)y) \geq \alpha f(x) + (1-\alpha) f(y)$$

holds true for all scalars $0 \leq \alpha \leq 1$ and all $x, y \in C$.

Let $\mathcal{C}$ be the cone of all concave functions from $\mathbb{R}_+$ to $\mathbb{R}_+$ in the vector space $\mathbb{R}^{\mathbb{R}_+}$. Also, let $L = \mathcal{C} - \mathcal{C}$ be the vector subspace generated by $\mathcal{C}$ in $\mathbb{R}^{\mathbb{R}_+}$. We let $L_+ = \mathcal{C}$ and consider L ordered by the cone L_+.

(a) If $\mathcal{K} = \{f \in L\colon\ f(x) \geq 0 \text{ for all } x \in \mathbb{R}_+\}$, then show that L is a $\mathcal{K}$-lattice.

(b) Is $\mathcal{C}$ a lattice cone?

11. Show that if f is a $\mathcal{K}$-positive linear functional and A is a finite (nonempty) subset of a $\mathcal{K}$-lattice, then $\veebar A$ solves the following minimization problem:

$$\min\{f(x)\colon\ x \geq A\}.$$

Also show that, in this case, if f is $\mathcal{K}$-strictly positive, then $\veebar A$ is the only minimizer.

12. Let $L = \mathbb{R}^3$ and consider L ordered by the standard ice cream cone L_+ of $\mathbb{R}^3$, that is,

$$L_+ = \big\{(x, y, z) \in \mathbb{R}^3\colon\ z \geq \sqrt{x^2 + y^2}\,\big\}.$$

Also consider the linear functional $f: L \to \mathbb{R}$ defined by $f(x, y, z) = y + z$. Establish the following.

(a) The linear functional f is positive but not strictly positive.

(b) The minimization problem

$$\min\big\{f(x, y, z)\colon\ (x, y, z) \geq \{(0,0,0), (1,0,0)\}\,\big\}$$

has no solution.

(c) If $\mathcal{K}$ is a super cone of L_+ for which the linear functional f is also $\mathcal{K}$-positive, then L cannot be a $\mathcal{K}$-lattice.

[*Hint*: The proof of (b) goes as follows. First note that if $(x,y,z) \in L_+$, then $y+z \geq y+\sqrt{x^2+y^2} \geq 0$. Now if we let $\mathbf{x}_n = (1,-n,\sqrt{1+n^2})$, then $\mathbf{x}_n \geq (0,0,0)$ and $\mathbf{x}_n \geq (1,0,0)$. Hence, from $-n+\sqrt{1+n^2} \to 0$ it follows that $\inf\bigl\{(x,y,z)\colon\ (x,y,z) \geq \{(0,0,0),(1,0,0)\}\bigr\} = 0$. Next, suppose that for some $(x_0,y_0,z_0) \in \mathbb{R}^3$ we have $(x_0,y_0,z_0) \geq \{(0,0,0),(1,0,0)\}$ and $y_0+z_0=0$. It follows that $(x_0,-z_0,z_0) \geq \{(0,0,0),(1,0,0)\}$ and $x_0=0$. But then $(0,-z_0,z_0) \not\geq (1,0,0)$, which is a contradiction.]

13. For a closed super cone $\mathcal{K}$ of some ordered locally convex space L show that the following statements are equivalent.
(a) L is a $\mathcal{K}$-lattice.
(b) For each nonempty finite subset A of L there is a vector $z_A \in L$ such that given any $f \in \mathcal{K}'$ the vector z_A solves the minimization problem
$$\min\{f(x)\colon\ x \geq A\}.$$
Moreover, in this case, show that $z_A = \veebar A$.
[*Hint*: To see that (b) $\Longrightarrow$ (a), assume that some $y \in L$ satisfies $y \geq A$ and $x \not\geq z_A$ or $y - z_A \notin \mathcal{K}$. Since $\mathcal{K}$ is closed, there exist (according to the separation theorem) some $f \in L'$ and a constant $c \in \mathbb{R}$ such that $f(y-z_A) < c$ and $f(x) \geq 0$ for all $x \in \mathcal{K}$. It follows that $f \in \mathcal{K}'$ and $c \leq 0$. In particular, we have $f(y) < f(z_A)$, and so z_A is not a solution of the optimization problem $\min\{f(x)\colon\ x \geq A\}$, which is a contradiction.]

14. Show that the standard ice cream cone in $\mathbb{R}^3$ is not a $\mathcal{K}$-lattice cone for any super cone $\mathcal{K}$. [*Hint*: For any super cone $\mathcal{K}$ choose $x,y \in L_+$ and a linear functional f that is $\mathcal{K}$-strictly positive such that the minimization problem
$$\min\{f(z)\colon\ z \geq \{x,y\}\}$$
has no solution. Now invoke Exercise 11 above.]

15. Assume that L is an ordered vector space with a generating cone and that $\mathcal{K}$ is a super cone of L_+. Show that L is a Dedekind complete lattice if and only if every nonempty subset of L_+ has a $\mathcal{K}$-infimum.

16. Is there an example of a Dedekind complete cone L_+ that is not a lattice cone but is a $\mathcal{K}$-lattice for which $\mathcal{K}$ is not a Dedekind complete lattice cone?

17. Prove in detail Lemma 5.13.

18. Prove Corollary 5.14.

5.2. The Riesz–Kantorovich transform

In this section L will denote an ordered vector space with a generating cone and N an arbitrary ordered vector space. We shall let M denote the vector lattice of all functions from L_+ to N, i.e., $M = N^{L_+}$. We already know that under the pointwise algebraic and lattice operations M is a Riesz space; see Lemma 5.13. Recall also that M_+ is the cone of all functions $f\colon L_+ \to N$ satisfying $f(x) \geq 0$ for all $x \in L_+$. The vector ordering induced by M_+ will be denoted by $\succcurlyeq$. That is, $f \succeq g$ in M means $f(x) \geq g(x)$ for all $x \in L_+$. Also keep in mind that the pointwise infima and suprema will be denoted $\curlywedge$ and $\curlyvee$.

We now isolate some special functions of M. A function $f\colon L_+ \to N$ is said to be

(1) **additive** if $f(x) + f(y) = f(x + y)$ holds for all $x, y \in L_+$,
(2) **superadditive** if $f(x) + f(y) \leq f(x + y)$ holds for all $x, y \in L_+$,
(3) **subadditive** if $f(x) + f(y) \geq f(x + y)$ holds for all $x, y \in L_+$,
(4) **positively homogeneous** if $f(\alpha x) = \alpha f(x)$ holds for each $\alpha \geq 0$ and all $x \in L_+$,
(5) **monotone** if $x \leq y$ implies $f(x) \leq f(y)$,
(6) **upper order bounded** if it carries order intervals of the form $[0, x]$ to bounded from above subsets of N,
(7) **lower order bounded** if it carries order intervals of the form $[0, x]$ to bounded from below subsets of N,
(8) **order bounded** if it carries order intervals of the form $[0, x]$ to order bounded subsets of N,
(9) **sublinear** if it is subadditive and positively homogeneous,
(10) **superlinear** if it is superadditive and positively homogeneous.

It should be clear that the above definitions are also applicable to any function defined on an arbitrary cone of a vector space with values in N. Clearly, every additive function is both superadditive and subadditive but a superadditive or a subadditive function need not be additive.

If $f \in \mathcal{L}(L, N)$, i.e., if f is a linear operator, then $f|_{L_+}$ (the restriction of f to the positive cone L_+) is additive and positively homogeneous. In particular, $f|_{L_+}$ is superadditive and positively homogeneous. On the other hand, in case L_+ is generating and the ordered vector space N is Archimedean, and a function $f\colon L_+ \to \mathrm{N}_+$ is additive and positively homogeneous, then (according to part (3) of Lemma 1.26) the formula

$$\widehat{f}(x) = f(x_1) - f(x_2)\,,$$

where $x = x_1 - x_2$ with $x_1, x_2 \in L_+$, extends f uniquely to a positive operator from L to N. Therefore, we have the following important property that we shall invoke quite often in the sequel.

Lemma 5.17. *If the cone L_+ is generating and N is Archimedean, then the collection of all positive operators from L to N can be identified with the collection of all additive positively homogeneous functions from L_+ to N_+.*

Two simple but useful monotonicity properties are stated next.

Lemma 5.18. *For a function $f\colon L_+ \to N$ we have the following.*

(1) *If f is monotone, then it is upper order bounded.*

(2) *If f is positive and superadditive, then it is monotone—and thus upper order bounded.*

Proof. (1) Clearly, if f is monotone and $x \geq 0$, then for each $0 \leq y \leq x$ we have $f(0) \leq f(y) \leq f(x)$. This implies $f([0,x]) \subseteq [f(0), f(x)]$ and hence f is upper order bounded.

(2) Now suppose that f is supperadditive and positive and let $0 \leq y \leq x$, i.e., $x - y \in L_+$. Then we have

$$f(x) = f((x-y)+y) \geq f(x-y) + f(y) \geq f(y)\,,$$

so that f is monotone. ■

From now on we shall assume

- **the Riesz space N is Dedekind complete.**

Having added this assumption, we can now introduce the Riesz–Kantorovich transform of a finite collection of functions.

Definition 5.19. *For each $i = 1, 2, \ldots, m$ let $f_i\colon L_+ \to N$ be an upper order bounded function and let $f = (f_1, \ldots, f_m)$. Then the two functions $\mathcal{R}_f, \mathcal{R}_f^{=}\colon L_+ \to N$, defined for each $x \in L_+$ by*

$$\begin{aligned} \mathcal{R}_f(x) &= \sup\Big\{ \sum_{i=1}^m f_i(x_i)\colon\ x_i \in L_+ \text{ for each } i \text{ and } \sum_{i=1}^m x_i \leq x \Big\} \text{ and} \\ \mathcal{R}_f^{=}(x) &= \sup\Big\{ \sum_{i=1}^m f_i(x_i)\colon\ x_i \in L_+ \text{ for each } i \text{ and } \sum_{i=1}^m x_i = x \Big\}, \end{aligned}$$

are called the ***Riesz–Kantorovich transforms*** *of the m-tuple of functions $f = (f_1, \ldots, f_m)$.*

If clarity requires it, we shall denote $\mathcal{R}_f$ by $\mathcal{R}(f_1, \ldots, f_m)$ and $\mathcal{R}_f^{=}$ by $\mathcal{R}^{=}(f_1, \ldots, f_m)$. In case a distinction between these two functions is warranted, we shall call $\mathcal{R}_f$ the **Riesz–Kantorovich transform** (or the **R–K transform** in brief) and $\mathcal{R}_f^{=}$ the **exact Riesz–Kantorovich transform** of the m-tuple $f = (f_1, \ldots, f_m)$.

Clearly, $\mathcal{R}_f$ and $\mathcal{R}_f^{=}$ both belong to M. Moreover, it should be immediate that $\mathcal{R}_f$ always dominates $\mathcal{R}_f^{=}$, i.e., $\mathcal{R}_f \succeq \mathcal{R}_f^{=}$ is true, and that if one of the functions f_i is monotone, then $\mathcal{R}_f = \mathcal{R}_f^{=}$. Also a glance at the definitions of the Riesz–Kantorovich transform shows that

$$\mathcal{R}(f_1, \ldots, f_m) = \mathcal{R}^{=}(0, f_1, \ldots, f_m)\,.$$

Lemma 5.20. *If $f = (f_1, \dots, f_m)$ is an m-tuple of upper order bounded functions from L_+ to N, then their Riesz–Kantorovich transform $\mathcal{R}_f$ is a monotone function from L_+ to N.*

Proof. Assume that $\mathcal{R}_f$ is the Riesz–Kantorovich transform of an m-tuple of functions $f = (f_1, \dots, f_m)$ and let $0 \leq x \leq y$. If $\sum_{i=1}^m x_i \leq x$ holds with $x_i \geq 0$ for each i, then $\sum_{i=1}^m x_i \leq y$, and so $\sum_{i=1}^m f_i(x_i) \leq \mathcal{R}_f(y)$. This implies $\mathcal{R}_f(x) \leq \mathcal{R}_f(y)$. ■

We turn to an important property of the Riesz–Kantorovich transform.

Lemma 5.21. *The Riesz–Kantorovich transform $\mathcal{R}_f$ of an arbitrary m-tuple $f = (f_1, \dots, f_m)$ of superadditive, positively homogeneous, upper order bounded functions (from L_+ to N) is positive, superadditive, positively homogeneous, and upper order bounded.*

Proof. Let $f = (f_1, f_2, \dots, f_m)$ be an m-tuple of functions as above. We shall establish that the Riesz–Kantorovich transform $\mathcal{R}_f$ is superadditive and positively homogeneous. Start by observing that since each f_i is upper order bounded and satisfies $f_i(0) = 0$, the Riesz–Kantorovich transform $\mathcal{R}_f$ takes values in N_+, i.e., $\mathcal{R}_f$ is positive.

The positive homogeneity of $\mathcal{R}_f$ is a consequence of the fact that each f_i is positively homogeneous and that for each $\alpha > 0$ and $x \in L_+$, a vector $(x_1, \dots, x_m) \in L_+^m$ satisfies $\sum_{i=1}^m x_i \leq x$ if and only if $\sum_{i=1}^m \alpha x_i \leq \alpha x$.

For the superadditivity of $\mathcal{R}_f$ fix $x, y \in L_+$, and let $(x_1, \dots, x_m) \in L_+^m$ satisfy $\sum_{i=1}^m x_i \leq x$ and $(y_1, \dots, y_m) \in L_+^m$ satisfy $\sum_{i=1}^m y_i \leq y$. Then, we have $(x_1 + y_1, \dots, x_m + y_m) \in L_+^m$ and $\sum_{i=1}^m (x_i + y_i) \leq x + y$. Now the superadditivity of each f_i yields

$$\begin{aligned}
\big[f_1(x_1) + \cdots + f_m(x_m)\big] &+ \big[f_1(y_1) + \cdots + f_m(y_m)\big] \\
&= \big[f_1(x_1) + f(y_1)\big] + \cdots + \big[f(x_m) + f(y_m)\big] \\
&\leq f_1(x_1 + y_1) + \cdots + f_m(x_m + y_m) \\
&\leq \mathcal{R}_f(x + y).
\end{aligned}$$

This easily implies $\mathcal{R}_f(x) + \mathcal{R}_f(y) \leq \mathcal{R}_f(x + y)$. Therefore, $\mathcal{R}_f$ is superadditive. The function $\mathcal{R}_f$ is also upper order bounded because it is monotone (see Lemma 5.20). ■

The Riesz–Kantorovich transform is dominated by all superadditive monotone functions that dominate the underlying functions

Lemma 5.22. *Let $f = (f_1, f_2, \dots, f_m)$ be an m-tuple of upper order bounded functions from L_+ to N and let $g \colon L_+ \to N$. If $g \succeq f_i$ for all i and g*

is superadditive and monotone, then $g \succeq \mathcal{R}_f$. *In particular, if each* f_i *is superadditive, positively homogeneous and upper order bounded, then*

$$\mathcal{R}_f = \bigwedge \big\{ g \in M_+ \colon\ g \text{ is superadditive, positively homogeneous and } g \succeq \{f_1, \ldots, f_m\} \big\} .$$

Proof. Take $x \in L_+$ and $x_1, x_2, \ldots, x_m \in L_+$ such that $\sum_{i=1}^m x_i \le x$. Then we have

$$g(x) \ge g\Big(\sum_{i=1}^n x_i\Big) \ge \sum_{i=1}^m g(x_i) \ge \sum_{i=1}^m f_i(x_i) .$$

This implies $g(x) \ge \mathcal{R}_f(x)$. For the last statement notice that if we let

$$h = \bigwedge \big\{ g \in M_+ \colon\ g \text{ is superadditive, positively homogeneous and } g \succeq \{f_1, \ldots, f_m\} \big\} ,$$

then h exists (keep in mind that this is the pointwise infimum) and is superadditive. So, $h \succeq \mathcal{R}_f$. Since $\mathcal{R}_f \colon L_+ \to N_+$ is also superadditive, positively homogeneous, and dominates each f_i, we conclude that $h = \mathcal{R}_f$. ■

The Riesz–Kantorovich transform satisfies the associative law on the space of upper order bounded, superadditive and positively homogeneous functions.

Lemma 5.23. *If the functions* $f, g, h \colon L_+ \to N$ *are upper order bounded, superadditive and positively homogeneous, then we have*

$$\mathcal{R}(\mathcal{R}(f,g),h) = \mathcal{R}(f,\mathcal{R}(g,h)) = \mathcal{R}(f,g,h) .$$

Proof. It suffices to prove that

$$\mathcal{R}(\mathcal{R}(f,g),h) = \mathcal{R}(f,g,h) .$$

To this end, fix an arbitrary $x \in L_+$. Assume that $x_1, x_2, x_3 \in L_+$ satisfy $x_1 + x_2 + x_3 \le x$. Clearly,

$$\mathcal{R}(\mathcal{R}(f,g),h)(x) \ge \mathcal{R}(f,g)(x_1 + x_2) + h(x_3) \ge f(x_1) + g(x_2) + h(x_3) ,$$

and from this we easily infer that $\mathcal{R}(\mathcal{R}(f,g),h)(x) \ge \mathcal{R}(f,g,h)(x)$. Since $x \in L_+$ is arbitrary, we see that

$$\mathcal{R}(\mathcal{R}(f,g),h) \succeq \mathcal{R}(f,g,h) .$$

For the reverse inequality, note first that $\mathcal{R}(f,g,h) \succeq \{f,g,h\}$. By Lemma 5.22 we have $\mathcal{R}(f,g,h) \succeq \mathcal{R}(f,g)$ and one more application of Lemma 5.22 shows that $\mathcal{R}(f,g,h) \succeq \mathcal{R}(\mathcal{R}(f,g),h)$, and the proof is finished. ■

We conclude this section with an interesting extenson of the classical Riesz–Kantorovich Theorem 1.59.

Definition 5.24. *Let L be an ordered vector space and let $\mathcal{C}$ be a subcone of L_+, i.e., $\mathcal{C} \subseteq L_+$. We say that L_+ has the* **decomposition property on $\mathcal{C}$** *(or simply the $\mathcal{C}$-**decomposition property**) if*

$$[0,x]+[0,y]=[0,x+y]$$

for all $x, y \in \mathcal{C}$, where for each $z \in L_+$ we let $[0,z]=\{a \in L_+\colon\ 0 \le a \le z\}$.

The proof of the next result indicates the importance of $\mathcal{K}$-lattices.

Theorem 5.25. *Let L be an ordered vector space, $\mathcal{C}$ a generating subcone of L_+, and N a Dedekind complete vector lattice. If the cone L_+ has the $\mathcal{C}$-decomposition property, then $\mathcal{L}_\mathrm{b}(L,N)$ is a vector lattice.*

Moreover, for any operator $T \in \mathcal{L}_\mathrm{b}(L,N)$ and any $x \in \mathcal{C}$ we have

$$T^+(x)=\mathcal{R}_T(x)\,.$$

Proof. Consider the super cone $\mathcal{K}$ of $\mathcal{L}_\mathrm{b}^+(L,N)$ defined by

$$\mathcal{K}=\{T \in \mathcal{L}_\mathrm{b}(L,N)\colon\ T(x) \ge 0 \text{ for all } x \in \mathcal{C}\}\,.$$

As usual, we denote by $\succeq$ the vector ordering induced on $\mathcal{L}_\mathrm{b}(L,N)$ by this super cone $\mathcal{K}$. First, we claim that $\mathcal{L}_\mathrm{b}(L,N)$ is a $\mathcal{K}$-lattice.

To see this, let $X=(T_1,T_2,\ldots,T_m)$ be an m-tuple of order bounded operators from L to N. We must show that $\veebar X$ exists. Replacing $X=(T_1,T_2,\ldots,T_m)$ by $(0,T_2-T_1,\ldots,T_m-T_1)$ (and using part (2) of Lemma 5.3), we see that we can suppose without loss of generality that $T_1=0$.

Start by noticing (as in the proof of Theorem 1.59) that the Riesz–Kantorovich transform $\mathcal{R}_X\colon L_+ \to N_+$ is additive on $\mathcal{C}$. Since $\mathcal{C}$ is generating and N is Archimedean, it follows from part (3) of Lemma 1.26 that the restriction of $\mathcal{R}_X$ to $\mathcal{C}$ has a unique $\mathcal{C}$-positive linear extension to all of L, which we shall denote by R, i.e., $R\colon L \to N$ is the (unique) $\mathcal{C}$-positive operator satisfying $R(x)=\mathcal{R}_X(x)$ for each $x \in \mathcal{C}$. We claim that $R=\veebar X$.

To this end, we shall show first that $R(x) \ge \mathcal{R}_X(x)$ holds for all $x \in L_+$. To this end, fix $x \in L_+$. According to part (3) of Lemma 2.54 there exist $y \in \mathcal{C}$ and $0<\alpha<1$ such that $\alpha x+(1-\alpha)y \in \mathcal{C}$. But then we have

$$\begin{aligned}
\alpha\big[\mathcal{R}_X(x)-R(x)\big] &= \mathcal{R}_X(\alpha x)-R(\alpha x)\\
&= \mathcal{R}_X(\alpha x)+\big[\mathcal{R}_X((1-\alpha)y)-R((1-\alpha)y)\big]-R(\alpha x)\\
&= \big[\mathcal{R}_X(\alpha x)+\mathcal{R}_X((1-\alpha)y)\big]-R(\alpha x+(1-\alpha)y)\\
&\le \mathcal{R}_X(\alpha x+(1-\alpha)y)-R(\alpha x+(1-\alpha)y)\\
&= 0\,.
\end{aligned}$$

Since $\alpha>0$, we see that $\mathcal{R}_X(x)-R(x) \le 0$ or $R(x) \ge \mathcal{R}_X(x)$ for each $x \in L_+$. In particular, we have $R \ge T_i$ in $\mathcal{L}_\mathrm{b}(L,N)$ for each i.

Next, suppose that an operator $S \in \mathcal{L}_b(L, N)$ satisfies $S \geq T_i$ in $\mathcal{L}_b(L, N)$ for each i. This implies $S(x) \geq \mathcal{R}_X(x)$ for all $x \in L_+$. In particular, for all $x \in \mathcal{C}$ we have $R(x) = \mathcal{R}_X(x) \leq S(x)$, i.e., $S \succeq R$. This shows that $\veebar X = R$ and so $\mathcal{L}_b(L, N)$ is indeed a $\mathcal{K}$-lattice.

To complete the proof, we only need to show that for any three positive linear operators T_1, T_2, T_3 we have $(T_1 \veebar T_2) \veebar T_3 = T_1 \veebar (T_2 \veebar T_3)$; see Theorem 5.10. By Lemma 5.23, we know that $(T_1 \veebar T_2) \veebar T_3 = T_1 \veebar (T_2 \veebar T_3)$ holds on $\mathcal{C}$. Since $\mathcal{C}$ is generating, it follows that $(T_1 \veebar T_2) \veebar T_3 = T_1 \veebar (T_2 \veebar T_3)$ holds true on L. ∎

As an application of the preceding result we have the following version of Andô's Theorem 2.47.

Corollary 5.26. *For an ordered Banach space L with a closed, generating and norm normal cone, the following statements are equivalent.*

(1) *L has the Riesz decomposition property.*
(2) *L' has the Riesz decomposition property.*
(3) *There exists a generating subcone $\mathcal{C}$ of L_+ such that L_+ has the $\mathcal{C}$-decomposition property.*
(4) *There exists a generating subcone $\mathcal{C}$ of L'_+ such that L'_+ has the $\mathcal{C}$-decomposition property.*
(5) *L' is a Riesz space.*

We conclude the section with two open questions regarding an ordered vector space L and a Dedekind complete Riesz space N.

(1) Suppose that L_+ has the $\mathcal{C}$-decomposition property for some generating subcone $\mathcal{C}$ of L_+. Does L have the decomposition property?
(2) Assume that there exists a generating subcone $\mathcal{C}$ of L_+ such that for all order bounded operators $T \colon L \to N$ the Riesz–Kantorovich transform $\mathcal{R}_T$ of T is additive on $\mathcal{C}$. Does it necessarily follow that $\mathcal{R}_T$ is additive on L_+?

Exercises

1. Let $\mathcal{F}$ be a nonempty bounded from below subset of $M = N^{L_+}$, where N is assumed Dedekind complete. Also define the function $g \colon L_+ \to N$ for each $x \in L_+$ by

$$g(x) = \inf_M \{f(x) \colon \ f \in \mathcal{F}\}.$$

Establish the following.
(a) If each $f \in \mathcal{F}$ is concave, then g is concave.
(b) If each $f \in \mathcal{F}$ is positively homogeneous, then so is g.

(c) If each $f \in \mathcal{F}$ is monotone, then g is monotone.
(d) If each $f \in \mathcal{F}$ is superadditive, then g is likewise superadditive.
(e) If each $f \in \mathcal{F}$ is superlinear, then so is g.

2. If $f = (f_1, \ldots, f_m)$ is an m-tuple of upper order bounded functions (from L_+ to a Dedekind complete Riesz space N), then show that $\mathcal{R}_f$ dominates $\mathcal{R}_f^{=}$, i.e., $\mathcal{R}_f \geq \mathcal{R}_f^{=}$. Also, show that if one of the f_i is monotone, then equality holds, that is, $\mathcal{R}_f = \mathcal{R}_f^{=}$.

3. For any m-tuple $f = (f_1, \ldots, f_m)$ of upper order bounded functions (from L_+ to a Dedekind complete Riesz space N) show that

$$\mathcal{R}(f_1, \ldots, f_m) = \mathcal{R}^{=}(0, f_1, \ldots, f_m).$$

4. Let $L = \mathbb{R}^3$ and $N = \mathbb{R}$, and consider the three linear functionals $f, g, h\colon L \to \mathbb{R}$ defined by

$$f(x, y, z) = x, \quad g(x, y, z) = y, \quad \text{and} \quad h(x, y, z) = z.$$

Compute the Riesz–Kantorovich transform $\mathcal{R}(f, g, h)$ in the following three cases.

(a) The cone of L is the standard 45° ice cream cone, that is,

$$L_+ = \Big\{(x, y, z) \in \mathbb{R}^3\colon\ z \geq \sqrt{x^2 + y^2}\,\Big\}.$$

(b) The cone of L is the polyhedral subcone of $\mathbb{R}^3_+$ given by

$$L_+ = \{(x, y, z) \in L\colon\ x \geq 0,\ y \geq 0,\ z \geq 0 \text{ and } x + y - z \geq 0\}.$$

(c) The cone of L is the positive orthant, i.e., $L_+ = \mathbb{R}^3_+$.

5. Let L be an ordered vector space and let $\mathcal{C}$ be a subcone of L_+. Show that L_+ has the $\mathcal{C}$-decomposition property if and only if for any two nonempty finite subsets A and B of $\mathcal{C}$ satisfying $A \leq B$ there is an intermediate point between them.

6. Assume that there exists a generating subcone $\mathcal{C}$ of L_+ such that for any operator $T \in \mathcal{L}_b(L, N)$ the Riesz–Kantorovich functional $\mathcal{R}_T$ is additive on $\mathcal{C}$. Show that $\mathcal{L}_b(L, N)$ is a Riesz space.

7. Prove Corollary 5.26.

8. Show that if for every collection $X = \{T_1, T_2, \ldots, T_m\}$ of order bounded linear operators from L to a Dedekind complete Riesz space N the Riesz–Kantorovich transform $\mathcal{R}_T$ is additive on a fixed generating subcone $\mathcal{C}$ of L_+, then $\mathcal{L}_b(L, N)$ is a Riesz space.

5.3. The order extension of $\mathcal{L}_b(L, N)$

In this section, L is again an ordered vector space whose cone L_+ is generating and N is a Dedekind complete vector lattice. As before, we let $M = N^{L_+}$, the Dedekind complete vector lattice of all functions from L_+ to N. Once again, the pointwise ordering of functions in M is denoted $\succeq$, and the pointwise suprema and infima are denoted $\curlyvee$ and $\curlywedge$, respectively.

As usual, the positive cone in this vector ordering is denoted M_+. Always keep in mind that M is a Dedekind complete Riesz space; see Lemma 5.13.

We are now ready to define the notion of the extension cone in M.

Definition 5.27. *The collection of all functions $f\colon L_+ \to N_+$ that are superadditive and positively homogeneous functions is called the* ***extension cone*** *of M and is denoted $\mathcal{L}_+^\diamond(L, N)$. That is,*

$$\mathcal{L}_+^\diamond(L, N) = \big\{ f \in M_+ \colon\ f \ \text{ is superadditive and positively homogeneous} \big\}\,.$$

It should be clear that $\mathcal{L}_+^\diamond(L, N) \subseteq M_+$. It is also easy to show that the extension cone is indeed a cone of M.

Lemma 5.28. *The collection $\mathcal{L}_+^\diamond(L, N)$ is a cone of M.*

Proof. Obviously $\alpha\mathcal{L}_+^\diamond(L, N) \subseteq \mathcal{L}_+^\diamond(L, N)$ holds true for all $\alpha \geq 0$. Now if $f, h \in \mathcal{L}_+^\diamond(L, N)$, then $f + h \in M_+$ and $f + h$ is positive and positively homogeneous. Furthermore, for any $x, y \in L_+$ we have

$$\begin{aligned} [f+h](x+y) &= f(x+y) + h(x+y) \\ &\geq f(x) + h(x) + f(y) + h(y) \\ &= [f+h](x) + [f+h](y)\,, \end{aligned}$$

from which we conclude that $f + g$ is superadditive and that $\mathcal{L}_+^\diamond(L, N)$ is a wedge. Because $\mathcal{L}_+^\diamond(L, N)$ is a subset of the cone M_+, it must be the case that it is also a cone. ■

We continue with some notation. The vector space generated by the cone $\mathcal{L}_+^\diamond(L, N)$ in M will be denoted by $\mathcal{L}^\diamond(L, N)$, i.e.,

$$\mathcal{L}^\diamond(L, N) = \mathcal{L}_+^\diamond(L, N) - \mathcal{L}_+^\diamond(L, N)\,.$$

The vector space $\mathcal{L}^\diamond(L, N)$ will be considered ordered by the cone $\mathcal{L}_+^\diamond(L, N)$. The vector ordering induced by $\mathcal{L}_+^\diamond(L, N)$ will be designated by $\geq$. In other words, $f \geq g$ in $\mathcal{L}^\diamond(L, N)$ means $f - g \in \mathcal{L}_+^\diamond(L, N)$, i.e., $f - g$ is positive, superadditive and positively homogeneous.

The restriction of the cone M_+ to $\mathcal{L}^\diamond(L, N)$ will be denoted by $\mathcal{K}$, i.e.,

$$\mathcal{K} = \mathcal{L}^\diamond(L, N) \cap M_+\,.$$

Clearly, $\mathcal{L}_+^\diamond(L, N) \subseteq \mathcal{K}$, i.e., $\mathcal{K}$ is a super cone of $\mathcal{L}_+^\diamond(L, N)$. As usual, we shall denote the order induced by $\mathcal{K}$ on $\mathcal{L}^\diamond(L, N)$ by $\succeq$. That is, $f \succeq g$ holds in $\mathcal{L}^\diamond(L, N)$ if and only if $f - g \in \mathcal{K}$, which means $f(x) \geq g(x)$ for all $x \in L_+$.

It turns out that the ordered vector space $\mathcal{L}^\diamond(L, N)$ is a $\mathcal{K}$-lattice.

Lemma 5.29. *The vector space $\mathcal{L}^\diamond(L,N)$ is a Dedekind complete $\mathcal{K}$-lattice.*

Moreover, for any nonempty subset A of $\mathcal{L}^\diamond(L,N)$ that is bounded from above we have

$$\veebar A = \bigwedge\{f \in \mathcal{L}^\diamond(L,N) \colon\ f \geq A\}.$$

Proof. Let X be a nonempty subset of $\mathcal{L}^\diamond_+(L,N)$ and let f be the pointwise infimum of X. Since N is Dedekind complete, the function f exists and obviously $f \in \mathcal{K}$. Clearly, f is super additive and positively homogeneous, and thus $f \in \mathcal{L}^\diamond_+(L,N)$. Now a glance at Lemma 5.12 guarantees that $\mathcal{L}^\diamond(L,N)$ is a $\mathcal{K}$-lattice. The last part also follows from Lemma 5.12. ∎

Our task now is to embed the space of order bounded operators $\mathcal{L}_\mathrm{b}(L,N)$ into the ordered vector space $\mathcal{L}^\diamond(L,N)$.

Lemma 5.30. *The mapping $T \mapsto T|_{L_+}$ is a one-to-one operator from $\mathcal{L}_\mathrm{b}(L,N)$ to $\mathcal{L}^\diamond(L,N)$ [and so identifying $\mathcal{L}_\mathrm{b}(L,N)$ with its image in $\mathcal{L}^\diamond(L,N)$ we can consider $\mathcal{L}_\mathrm{b}(L,N)$ as a vector subspace of $\mathcal{L}^\diamond(L,N)$].*

Proof. Since L_+ is generating, it is straightforward to verify that the mapping $T \mapsto T|_{L_+}$ is indeed a one-to-one operator from $\mathcal{L}_\mathrm{b}(L,N)$ to $\mathcal{L}^\diamond(L,N)$.

Now fix $T \in \mathcal{L}_\mathrm{b}(L,N)$ and let $S = T|_{L_+}$. Since S is order bounded from above, it follows that $\mathcal{R}_S \in \mathcal{L}^\diamond_+(L,N)$ and clearly $\mathcal{R}_S - S \in \mathcal{L}^\diamond_+(L,N)$. Finally, note that $S = \mathcal{R}_S - (\mathcal{R}_S - S)$. ∎

From now on we shall consider the space $\mathcal{L}_\mathrm{b}(L,N)$ as a subspace of $\mathcal{L}^\diamond(L,N)$. We are now ready to introduce the order extension of $\mathcal{L}_\mathrm{b}(L,N)$.

Definition 5.31. *The **order extension** of $\mathcal{L}_\mathrm{b}(L,N)$, denoted $\mathcal{O}(L,N)$, is the $\mathcal{K}$-sublattice generated by $\mathcal{L}_\mathrm{b}(L,N)$ in the $\mathcal{K}$-lattice $\mathcal{L}^\diamond(L,N)$.*

Recall that the positive cone of $\mathcal{O}(L,N)$ is the extension cone $\mathcal{L}^\diamond_+(L,N)$ and the ordering induced by this cone is denoted $\geq$. Furthermore, the pointwise ordering is denoted $\succeq$ and its positive cone is $\mathcal{K} = \mathcal{O}(L,N) \cap M_+$.

The $\mathcal{K}$-lattice operations in $\mathcal{O}(L,N)$ enjoy the following property.

Lemma 5.32. *If $T = (T_1, T_2, \ldots, T_m)$ is an m-tuple of operators in $\mathcal{L}_\mathrm{b}(L,N)$, then*

$$\veebar\{0, T_1, T_2, \ldots, T_m\} = \mathcal{R}_T$$

in the $\mathcal{K}$-lattice $\mathcal{O}(L,N)$. If some T_i is positive, then

$$\veebar\{T_1, T_2, \ldots, T_m\} = \mathcal{R}_T^{=}.$$

Proof. We only need to prove the first claim and that $\veebar\{0,T\} = \mathcal{R}_T$. We know from Lemma 5.29 that

$$\veebar T = \bigwedge\{f \in \mathcal{L}^\diamond_+(L,N) \colon\ f \geq \{T_1, T_2, \ldots, T_m\}\}.$$

We also know from Lemma 5.22 that

$$\mathcal{R}_T = \curlywedge\{f \in \mathcal{L}_+^\diamond(L,N):\ f \succeq \{T_1, T_2, \ldots, T_m\}\}\,.$$

So, we only need to show that $\mathcal{R}_T \geq \{T_1, T_2, \ldots, T_m\}$. That is, we must establish that for any T_i we have $\mathcal{R}_T - T_i \in \mathcal{L}_+^\diamond(L,N)$. However, this is obviously the case because each T_i is additive and positively homogeneous on L_+. ■

We are now ready to state and prove the main result of this section.

Theorem 5.33. *If $\mathcal{L}_b(L,N) = \mathcal{L}_r(L,N)$ holds true, then he following statements are equivalent.*

(1) *The order extension $\mathcal{O}(L,N)$ of $\mathcal{L}_b(L,N)$ is a vector lattice.*

(2) *For each operator $T \in \mathcal{L}_b(L,N)$ its Riesz–Kantorovich transform $\mathcal{R}_T$ is additive.*

(3) *The order extension $\mathcal{O}(L,N)$ of $\mathcal{L}_b(L,N)$ coincides with $\mathcal{L}_b(L,N)$, i.e., $\mathcal{O}(L,N) = \mathcal{L}_b(L,N)$.*

Proof. (1) $\Longrightarrow$ (2) Assume that $\mathcal{O}(L,N)$ is a vector lattice and suppose by way of contradiction that for some $T \in \mathcal{L}_b(L,N)$ the Riesz–Kantorovich transform $\mathcal{R}_T$ is superadditive but not additive. Let $S \in \mathcal{L}_b(L,N)$ be an operator that dominates T and zero in $\mathcal{L}_b(L,N)$. (The existence of such an operator follows from our assumption $\mathcal{L}_b(L,N) = \mathcal{L}_r(L,N)$.) Notice that $S - \mathcal{R}_T$ is not superadditive. So, $S - \mathcal{R}_T \notin \mathcal{L}_+^\diamond(L,N)$. This implies that $\veebar\{S, \mathcal{R}_T\} \neq S$. However, $\veebar$ is associative because $\mathcal{O}(L,N)$ is a vector lattice. Therefore, using Lemma 5.32, we have

$$\mathcal{R}_T \veebar S = (0 \veebar T) \veebar S = \veebar\{0, T, S\} = \mathcal{R}_{(0,T,S)} = S,$$

which is a contradiction. This shows that (1) implies (2).

(2) $\Longrightarrow$ (3) If (2) holds, then $\mathcal{L}_b(L,N)$ is a vector lattice. Furthermore, for any set of order bounded operators $T = \{0, T_1, \ldots, T_m\}$ we have

$$\vee T = \mathcal{R}_T = \veebar T\,,$$

which tells us that $\mathcal{L}_b(L,N) = \mathcal{O}(L,N)$.

(3) $\Longrightarrow$ (1) If (3) holds, then it must be the case that for each $T \in \mathcal{L}_b(L,N)$ the Riesz–Kantorovich transform $\mathcal{R}_T$ is additive. Therefore, $\mathcal{L}_b(L,N)$ is a vector lattice. Now because $M_+ \cap \mathcal{L}_b(L,N) = \mathcal{L}_+^\diamond(L,N) \cap \mathcal{L}_b(L,N)$, we see that $\mathcal{L}_b(L,N)$ is a vector lattice under the order induced by $\mathcal{L}_+^\diamond(L,N)$. This implies that $\mathcal{O}(L,N)$ is a vector lattice under the order induced by the extension cone $\mathcal{L}_+^\diamond(L,N)$. ■

As one of the consequences of the preceding result we have the following version of Andô's Theorem 2.47.

Corollary 5.34. *For an ordered Banach space L with a closed, generating and norm normal cone, the following statements are equivalent.*

(1) *L has the decomposition property.*
(2) *$\mathcal{O}(L,\mathbb{R})$ is a vector lattice.*
(3) *The norm dual L' is a vector lattice.*
(4) *$L' = \mathcal{O}(L,\mathbb{R})$.*

One important conclusion of Corollary 5.34 is the following: If an ordered Banach space with a closed, generating and norm normal cone does not have the decomposition property, then $\mathcal{O}(L,\mathbb{R})$ is a $\mathcal{K}$-lattice but never a Riesz space. This provides many examples of $\mathcal{K}$-lattices that are not Riesz spaces. From this perspective, we can say that $\mathcal{K}$-lattices are "seldom" Riesz spaces—which means, of course, that in an arbitrary $\mathcal{K}$-lattice its join operation $\veebar$ "most likely" does not satisfy the associative law.

We conclude the section with a major open problem.

- *Does there exist a pair (L,N), where L is an ordered vector space with a generating cone and N is a Dedekind complete vector lattice, such that $\mathcal{L}_b(L,N)$ is a vector lattice but $\mathcal{L}_b(L,N) \neq \mathcal{O}(L,N)$?*

We conjecture that no such pair exists.

Exercises

1. Let $f = (f_1, f_2, \ldots, f_m)$ be functions in $\mathcal{L}_+^\diamond(L,N)$ that are not additive. Show that it is not necessarily the case that $\veebar f = \mathcal{R}_f$. [*Hint*: If this is the case, then $\veebar$ must be associative.]

2. Show that if $\mathcal{O}(L,N)$ is a vector lattice, then for every $T \in \mathcal{L}_r(L,N)$ the pointwise infimum

$$Q = \curlywedge\{S \in \mathcal{L}_r(L,N) \colon\; S \succeq \{0,T\}\}$$

is additive. In particular $\mathcal{L}_r(L,N)$ is a vector lattice. [*Hint*: Using Q, follow the proof of Theorem 5.33.]

3. Assume that for every $T \in \mathcal{L}_b(L,N)$ and any $x \in L_+$ there exists a linear operator $S \succeq \{0,T\}$ such that $S(x) = \mathcal{R}_T(x)$. Show that if $\mathcal{L}_b(L,N)$ is a vector lattice, then $\mathcal{O}(L,N)$ is also a vector lattice. [*Hint:* In this case the Riesz–Kantorovich transform of T satisfies

$$\mathcal{R}_T = \curlywedge\{S \in \mathcal{L}_b(L,N) \colon S \succeq \{0,T\}\} = T^+.]$$

4. Give an example of a function $f \in \mathcal{L}_+^\diamond(L,N)$ such that $\veebar\{0,f\} \neq \mathcal{R}_f$. [*Hint*: Consider the function $S - [0 \veebar T]$ from the proof of Theorem 5.33.]

5. Suppose that for every $f \in \mathcal{L}_+^\diamond(L,N)$ we have $f \veebar 0 = \mathcal{R}_f$. Show that $\mathcal{O}(L,N)$ is a vector lattice.

6. Give an example of an infinite dimensional subcone $\mathcal{C}$ of $C[0,1]^+$ such that

(a) the vector space $V = \mathcal{C} - \mathcal{C}$ generated by $\mathcal{C}$ is not a vector lattice,

(b) the cone $\mathcal{C}$ is a $\mathcal{K}$-lattice cone of V, where $\mathcal{K} = V \cap C[0,1]^+$.

7. Give an example of a lattice-subspace H of the ordered vector space $\mathcal{O}(L, N)$ for which the $\mathcal{K}$-sublattice generated by H is not equal to H.

8. Show that if an ordered vector space with a generating cone L has the decomposition property, then $\mathcal{O}(L, N) = \mathcal{L}_b(L, N)$.

Also, give an example of a space L that does not have the decomposition property but for which $\mathcal{O}(L, N) = \mathcal{L}_b(L, N)$. [*Hint*: You can construct a finite dimensional example for which $\mathcal{R}_T$ is always additive for each $T \in \mathcal{L}_b(L, N)$ and L does not have decomposition property.]

Chapter 6

The order extension of L'

This chapter specializes the theory of $\mathcal{K}$-lattices and the order extension of $\mathcal{L}_b(L, N)$ introduced in the previous chapter to the case where L is an ordered topological vector space and $N = \mathbb{R}$.

We have in mind the usual linear programming problem that can be stated in the following framework. Let L be an ordered Hausdorff locally convex space and let N be an ordered topological vector space. Suppose that $T\colon L \to N$ is a linear operator such that its pull-back wedge

$$\mathcal{C} = \{x \in L\colon\ Tx \geq 0\}$$

as well as its dual wedge

$$\mathcal{C}' = \{f \in L'\colon\ f(x) \geq 0 \text{ for all } x \in \mathcal{C}\}$$

are both closed and generating cones.

Let $X = \{x_1, x_2, \ldots, x_m\}$ be a finite collection of vectors in L. For $f \in L'$ consider the minimization problem

$$\inf\{f(y)\colon\ y \in L \text{ and } Ty \geq T(X)\}\,.$$

It is easy to see that this minimization problem has the order theoretic version

$$\inf\{f(y)\colon\ y \in L \text{ and } y \geq_{\mathcal{C}} X\}\,.$$

For each f that is positive on $\mathcal{C}$, i.e., $f \in \mathcal{C}'$, this problem has a real value, say $v_X(f)$. Therefore, we can define a value function $v_X\colon \mathcal{C}' \to \mathbb{R}$.

Since we assume that the cone $\mathcal{C}'$ is generating, each vector $x \in \mathcal{C}$ represents a unique function $x\colon \mathcal{C}' \to \mathbb{R}$ by letting $x(f) = f(x)$. But v_X need not be additive on $\mathcal{C}'$ and therefore may not be represented by a vector in L_+.

However, from

$$v_X(f) = \inf\{x(f) \in L\colon\ x \geq_{\mathcal{C}} X\}$$

for each $f \in \mathcal{C}'$, we see that $v_X\colon \mathcal{C}' \to \mathbb{R}$ is necessarily a superadditive, positively homogeneous, and upper semicontinuous function on $\mathcal{C}'$.

The main purpose of this chapter is to provide a $\mathcal{K}$-lattice order extension of $(L, \mathcal{C})$ in which

$$\underline{\vee} X = v_X$$

for any choice of X.

For convenience we shall work in dual spaces. So suppose that L is an ordered topological vector space such that L'_+ and L_+ are generating cones. Let $M = \mathbb{R}^{L_+}$ be the vector space of all functions from L_+ to $\mathbb{R}$ ordered using the pointwise ordering $\succeq$. That is, M_+ is the cone of all functions $f\colon L_+ \to \mathbb{R}_+$. The objective here is to provide a $\mathcal{K}$-lattice order extension of the topological dual L' of an ordered topological vector space L in which

$$\underline{\vee} f = M\text{-}\inf\{g \in L'\colon\ g \succeq f\}$$

for any finite collection $f = \{f_1, f_2, \ldots, f_m\}$ of continuous linear functionals. The order extension we introduce here is an ordered vector space that is a $\mathcal{K}$-lattice which includes the solutions of such minimization problems. Therefore, the chapter brings to the theory of linear minimization the notions of $\mathcal{K}$-lattices and more generally the theory of ordered vector spaces. This provides very rich functional analytic structures for studying linear minimization.

We also want to explore when in this order extension the identity between the Riesz–Kantorovich transform and the join operation $\underline{\vee}$ holds. That is, when do we have $\mathcal{R}_f = \underline{\vee} f$? This is an important duality result that should be familiar to anyone who has studied linear optimization theory. That is, for any $f \in L'$ and $x \in L_+$ we have

$$\sup\{f(y)\colon\ y \in [0, x]\} = \inf\{g(x)\colon\ g \in L'_+ \text{ and } g \succeq f\}.$$

We show that this is indeed the case when $\mathcal{R}_f$ is upper semicontinuous. Furthermore, we introduce the generalized Riesz–Kantorovich functional that is computed in the double dual of L, which is needed because we are working with very general spaces.

The main pressing question arising from this chapter is the following: *Can we extend the results here to the case of linear operators from L to a Dedekind complete vector lattice N?*

6.1. The extension of L'

Let L be an ordered topological vector space. Throughout this section and, in fact, throughout this chapter we make the following assumptions.

(a) *The cone L_+ is generating.*

(b) *L is equipped with a Hausdorff locally convex topology τ for which L_+ is τ-closed and every order interval is τ-bounded.*

(c) *The dual cone L'_+ is generating, i.e., $L' = L'_+ - L'_+$.*[1]

As usual, L' denotes the topological dual of (L,τ) and L'_+ denotes the collection of all positive τ-continuous linear functionals on L, i.e.,

$$L'_+ = \{ f \in L' \colon\ f(x) \geq 0 \text{ for each } x \in L_+ \} .$$

Since we assume (from the outset) that the cone L_+ is generating, it follows that L'_+ is indeed a $\sigma(L',L)$-closed cone of L'. Moreover, since the order intervals of L are τ-bounded, it should be clear that any $f \in L'$ is order bounded. That is, we have $L' \subseteq L^\sim$.

Now consider the Dedekind complete Riesz space $M = \mathbb{R}^{L_+}$ ordered, as usual, by the pointwise ordering $\succeq$ and, of course, its positive cone is M_+. The infima and suprema of this vector ordering are denoted by the pointwise lattice operations $\curlywedge$ and $\curlyvee$, respectively.

We now isolate and name the collection of all upper τ-semicontinuous functions.[2]

Definition 6.1. *The* ***super topological dual cone*** *$L^{\textcircled{\prime}}_+$ of L is the collection of all superadditive, positively homogeneous, and upper τ-semicontinuous functions from L_+ to $\mathbb{R}_+$. That is,*

$$L^{\textcircled{\prime}}_+ = \{ f \in \mathcal{L}^{\diamond}_+(L,\mathbb{R}) \colon\ f \text{ is upper } \tau\text{-semicontinuous} \} .$$

Notice that $L^{\textcircled{\prime}}_+$ is a subcone of M_+. We denote by $L^{\textcircled{\prime}}$ the vector subspace generated by $L^{\textcircled{\prime}}_+$, i.e.,

$$L^{\textcircled{\prime}} = L^{\textcircled{\prime}}_+ - L^{\textcircled{\prime}}_+ .$$

As usual, the vector ordering induced on $L^{\textcircled{\prime}}$ by the cone $L^{\textcircled{\prime}}_+$ will be denoted by $\geq$ and the one by the cone $L^{\textcircled{\prime}} \cap M_+$ will be denoted by $\succeq$.

---

[1] Clearly, an ordered Banach space satisfies (a), (b), and (c) if and only if its cone is closed, generating, and norm normal; see Theorems 2.40.

[2] Recall that a function $f\colon X \to \mathbb{R}$ on a topological space is said to be **upper semicontinuous** if for each $c \in \mathbb{R}$ the set $\{x \in X \colon\ f(x) \geq c\}$ is closed. It is well known that a function $f\colon X \to \mathbb{R}$ is upper τ-semicontinuous if and only if $x_\alpha \to x$ in X implies $\limsup_\alpha f(x_\alpha) \leq f(x)$; see [**12**, Lemma 2.39, p. 43].

Next, notice that for any family of functions $\{f_i\}_{i\in I}$, where $f_i\colon L_+ \to \mathbb{R}$, and any $c \in \mathbb{R}$ we have

$$\big\{x \in L_+\colon\ \inf_{i\in I} f_i(x) \geq c\big\} = \bigcap_{i\in I} \big\{x \in L_+\colon\ f_i(x) \geq c\big\}\,.$$

This easily implies that whenever the pointwise infimum of a family of upper τ-semicontinuous functions defines a real-valued function, then it must be an upper τ-semicontinuous function (that is also superadditive and positively homogeneous if each f_i is likewise so).

Therefore, we observe the following from Lemma 5.12.

Lemma 6.2. *The vector space $L^{\oslash}$ is a Dedekind complete $\mathcal{K}$-lattice, where*

$$\mathcal{K} = L^{\oslash} \cap M_+\,.$$

Furthermore, for any finite set of functions $f = \{f_1, f_2, \ldots, f_m\}$ in $L^{\oslash}$ we have

$$\veebar f = \bigwedge\{g \in L^{\oslash}\colon\ g \geq f\}\,.$$

If $f \in L'_+$, then $f|_{L_+}$ (the restriction of f to the positive cone L_+) is additive, positively homogeneous, and continuous and so $f|_{L_+} \in L^{\oslash}_+$. Now if $f \in L'$, then there exist $g, h \in L'_+$ such that $f = g - h$. This implies that $f|_{L_+} = g|_{L_+} - h|_{L_+} \in L^{\oslash}_+ - L^{\oslash}_+ = L^{\oslash}$. Moreover, it is easy to see that the mapping $f \mapsto f|_{L_+}$, from L' to $L^{\oslash}$, is a one-to-one linear operator. This means that by identifying each $f \in L'$ with its restriction on L_+, we can consider L' as a vector subspace of $L^{\oslash}$. In the sequel, under this identification, we shall consider L' as a vector subspace of $L^{\oslash}$.

Our next objective is to characterize the join operation $\veebar f$ where $f = \{f_1, f_2, \ldots, f_m\}$ is a finite collection of linear functionals in L'. To do this, we need a useful characterization of the domination property amongst positively homogeneous functions.

Let us say that a nonempty subset $\mathcal{B}$ of a cone $\mathcal{C}$ in a vector space is a **pseudo-base** if for each $x \in \mathcal{C}$ there exist $\lambda \geq 0$ and $b \in \mathcal{B}$ (not necessarily uniquely determined) such that $x = \lambda b$. Clearly, a base is a pseudo-base but a pseudo-base need not be a base—in fact, it need not even be a convex set. For instance, the nonconvex set $\mathcal{B} = \big\{(x,y) \in \mathbb{R}^2_+\colon\ x^2 + y^2 = 1\big\}$ is a pseudo-base for the standard cone $\mathbb{R}^2_+$.

Lemma 6.3. *Let $f, g\colon \mathcal{C} \to \mathbb{R}$ be two positively homogeneous functions defined on a cone $\mathcal{C}$ of a vector space and let $\mathcal{B}$ be any pseudo-base of $\mathcal{C}$. Then f $\mathcal{C}$-dominates g, i.e., $f(x) \geq g(x)$ for all $x \in \mathcal{C}$, if and only if $f(b) \geq g(b)$ for all $b \in \mathcal{B}$.*

As an application of Lemma 6.3, let $\mathbb{R}^2_+$ be the standard cone of $\mathbb{R}^2$ and note that the set $\mathcal{B} = \big\{(x,y) \in \mathbb{R}^2_+\colon\ x^2 + y^2 = 1\big\}$ is a pseudo-base of $\mathbb{R}^2_+$.

Now consider the positively homogeneous functions $f, g\colon \mathbb{R}^2_+ \to \mathbb{R}$ defined by $f(0,0) = g(0,0) = 0$ and for each $(x,y) \in \mathbb{R}^2_+ \setminus \{(0,0)\}$ by

$$f(x,y) = \tfrac{3(x^2+y^2)}{x+y} \quad \text{and} \quad g(x,y) = \tfrac{x^3+y^3}{x^2+y^2}\,.$$

Note that f and g are indeed positively homogeneous functions and that for each $(x,y) \in \mathcal{B}$ we have $(x^3+y^3)(x+y) \le (x+y)^2 = x^2+y^2+2xy \le 3$. This implies that if $(x,y) \in \mathcal{B}$, then $f(x,y) = \frac{3}{x+y} \ge x^3+y^3 = g(x,y)$. Consequently, $f \ge g$ is true.

The following domination result regarding positively homogeneous functions is very useful.

Lemma 6.4. *Assume that $f, g\colon \mathcal{C} \to \mathbb{R}_+$ are positively homogeneous functions defined on a cone of a vector space, and assume that $g(x_0) > 0$ holds for some $x_0 \in \mathcal{C}$. Then the following statements are equivalent.*

(a) *The function f $\mathcal{C}$-dominates g, i.e., $f(x) \ge g(x)$ for each $x \in \mathcal{C}$.*

(b) *If $x \in \mathcal{C}$ satisfies $g(x) = g(x_0)$, then $f(x) \ge g(x_0)$.*

Proof. Let $\mathcal{B} = \big\{x \in \mathcal{C}\colon\ g(x) = g(x_0)\big\} \bigcup \big\{x \in \mathcal{C}\colon\ g(x) = 0\big\}$. Now if $g(x) > 0$, then letting $b = \frac{g(x_0)}{g(x)} x \in \mathcal{B}$, we get $x = \frac{g(x)}{g(x_0)} b$. This implies that $\mathcal{B}$ is a pseudo-base of $\mathcal{C}$. Now notice that (b) simply states that $f(b) \ge g(b)$ for all $b \in \mathcal{B}$, which is necessary and sufficient for f to $\mathcal{C}$-dominate g. ■

We now establish the following important result.

Lemma 6.5. *Assume that a functional $f\colon L_+ \to \mathbb{R}$ is super additive and positively homogenous and dominates some $\phi \in L'$, i.e., $f \succeq \phi$. Then f is upper τ-semicontinuous if and only if for each $x \in L_+$ we have*

$$f(x) = \inf\big\{h(x)\colon\ h \in L' \text{ and } h \succeq f\big\}\,.$$

Proof. Note first that replacing f by $f - \phi$, we can assume without loss of generality that $\phi = 0$, i.e., $f \in \mathcal{L}^{\diamond}_+(L, \mathbb{R})$. If for each $x \in L_+$ we have

$$f(x) = \inf\big\{h(x)\colon\ h \in L' \text{ and } h \ge f\big\}\,,$$

then f is upper τ-semicontinuous. For the converse suppose that f is upper τ-semicontinuous. Let $x_0 \in L_+$ and fix $\epsilon > 0$. We must establish that there exists some $h \in L'$ satisfying $h \succeq f$ and $f(x_0) \le h(x_0) < f(x_0) + \epsilon$.

To this end, assume first that $f(x_0) > 0$ is also true. Put $\alpha = \frac{f(x_0)}{f(x_0)+\epsilon}$ and notice that $0 < \alpha < 1$. By the positive homogeneity of f we get $f(\alpha x_0) = \alpha f(x_0) < f(x_0)$. That is, $\alpha x_0 \notin U_f^{x_0}$, where $U_f^{x_0}$ is the upper contour set of f at x_0, i.e., $U_f^{x_0} = \{x \in L_+\colon f(x) \ge f(x_0)\}$. Since the set $U_f^{x_0}$ is τ-closed and convex, there exists by the Separation Theorem (see Corollary 8.18) a τ-continuous linear functional g and a real number η satisfying $g(\alpha x_0) < \eta$

and $g(x) > \eta$ for each $x \in U_f^{x_0}$. The comprehensiveness of the set $U_f^{x_0}$ (see Lemma 1.32) guarantees that g is also a positive linear functional. This implies that $\eta > g(\alpha x_0) \geq 0$. Letting $h = \frac{f(x_0)}{\eta} g \in L'$, we observe that $h(x) = \frac{f(x_0)}{\eta} g(x) > f(x_0)$ for each $x \in U_f^{x_0}$. By Lemma 6.4 we get $h \geq f$. Finally, notice that from $h(\alpha x_0) = \frac{f(x_0)}{\eta} g(\alpha x_0) < f(x_0)$ it follows that $h(x_0) < \frac{f(x_0)}{\alpha} = f(x_0) + \epsilon$. This completes the proof when $f(x_0) > 0$.

Now we consider the case $f(x_0) = 0$. If $f = 0$, then f is linear and the conclusion holds true trivially. Thus, it remains to consider the case where $f(y) > 0$ holds for some $y \in L_+$. For each $0 < \alpha < 1$ let $z_\alpha = \alpha x_0 + (1-\alpha)y$. Since f is superadditive and positively homogeneous, it must be the case that $f(z_\alpha) > 0$ for each $0 < \alpha < 1$. From $\lim_{\alpha\uparrow 1} z_\alpha = x_0$ and the upper τ-semicontinuity of f, it follows that $0 \leq \limsup_{\alpha\uparrow 1} f(z_\alpha) \leq f(x_0) = 0$. This implies $\lim_{\alpha\uparrow 1} f(z_\alpha) = 0$. In particular, there exists some $\frac{1}{2} \leq \alpha_0 < 1$ such that $0 < f(z_\alpha) < \frac{\epsilon}{2}$ for all $\alpha_0 \leq \alpha < 1$. According to the preceding case, for each $\alpha_0 \leq \alpha < 1$ there exists some $h_\alpha \in L'$ satisfying $h_\alpha \geq f \geq 0$ and $0 < f(z_\alpha) \leq h_\alpha(z_\alpha) < \frac{\epsilon}{2}$. Now observe that for each $\alpha_0 \leq \alpha < 1$ we have

$$\tfrac{1}{2} h_\alpha(x_0) \leq \alpha h_\alpha(x_0) \leq \alpha h_\alpha(x_0) + (1-\alpha) h_\alpha(y) = h_\alpha(z_\alpha) < \tfrac{\epsilon}{2},$$

and consequently $f(x_0) = 0 \leq h_\alpha(x_0) < \epsilon$ holds for all $\alpha_0 \leq \alpha < 1$, and the proof is finished. ■

We are now ready to define the super topological dual of L.

Definition 6.6. *The $\mathcal{K}$-sublattice generated by L' in $L^{\textcircled{\tiny T}}$ is called the **super topological dual** of L and is denoted $\mathcal{O}'(L)$.*

Immediate consequences of Lemma 6.5 follow.

Lemma 6.7. *For any finite collection of functions $f = \{f_1, f_2, \ldots, f_m\}$ in L' we have*

$$\veebar f = \bigwedge \{g \in L' \colon\ g \succeq f\}.$$

Corollary 6.8. *If L' is a vector lattice, then $L' = \mathcal{O}'(L)$ and the join operations $\veebar$ and $\vee$ are identical.*

We now establish a partial converse of this corollary.

Lemma 6.9. *If $\mathcal{O}'(L)$ is a vector lattice, then for any finite collection of linear functionals $f = \{f_1, f_2, \ldots, f_m\}$ in L' its $\mathcal{K}$-supremum*

$$\veebar f = \bigwedge \{g \in L' \colon g \succeq f\}$$

is additive.

Proof. Suppose by way of contradition that $\veebar f$ is not additive. Take a function $g \in L'$ satisfying $g \succeq \veebar f$. Notice that $g - \veebar f$ is not in $L^{\mathcal{O}}_{+}$. Therefore, $h = \veebar\{g, \veebar f\} \neq g$. However, since $\mathcal{O}'(L)$ is a vector lattice, we see that $h = \veebar\{g, f_1, f_2, \ldots, f_m\}$. But then this implies, of course, that $h = g$, which is a contradiction. ∎

We conclude this section with the main theorem.

Theorem 6.10. *Assume that $L^{\sim} = L'$. Then $L' = \mathcal{O}'(L)$ holds true if and only if L' is a vector lattice.*

Furthermore, whenever $L^{\sim} = L' = \mathcal{O}'(L)$ is true, then the join operations $\veebar$ and $\vee$ are identical.

Exercises

1. Show that a function $f\colon X \to \mathbb{R}$ on a topological space is upper semicontinuous if and only if $x_\alpha \to x$ in X implies $\limsup_\alpha f(x_\alpha) \leq f(x)$.
2. Show that the sum of an upper semicontinuous function and a continuous function is upper semicontinuous. Is the sum of two upper semicontinuous functions an upper semicontinuous function?
3. Let $\mathcal{K}$ be a cone in a vector space X and for any function $f\colon \mathcal{K} \to [0, \infty)$ put
$$\mathcal{K}_f = \{(x, \lambda) \in X \times \mathbb{R}\colon\ 0 \leq \lambda \leq f(x)\}.$$
For a homogeneous and superadditive function f show that
   (a) the set $\mathcal{K}_f$ is a cone of $X \times \mathbb{R}$,
   (b) the cone $\mathcal{K}_f$ is closed for some Hausdorff linear topology τ on X if and only if $f\colon \mathcal{K} \to [0, \infty)$ is upper τ-semicontinuous.

   [*Hint*: Assume that f is upper τ-semicontinuous and let $(x_\alpha, \lambda_\alpha) \to (x, \lambda)$ in $X \times \mathbb{R}$. From $0 \leq \lambda_\alpha \leq f(x_\alpha)$ and the upper τ-semicontinuity of f we get $\lambda = \lim \lambda_\alpha \leq \limsup f(x_\alpha) \leq f(x)$. This shows that $(x, \lambda) \in \mathcal{K}_f$ and so $\mathcal{K}_f$ is closed in $\mathcal{K}_f \times \mathbb{R}$. Conversely, if $\mathcal{K}_f$ is closed in $X \times \mathbb{R}$ and a net $\{x_\alpha\} \subseteq \mathcal{K}$ satisfies $x_\alpha \xrightarrow{\tau} x$ in X and $f(x_\alpha) \geq c$ for all α and some fixed $c > 0$, then from $(x_\alpha, c) \in \mathcal{K}_f$, $(x_\alpha, c) \to (x, c)$ and the closedness of $\mathcal{K}_f$ we get $(x, c) \in \mathcal{K}_f$. This shows that $f(x) \geq c$ and from this we get that f is upper τ-semicontinuous.]
4. Show that if $\mathcal{O}'(L)$ is a vector lattice, then it is not necessarily the case that L' is a vector lattice. [*Hint*: Begin with a Banach lattice N and choose a dense subspace L' of N' that is not a vector lattice. Consider the space $L = N$ with dual L'. Is $\mathcal{O}'(L)$ the Riesz space generated by L' in N?]
5. Show an example where $\veebar f$ in $\mathcal{O}'(L)$ is not the same as $\mathcal{R}_f$. [*Hint*: Take an example in which $L^{\sim}$ is a vector lattice but L' is a lattice-subspace that is not a Riesz subspace of $L^{\sim}$. Then apply Lemma 6.9.]
6. Prove Lemma 6.2.

7. Prove Lemma 6.3.

8. Prove Corollary 6.8.

9. Prove Theorem 6.10.

6.2. Generalized Riesz–Kantorovich functionals

As in the previous section, recall that L will denote an ordered vector space that satisfies the following three properties.

(a) *The cone L_+ is generating.*

(b) *L is equipped with a Hausdorff locally convex topology τ for which L_+ is τ-closed and every order interval is τ-bounded.*

(c) *The dual cone L'_+ is generating in L'.*

Also recall that for any m-tuple $f = (f_1, f_2, \ldots, f_m)$ of order upper bounded functions, from L_+ to $\mathbb{R}$, its Riesz–Kantorovich transform is the function $\mathcal{R}_f \colon L_+ \to \mathbb{R}$ defined for $x \in L_+$ by

$$\mathcal{R}_f(x) = \sup\Big\{\sum_{i=1}^{m} f_i(x_i)\colon\ x_i \in [0,x] \text{ for each } i \text{ and } \sum_{i=1}^{m} x_i \le x\Big\}.$$

The exact Riesz–Kantorovich transform of f is defined for each $x \in L_+$ by

$$\mathcal{R}_f^{=}(x) = \sup\Big\{\sum_{i=1}^{m} f_i(x_i)\colon\ x_i \in [0,x] \text{ for each } i \text{ and } \sum_{i=1}^{m} x_i = x\Big\}.$$

When each f_i is an order bounded linear functional, i.e., $f_i \in L^{\sim}$, we shall call $\mathcal{R}_f$ the **Riesz–Kantorovitch functional** of f and $\mathcal{R}_f^{=}$ the **exact Riesz–Kantorovitch functional**.

Lemma 6.11. *Assume that $f_1, \ldots, f_m \in L'$. Then for the $(m+1)$-tuple $f = (0, f_1, f_2, \ldots, f_m)$ the following statements are equivalent.*

(1) *The Riesz–Kantorovich functional $\mathcal{R}_f$ is upper semicontinuous.*

(2) *In the $\mathcal{K}$-lattice $\mathcal{O}'(L)$ we have $\veebar f = \mathcal{R}_f$. That is,*

$$\mathcal{R}_f = \bigwedge\{g \in L'\colon\ g \succeq \{0, f_1, \ldots, f_m\}\}.$$

Proof. (1) $\Longrightarrow$ (2) Assume that $\mathcal{R}_f$ is upper semicontinuous. That is, $\mathcal{R}_f \in \mathcal{O}'(L)$. Clearly, $\mathcal{R}_f \succeq f_i$ for each i and $\mathcal{R}_f \succeq 0$. Since each f_i is linear and continuous, it must be the case that $\mathcal{R}_f - f_i$ is upper semicontinuous, superadditive, positive homogeneous, and so $\mathcal{R}_f - f_i \in L_+^{\mathcal{O}}$, i.e., $\mathcal{R}_f \geq f_i$ for each i and $\mathcal{R}_f \geq 0$. Therefore, $\mathcal{R}_f \succeq \veebar\{0, f_1, \ldots, f_m\}$. On the other hand, it is easy to see that $\veebar\{0, f_1, \ldots, f_m\} \succeq \mathcal{R}_f$ is also true. Therefore, $\mathcal{R}_f = \veebar\{0, f_1, \ldots, f_m\}$.

(2) $\Longrightarrow$ (1) This is obvious. ∎

We now turn our attention to the **algebraic dual** $(L')^*$ of L'. Recall that every vector $x \in L$ can be considered as a vector of $(L')^*$ by letting $x(f) = f(x)$ for each $f \in L'$. Also, if $\phi, \psi \in (L')^*$, then $\phi \geq \psi$ means that $\phi(f) \geq \psi(f)$ holds for all $f \in L'_+$. Thus, for each $x \in L_+$, besides the order interval $[0, x]$ in L, there is another order interval associated with x in $(L')^*$. This is the order interval

$$[\![0, x]\!] = \{\phi \in (L')^*\colon\ 0 \leq \phi \leq x\}.$$

Clearly $[0, x] \subseteq [\![0, x]\!]$ holds for each $x \in L_+$.[3]

Lemma 6.12. *The order intervals of $(L')^*$ are $\sigma((L')^*, L')$-compact.*

Proof. Notice that the vector space $(L')^*$ equipped with the topology $\sigma((L')^*, L')$ is an ordered topological vector space whose order intervals are $\sigma((L')^*, L')$-closed and $\sigma((L')^*, L')$-bounded—and hence they are also $\sigma((L')^*, L')$-compact. ∎

We continue with the introduction of the notion of the generalized Riesz–Kantorovich functional.

Definition 6.13. *If $f = (f_1, \ldots, f_m)$ is an m-tuple of τ-continuous linear functionals on L, then the **generalized Riesz–Kantorovich functional** $\overline{\mathcal{R}}_f\colon L_+ \to [0, \infty]$ is defined for each $x \in L_+$ by*

$$\overline{\mathcal{R}}_f(x) = \sup\Big\{\sum_{i=1}^{m} \phi_i(f_i)\colon\ \phi_i \in [\![0, x]\!] \text{ for each } i \text{ and } \sum_{i=1}^{m} \phi_i \leq x\Big\}.$$

Clearly, $\overline{\mathcal{R}}_f$ is positive, superadditive, and positively homogeneous. Therefore, if $\overline{\mathcal{R}}_f$ is also real-valued, then $\overline{\mathcal{R}}_f \in \mathcal{L}_+^{\diamond}(L, \mathbb{R})$.

Two basic properties of the generalized Riesz–Kantorovich functionals are described in the next result.

Lemma 6.14. *For an m-tuple $f = (f_1, \ldots, f_m)$ of τ-continuous linear functionals on L we have the following.*

1. *$\overline{\mathcal{R}}_f$ dominates $\mathcal{R}_f$, i.e., $\overline{\mathcal{R}}_f(x) \geq \mathcal{R}_f(x)$ holds for all $x \in L_+$.*
2. *A linear functional $g \in L'_+$ dominates $\overline{\mathcal{R}}_f$ if and only if $g \succeq f_i$ for each i (in which case we have $\overline{\mathcal{R}}_f \in \mathcal{L}_+^{\diamond}(L, \mathbb{R})$).*

---

[3] Since L'_+ is a generating cone of L', it follows that the collection

$$(L')^*_+ = \{\phi \in (L')^*\colon\ \phi(f) \geq 0 \text{ for all } f \in L'_+\}$$

of all positive linear functionals on L' is indeed a cone of $(L')^*$.

Proof. (1) This follows immediately from the inclusion $[0,x] \subseteq [\![0,x]\!]$.

(2) Fix $x \in L_+$ and let $g \in L'_+$ satisfy $g \succeq f_i$ for each i. Now assume that the vector $(\phi_1,\ldots,\phi_m)$ satisfies $\phi_i \in [\![0,x]\!]$ for each i and $\sum_{i=1}^m \phi_i \leq x$. Since g dominates each f_i, we get

$$\sum_{i=1}^m \phi_i(f_i) \leq \sum_{i=1}^m \phi_i(g) = \Big[\sum_{i=1}^m \phi_i\Big](g) \leq x(g) = g(x) < \infty\,.$$

This implies that $\overline{\mathcal{R}}_f(x) \leq g(x) < \infty$. On the other hand, if $\overline{\mathcal{R}}_f(x) \leq g(x)$ for all $x \in L_+$, then it is easy to see that $g \succeq f_i$ also holds for all i. ■

Whenever $\mathcal{R}_f$ is upper τ-semicontinuous, it coincides with $\overline{\mathcal{R}}_f$.

Lemma 6.15. *If $f = (f_1,\ldots,f_m)$ is an m-tuple of τ-continuous linear functionals on L and $\mathcal{R}_f$ is upper τ-semicontinuous, then $\mathcal{R}_f = \overline{\mathcal{R}}_f$.*

Proof. Take an arbitrary $x \in L_+$. Clearly, $0 \leq \mathcal{R}_f(x) < \infty$. We claim that $\overline{\mathcal{R}}_f(x) = \mathcal{R}_f(x)$. To see this, suppose by way of contradiction that $\overline{\mathcal{R}}_f(x) > \mathcal{R}_f(x)$. According to Lemma 6.5, there exists some $g \in L'_+$ such that $g \succeq \mathcal{R}_f$ and $\overline{\mathcal{R}}_f(x) > g(x)$. However, since $g \succeq f_i$ for each i, it follows from Lemma 6.14(2) that $\overline{\mathcal{R}}_f(x) \leq g(x)$, which is a contradiction. Therefore, $\overline{\mathcal{R}}_f(x) = \mathcal{R}_f(x)$. ■

Recall that if L is a locally convex-solid Riesz space, then the order intervals in L are $\sigma(L'',L')$-dense in the order intervals of L''; see [**15**, Lemma 3.56, p. 176]. An analogous property holds true whenever the Riesz–Kantorovich functionals of functions in L' are upper τ-semicontinuous.

Theorem 6.16. *If $\mathcal{R}_f$ is upper τ-semicontinuous for every $f \in L'$, then each order interval $[0,x]$ is $\sigma((L')^*,L')$-dense in $[\![0,x]\!]$.*

Proof. Fix $x_0 \in L_+$ and suppose by way of contradiction that there exists some $\phi \in [\![0,x_0]\!]$ satisfying $\phi \notin \overline{[0,x_0]}$, the $\sigma((L')^*,L')$-closure of $[0,x_0]$. Since $\sigma((L')^*,L')$ is locally convex, it follows from the Separation Theorem (see Corollary 8.18) that there exists a linear functional $f \in L'$ and a constant c such that $\phi(f) > c$ and $f(x) \leq c$ for all $x \in [0,x_0]$. Therefore, $\mathcal{R}_f(x_0) \leq c$. On the other hand, by Lemma 6.15 we have $\mathcal{R}_f(x_0) = \overline{\mathcal{R}}_f(x_0) \geq \phi(f) > c$, a contradiction. This contradiction establishes that $[0,x]$ is $\sigma((L')^*,L')$-dense in $[\![0,x]\!]$. ■

The continuity of certain Riesz–Kantorovich functionals at zero is equivalent to the identity $L^r = L^\sim$.

Theorem 6.17. *Every order bounded linear functional on L is regular (that is, $L^r = L^\sim$) if and only if for every $f \in (L^\sim)^m$ its Riesz–Kantorovich functional $\mathcal{R}_f$ is continuous at zero for some locally convex topology on L.*

Proof. Assume first that $L^r = L^\sim$. From $L' \subseteq L^\sim$, we know that $\sigma(L, L^\sim)$ is a Hausdorff locally convex topology on L. Suppose by way of contradiction that for some m-tuple $f = (f_1, \ldots, f_m)$ of linear functionals of $L^\sim$ the Riesz–Kantorovich functional $\mathcal{R}_f$ is not $\sigma(L, L^\sim)$-continuous at zero. This means that there exist a net $\{x_\alpha\} \subseteq L_+$ and some $\epsilon > 0$ such that $x_\alpha \xrightarrow{\sigma(L,L^\sim)} 0$ while $\mathcal{R}_f(x_\alpha) \geq \epsilon > 0$ holds for every α. Since $L^r = L^\sim$, there exists some positive linear functional g on L that dominates each f_i, and therefore $\mathcal{R}_f(x) \leq g(x)$ for every $x \in L_+$. This implies that $g(x_\alpha) \geq \epsilon > 0$ holds for each α. This contradicts the fact that $g \in L^\sim$ and $x_\alpha \xrightarrow{\sigma(L,L^\sim)} 0$.

For the converse, suppose that for every $f \in (L^\sim)^m$ the Riesz–Kantorovich functional $\mathcal{R}_f$ is continuous at zero for some locally convex topology τ_f on L. Since $L^r \subseteq L^\sim$ is always true, it suffices to show that for any $h \in L^\sim$ there exists a positive linear functional g such that $g \succeq h$. To this end, fix $h \in L^\sim$. If $0 \succeq hd$, then we are done. So, we can assume that there exists some $x \in L_+$ such that $h(x) > 0$. Clearly, $\mathcal{R}_h(x) > 0$. Since $\mathcal{R}_h$ is τ_h-continuous at zero, it follows that zero does not belong to the τ_h-closure of the convex set $U^x_{\mathcal{R}_h} = \{y \in L_+ \colon \mathcal{R}_h(y) \geq \mathcal{R}_h(x)\}$.[4] By the Separation Theorem, there exists a nonzero τ_h-continuous linear functional f and a real number $c > 0$ such that $z \in U^x_{\mathcal{R}_h}$ implies $f(z) \geq c$. Since the set $U^x_{\mathcal{R}_h}$ is also comprehensive, it must be the case that $f \succeq 0$; and so $f \in L^\sim$. From $\frac{\mathcal{R}_h(x)}{c} f(y) \geq \mathcal{R}_h(x)$ for all $y \in U^x_{\mathcal{R}_h}$ and Lemma 6.4, we see that the positive linear functional $\frac{\mathcal{R}_h(x)}{c} f$ dominates $\mathcal{R}_h$, and so it also dominates the linear functional h. ■

The next result presents a condition under which the suprema of certain m-tuples of continuous linear functionals satisfy the Riesz–Kantorovich formula. The proof of this theorem is left as an excercise.

Theorem 6.18. *Let $f = (f_1, \ldots, f_m)$ be an m-tuple of τ-continuous linear functionals such that the supremum $g = \sup\{0, f_1, \ldots, f_m\}$ exists in L'. Then*

(1) $g = \mathcal{R}_f$ *if and only if* $\mathcal{R}_f \in L^{\textcircled{r}}$,

(2) $g = \overline{\mathcal{R}}_f$ *if and only if* $\overline{\mathcal{R}}_f \in L^{\textcircled{r}}$.

It is well known that for a **Fréchet lattice** (i.e., a locally convex-solid Riesz space whose topology is completely metrizable) we have $L' = L^r = L^\sim$; see [**12**, Theorem 9.11, p. 352].[5] The next result informs us that we can do without the lattice property.

[4] To see this, consider the neighborhood $W = \{z \in L \colon \mathcal{R}_h(z) < \mathcal{R}_h(x)\}$ of zero and note that $W \cap U^x_{\mathcal{R}_h} = \emptyset$.

[5] It is also true that a given Riesz space admits at most one completely metrizable locally solid topology; see [**12**, Corollary 9.10, p. 352]. In particular, a Riesz space admits at most one (up to an equivalence) lattice norm under which it is a Banach lattice.

Theorem 6.19. *If the topology τ is metrizable and either*

(i) *the positive cone L_+ has a τ-interior point or else*

(ii) *the locally convex topology τ is complete and metrizable,*

then each function in $\mathcal{L}_+^\diamond(L,\mathbb{R})$ is continuous at zero.

In particular, in this case, we have $L' = L^r = L^\sim$.

Proof. Fix a function $f \in \mathcal{L}_+^\diamond(L,\mathbb{R})$, and assume by way of contradiction that f is not τ-continuous at zero. This means that there exist a sequence $\{x_n\}$ of L_+ and some $\epsilon > 0$ satisfying $x_n \xrightarrow{\tau} 0$ and $f(x_n) \geq \epsilon > 0$ for each n. By Lemma 2.30, we can suppose (by passing to a subsequence if necessary) that $0 \leq x_n \leq \frac{1}{n}u$ holds for each n and some $u \in L_+$. But then, the monotonicity and homogeneity of f imply

$$\tfrac{1}{n}f(u) = f(\tfrac{1}{n}u) \geq f(x_n) \geq \epsilon > 0$$

for every n, which is impossible. This contradiction shows that f is continuous at zero.

The preceding conclusion implies that the Riesz–Kantorovich functional of any m-tuple of functionals in $L^\sim$ is continuous at zero. So, by Theorem 6.17, we have $L^r = L^\sim$. On the other hand, a glance at Corollary 2.31 shows that $L^r \subseteq L'$. Since $L' \subseteq L^\sim$ is also the case, we infer that indeed $L' = L^r = L^\sim$ holds true, as desired. ■

Sufficient conditions for the upper semicontinuity of the Riesz–Kantorovich functionals are presented next.

Lemma 6.20. *Assume that the locally convex topology τ is complete and metrizable. Then for an m-tuple $f = (f_1, \ldots, f_m)$ of order bounded linear functionals on L we have the following.*

(1) *The generalized Riesz–Kantorovich functional $\overline{\mathcal{R}}_f$ is real-valued and upper τ-semicontinuous.*

(2) *If the order intervals in L are also weakly compact, then*
   (a) *$\mathcal{R}_f = \overline{\mathcal{R}}_f$,*
   (b) *the Riesz–Kantorovich functional $\mathcal{R}_f$ is upper τ-semicontinuous, and*
   (c) *$[0,x] = [\![0,x]\!]$ holds for each $x \in L_+$.*

Proof. Notice first that (according to Theorem 6.19) we have $L^\sim = L' = L^r$. This implies that each linear functional f_i is τ-continuous and that the generalized Riesz–Kantorovich functional $\overline{\mathcal{R}}_f$ is real-valued.

(1) Take a sequence $\{x_n\} \subseteq L_+$ satisfying $x_n \xrightarrow{\tau} x$ for some $x \in L_+$ and suppose that $\overline{\mathcal{R}}_f(x_n) \geq c$ for all n. We need to show that $\overline{\mathcal{R}}_f(x) \geq c$.

From Lemma 2.30, we can assume that $\{x_n\}$ is order bounded. That is, we can assume that there exists some $e \in L_+$ such that $\{x_n\} \subseteq [0,e] \subseteq [\![0,e]\!]$. Next, fix $\epsilon > 0$ and for each n pick some m-tuple $(\phi_1^n, \ldots, \phi_m^n) \in [\![0,e]\!]^m$ satisfying $\sum_{i-1}^m \phi_i^n \leq x_n$ and

$$\sum_{i=1}^m \phi_i^n(f_i) > \overline{\mathcal{R}}_f(x_n) - \epsilon \geq c - \epsilon\,. \tag{$\star$}$$

Now, by Lemma 6.12, the order interval $[\![0,e]\!]$ is $\sigma((L')^*, L')$-compact. So, there exists a subnet $\{(\phi_1^{n_\alpha}, \ldots, \phi_m^{n_\alpha})\}$ of the sequence $\{(\phi_1^n, \ldots, \phi_m^n)\}$ such that $\phi_i^{n_\alpha} \xrightarrow[\alpha]{\sigma((L')^*,L')} \phi_i \in [\![0,e]\!]$ holds in $(L')^*$ for each $i = 1, \ldots, m$. Clearly, $\sum_{i=1}^m \phi_i \leq x$.

Next, using $(\star)$ and the $\sigma((L')^*, L')$-continuity of each f_i, we see that

$$c - \epsilon \leq \limsup_\alpha \Big[\sum_{i=1}^m \phi_i^{n_\alpha}(f_i)\Big] \leq \sum_{i=1}^m \limsup_\alpha \phi_i^{n_\alpha}(f_i) = \sum_{i=1}^m \phi_i(f_i) \leq \overline{\mathcal{R}}_f(x)\,.$$

Since $\epsilon > 0$ is arbitrary, we get $\overline{\mathcal{R}}_f(x) \geq c$, as desired.

(2) In this case, we can modify the preceding proof to show that $\mathcal{R}_f$ is also upper τ-semicontinuous. The rest of the proof follows from Lemma 6.15 and Theorem 6.16. ■

The next result is a simple consequence of Lemma 6.20 and Theorem 6.18.

Theorem 6.21. *Assume again that the locally convex topology τ is complete and metrizable. If for some m-tuple of order bounded linear functionals $f = (f_1, \ldots, f_m)$ the supremum $g = \sup\{0, f_1, \ldots, f_m\}$ exists in $L^\sim$, then for each $x \in L_+$ we have*

$$g(x) = \sup\Big\{\sum_{i=1}^m \phi_i(f_i)\colon\ \phi_i \in [\![0,x]\!] \text{ for each } i \text{ and } \sum_{i=1}^m \phi_i \leq x\Big\}.$$

If, in addition, the order intervals in L are weakly compact, then

$$g(x) = \sup\Big\{\sum_{i=1}^m f_i(x_i)\colon\ x_i \in [0,x] \text{ for each } i \text{ and } \sum_{i=1}^m x_i \leq x\Big\}$$

holds for each $x \in L_+$.

We close this section with a result that has many implications in economic theory.

Theorem 6.22. *Assume once more that the locally convex topology τ is complete and metrizable. If $f = (f_1, \ldots, f_m)$ is an m-tuple of order bounded linear functionals, then for each $x \in L_+$ we have*

$$\inf\big\{g(x)\colon g \in L'_+ \text{ and } g \succeq f_i \text{ for each } i = 1, \ldots, m\big\} = \overline{\mathcal{R}}_f(x)\,.$$

If, in addition, the order intervals in L are weakly compact, then

$$\inf\{g(x)\colon\ g \in L'_+ \text{ and } g \succeq f_i \text{ for each } i = 1, \ldots, m\} = \mathcal{R}_f(x)\,.$$

Proof. From Theorem 6.19 we know that $L' = L^r = L^\sim$. By Lemma 6.20 we also know that $\overline{\mathcal{R}}_f$ is upper τ-semicontinuous. Therefore, from Lemma 6.5 and Theorem 6.19 we have

$$\overline{\mathcal{R}}_f(x) = \inf\left\{g(x)\colon\ g \in L_+{}' \text{ and } g \geq \overline{\mathcal{R}}_f\right\}$$

for each $x \in L_+$. Furthermore, Lemma 6.14(2) guarantees that some $g \in L_+^\sim$ satisfies $g \geq \overline{\mathcal{R}}_f$ if and only if $g \geq f_i$ for each i. This completes the proof of the first part of the theorem. Now if the order intervals in L are weakly compact, then by Lemma 6.20(2) we have $\mathcal{R}_f = \overline{\mathcal{R}}_f$ and $[0, x] = [\![0, x]\!]$, and the proof is finished. ■

Exercises

1. Prove Theorem 6.18.
2. Assume that for some m-tuple $f = (f_1, \ldots, f_m)$ of τ-continuous linear functionals the supremum $g = \sup\{0, f_1, \ldots, f_m\}$ exists in L'. Show that
   (a) $g = \mathcal{R}_f$ if and only if $\mathcal{R}_f$ is τ-continuous,
   (b) $g = \overline{\mathcal{R}}_f$ if and only if $\overline{\mathcal{R}}_f$ is τ-continuous.
3. Let $\mathcal{K}$ be a closed and generating cone of some Euclidean space $\mathbb{R}^\ell$ and let p_1, p_2 be two vectors in $\mathbb{R}^\ell$. Show that for each $x \in \mathcal{K}$ we have

$$\inf\{p \cdot x\colon\ p \geq_{\mathcal{K}'} \{0, p_1, p_2\}\} = \sup\{p_1 \cdot y + p_2 \cdot (x - y)\colon\ 0 \leq_{\mathcal{K}} y \leq_{\mathcal{K}} x\}\,.$$

6.3. When is the Riesz–Kantorovich functional additive?

The question of the title is of course (according to Lemma 1.26) equivalent to "When is the Riesz–Kantorovich functional linear?" Since we know the Riesz–Kantorovich functional is always superadditive, the question can also be posed as "When is the Riesz–Kantorovich functional subadditive?" The purpose of this section is to provide some answers to this question. In this section, L will simply denote an ordered vector space. All results in this section are taken from [**21**].

Before discussing the major results in this section, let us introduce some extra notation. For any natural number m and any $x \in L_+$ we define the following nonempty convex sets:

$$\mathcal{A}_x^m = \Big\{(y_1,\ldots,y_m)\in L_+^m\colon\ \sum_{i=1}^m y_i\le x\Big\} \quad\text{and}$$

$$\mathcal{F}_x^m = \Big\{(y_1,\ldots,y_m)\in L_+^m\colon\ \sum_{i=1}^m y_i = x\Big\}.$$

In terms of this notation, notice that the Riesz–Kantorovich functionals for each $x\in L_+$ are given by the formulas

$$\mathcal{R}_f(x) = \sup\Big\{\sum_{i=1}^m f_i(x_i)\colon\ (x_1,\ldots,x_m)\in\mathcal{A}_x^m\Big\} \quad\text{and}$$

$$\mathcal{R}_f^{=}(x) = \sup\Big\{\sum_{i=1}^m f_i(x_i)\colon\ (x_1,\ldots,x_m)\in\mathcal{F}_x^m\Big\}.$$

We also introduce the following terminology.

Definition 6.23. *Let $f_i\colon L_+\to(-\infty,\infty]$, $i=1,\ldots,m$, be m-functions. We shall say that the Riesz–Kantorovich functional $\mathcal{R}_f$ of the m-tuple of functions $f=(f_1,\ldots,f_m)$ is **exact** at a vector $x\in L_+$ if*

$$\mathcal{R}_f(x) = \max\Big\{\sum_{i=1}^m f_i(x_i)\colon\ (x_1,\ldots,x_m)\in\mathcal{A}_x^m\Big\}.$$

In other words, the Riesz–Kantorovich functional $\mathcal{R}_f$ is exact at some x in L_+ if there exists some $(x_1,\ldots,x_m)\in\mathcal{A}_x^m$ such that $\mathcal{R}_f(x)=\sum_{i=1}^m f_i(x_i)$.

*Similarly, the Riesz–Kantorovich functional $\mathcal{R}_f^{=}$ is **exact** at some vector $x\in L_+$ if $\mathcal{R}_f^{=}(x)=\max\{\sum_{i=1}^m f_i(x_i)\colon\ (x_1,\ldots,x_m)\in\mathcal{F}_x^m\}$.*

Regarding the two convex sets $\mathcal{A}_x^m$ and $\mathcal{F}_x^m$ and the "exactness" of the Riesz–Kantorovich functionals we have the following result—whose easy proof is omitted.

Lemma 6.24. *Assume that τ is a Hausdorff locally convex topology on L such that for some $x\in L_+$ the order interval $[0,x]$ is weakly compact (i.e., $\sigma(L,L')$-compact). Then we have the following.*

1. *For each m the convex sets $\mathcal{A}_x^m$ and $\mathcal{F}_x^m$ are both weakly compact subsets of $(L,\tau)^m$.*
2. *If $f_1,\ldots,f_m$ are τ-continuous linear functionals, i.e., $f_i\in L'$ for each i, then the Riesz–Kantorovich functionals $\mathcal{R}_f$ and $\mathcal{R}_f^{=}$ of the m-tuple $f=(f_1,\ldots,f_m)$ are both exact at the vector $x\in L_+$.*

Also, recall that for a finite number $f_1, \ldots, f_m$ of order bounded linear functionals on L we let

$$\bigvee_{i=1}^{m} f_i = \sup\{f_1, f_2, \ldots, f_m\} \text{ and } \Big[\bigvee_{i=1}^{m} f_i\Big]^+ = \sup\{0, f_1, f_2, \ldots, f_m\},$$

where the suprema are taken in $L^\sim$.[6] Now notice that if the ordered vector space L has the Riesz decomposition property and the cone L_+ is generating, then the Riesz–Kantorovich formulas can be rewritten as follows:

$$\begin{aligned} \Big[\bigvee_{i=1}^{m} f_i\Big](x) &= \sup\Big\{\sum_{i=1}^{m} f_i(y_i)\colon\ (y_1, y_2, \ldots, y_m) \in \mathcal{F}_x^m\Big\}, \\ \Big[\bigvee_{i=1}^{m} f_i\Big]^+(x) &= \sup\Big\{\sum_{i=1}^{m} f_i(y_i)\colon\ (y_1, y_2, \ldots, y_m) \in \mathcal{A}_x^m\Big\} \end{aligned}$$

for all $x \in L_+$. (See Theorem 1.59.) In other words, we have $\mathcal{R}_f^= = \bigvee_{i=1}^{m} f_i$ and $\mathcal{R}_f = \big[\bigvee_{i=1}^{m} f_i\big]^+$.

A useful case concerning the "exactness" of the Riesz–Kantorovich functionals is presented next.

Theorem 6.25. *Let L be an ordered vector space and assume that for the order bounded linear functionals $f_1, f_2, \ldots, f_m$ the supremum $\big[\bigvee_{i=1}^{m} f_i\big]^+$ exists in $L^\sim$. If for the m-tuple $f = (f_1, \ldots, f_m)$ the Riesz–Kantorovich formula $\mathcal{R}_f$ is exact at an order unit e of L, then we have*

$$\Big[\bigvee_{i=1}^{m} f_i\Big]^+(e) = \mathcal{R}_f(e).$$

In other words, if for an order unit e of L there is some $(x_1^*, x_2^*, \ldots, x_m^*)$ in $\mathcal{A}_e^m$ satisfying

$$\sum_{i=1}^{m} f_i(x_i^*) \ge \sum_{i=1}^{m} f_i(x_i)$$

for each $(x_1, x_2, \ldots, x_m) \in \mathcal{A}_e^m$, then

$$\Big[\bigvee_{i=1}^{m} f_i\Big]^+(e) = \mathcal{R}_f(e) = \sum_{i=1}^{m} f_i(x_i^*).$$

Proof. Let $g = \big[\bigvee_{i=1}^{m} f_i\big]^+$ and notice first that

$$g(e) \ge g\Big(\sum_{i=1}^{m} x_i^*\Big) = \sum_{i=1}^{m} g(x_i^*) \ge \sum_{i=1}^{m} f_i(x_i^*) \ge 0.$$

To finish the proof, we must verify that $g(e) \le \sum_{i=1}^{m} f_i(x_i^*)$ is also true.

[6] It should be observed that $\bigvee_{i=1}^{m} f_i$ and $\big(\bigvee_{i=1}^{m} f_i\big)^+$ are in general different linear functionals. For an example consider a linear functional $f < 0$ and note that $f \vee f = f < 0 = f^+$.

If $\sum_{i=1}^m f_i(x_i^*) = 0$, then it should be clear that $f_i \leq 0$ for each i. Indeed, since e is an order unit for each $x \geq 0$, there exists some $\lambda > 0$ with $\lambda x \leq e$ and so $\lambda f_i(x) = f_i(\lambda x) \leq \sum_{i=1}^m f_i(x_i^*) = 0$, from which it follows that $f_i(x) \leq 0$ for each i and all $x \geq 0$. Therefore, in this case, we have $g = 0$ so that $g(e) = \sum_{i=1}^m f_i(x_i^*) = 0$. Thus, we can assume $\sum_{i=1}^m f_i(x_i^*) > 0$. Let

$$Y = \Big\{(y_1, y_2, \ldots, y_m) \in L_+^m \colon \ \sum_{i=1}^m f_i(y_i) > \sum_{i=1}^m f_i(x_i^*)\Big\}.$$

Notice that $Y \subseteq L_+^m$ is nonempty (for instance, $2(x_1^*, x_2^*, \ldots, x_m^*) \in Y$) and convex. Furthermore, it follows from our hypothesis that Y is disjoint from $\mathcal{A}_e^m$. This implies that Y is also disjoint from the convex set

$$Z = \Big\{(z_1, z_2, \ldots, z_m) \in L^m \colon \ \sum_{i=1}^m z_i \leq e\Big\},$$

which in turn (in view of Exercise 13 of Section 1.1) has an internal point in L^m. By the Algebraic Separating Hyperplane Theorem 8.15 there exists a nonzero linear functional $(h_1, h_2, \ldots, h_m) \in (L^*)^m$ that separates Z and Y. That is, for all $(y_1, y_2, \ldots, y_m) \in Y$ and all $(z_1, z_2, \ldots, z_m) \in Z$ we have

$$\sum_{i=1}^m h_i(y_i) \geq \sum_{i=1}^m h_i(z_i). \tag{$\star$}$$

Since $(x_1^*, x_2^*, \ldots, x_m^*) \in Z$, it follows from $(\star)$ that

$$\sum_{i=1}^m h_i(y_i) \geq \sum_{i=1}^m h_i(x_i^*) \quad \text{for all } (y_1, y_2, \ldots, y_m) \in Y. \tag{$\star\star$}$$

Furthermore, since $\alpha(x_1^*, x_2^*, \ldots, x_m^*) \in Y$ for all $\alpha > 1$, it also follows from $(\star)$ that

$$\sum_{i=1}^m h_i(x_i^*) = \lim_{\alpha \downarrow 1} \sum_{i=1}^m h_i(\alpha x_i^*) \geq \sum_{i=1}^m h_i(z_i) \tag{$\star\star\star$}$$

holds true for each $(z_1, z_2, \ldots, z_m) \in Z$.

Next, we show that $h_1 = h_2 = \cdots = h_m$. Suppose by way of contradiction that there exists some $z \in L$ such that $h_1(z) > h_2(z)$. Then for some $\alpha > 1$ we have

$$h_1(x_1^* + z) + h_2(x_2^* - z) + \sum_{i=3}^m h_i(x_i^*) > \sum_{i=1}^m h_i(\alpha x_i^*).$$

But $(\alpha x_1^*, \alpha x_2^*, \ldots, \alpha x_m^*) \in Y$ and $(x_1^* + z, x_2^* - z, x_3^*, \ldots, x_m^*) \in Z$, which contradicts $(\star)$. By symmetry, we get $h_1 = h_2 = \cdots = h_m = h$.

Now we show that $h \geq 0$. Let $x \in L_+$ and note that $(e-\alpha x, 0, \ldots, 0) \in Z$ for all $\alpha > 0$. Therefore, from $(\star\star\star)$ we get $\sum_{i=1}^m h(x_i^*) \geq h(e) - \alpha h(x)$, or

$$h(x) \geq \frac{h(e) - \sum_{i=1}^m h(x_i^*)}{\alpha}$$

for each $\alpha > 0$. Letting $\alpha \to \infty$ yields $h(x) \geq 0$, i.e., $h \geq 0$, and so $h \in L^\sim$.

Furthermore, since $h \neq 0$, it must be the case that $h(e) > 0$ and since $(e, 0, \ldots, 0) \in Z$, it follows from $(\star\star\star)$ that $\sum_{i=1}^m h(x_i^*) > 0$. Let

$$\delta = \frac{\sum_{i=1}^m f_i(x_i^*)}{\sum_{i=1}^m h(x_i^*)} > 0\,, \quad \text{or} \quad \sum_{i=1}^m f_i(x_i^*) = \delta \sum_{i=1}^m h(x_i^*)\,.$$

We claim that $\delta h \geq f_i$ for $i = 1, 2, \ldots, m$. To see this, fix i and let $x \in L_+$. If $f_i(x) \leq 0$, then $\delta h(x) \geq 0 \geq f_i(x)$ is trivially true. Assume therefore that $f_i(x) > 0$ and let

$$\gamma = \frac{\sum_{i=1}^m f_i(x_i^*)}{f_i(x)} > 0\,, \quad \text{or} \quad \sum_{i=1}^m f_i(x_i^*) = \gamma f_i(x)\,.$$

It is clear that the vector $(0, \ldots, \alpha\gamma x, 0, \ldots, 0)$, where $\alpha\gamma x$ occupies the ith position, satisfies $(0, \ldots, \alpha\gamma x, 0, \ldots, 0) \in Y$ for all $\alpha > 1$, and from $(\star\star)$ we see that

$$\delta\gamma h(x) = \delta \lim_{\alpha\downarrow 1} h(\alpha\gamma x) \geq \delta \sum_{i=1}^m h(x_i^*) = \sum_{i=1}^m f_i(x_i^*) = \gamma f_i(x)\,,$$

or $\delta h(x) \geq f_i(x)$ for each $x \geq 0$. Thus, $\delta h \geq f_i$ for $i = 1, 2, \ldots, m$.

Consequently, $g \leq \delta h$. In particular, from $(\star\star\star)$ it follows that

$$g(e) \leq \delta h(e) \leq \delta \sum_{i=1}^m h(x_i^*) = \sum_{i=1}^m f_i(x_i^*)\,,$$

and the proof is finished. ■

Corollary 6.26. *Let L be an ordered vector space and assume that for an m-tuple $f = (f_1, \ldots, f_m)$ of order bounded linear functionals the supremum $\bigvee_{i=1}^m f_i$ exists in $L^\sim$. If the Riesz–Kantorovich functional $\mathcal{R}_{\overline{f}}$ is exact at some order unit e, then*

$$\Big[\bigvee_{i=1}^m f_i\Big](e) = \mathcal{R}_{\overline{f}}(e)\,.$$

In other words, if there exists some $(x_1^*, x_2^*, \ldots, x_m^*) \in \mathcal{F}_e^m$ satisfying

$$\sum_{i=1}^m f_i(x_i^*) \geq \sum_{i=1}^m f_i(x_i)$$

for each $(x_1, x_2, \dots, x_m) \in \mathcal{F}_e^m$, *then*

$$\Big[\bigvee_{i=1}^m f_i\Big](e) = \mathcal{R}_{\vec{f}}(e) = \sum_{i=1}^m f_i(x_i^*)\,.$$

Proof. We can assume without loss of generality that $m \geq 2$ holds. Let $g = \bigvee_{i=1}^m f_i$. Start by observing that $g - f_1 = \big[\bigvee_{i=2}^m (f_i - f_1)\big]^+$ and that for each $(x_2, \dots, x_m) \in \mathcal{A}_e^{m-1}$ we have

$$\begin{aligned}\sum_{i=2}^m (f_i - f_1)(x_i^*) &= \sum_{i=1}^m f_i(x_i^*) - f_1(e)\\ &\geq \sum_{i=2}^m f_i(x_i) + f_1\Big(e - \sum_{i=2}^m x_i\Big) - f_1(e)\\ &= \sum_{i=2}^m (f_i - f_1)(x_i)\,.\end{aligned}$$

Therefore, Theorem 6.25 yields

$$(g - f_1)(e) = \sum_{i=2}^m (f_i - f_1)(x_i^*)\,.$$

In particular, we have

$$g(e) = f_1\Big(e - \sum_{i=2}^m x_i^*\Big) + \sum_{i=2}^m f_i(x_i^*) = \sum_{i=1}^m f_i(x_i^*)\,,$$

as desired. ∎

A topological version of Theorem 6.25 is as follows.

Theorem 6.27. *Assume that an ordered vector space L with order units is equipped with a Hausdorff locally convex topology such that the order intervals of L are weakly compact. If for some m-tuple $f = (f_1, \dots, f_m)$ of continuous linear functionals the supremum $g = \bigvee_{i=1}^m f_i$ exists in $L^\sim$, then g coincides with the Riesz–Kantorovich functional $\mathcal{R}_{\vec{f}}$. That is, for each $x \in L_+$ we have*

$$\Big[\bigvee_{i=1}^m f_i\Big](x) = \mathcal{R}_{\vec{f}}(x) = \max\Big\{\sum_{i=1}^m f_i(y_i)\colon\ (y_1, y_2, \dots, y_m) \in \mathcal{F}_x^m\Big\}\,.$$

Proof. The exactness at each $x \in L_+$ of the Riesz–Kantorovich functional $\mathcal{R}_{\vec{f}}$ follows from Lemma 6.24. Suppose that τ is a Hausdorff locally convex topology on L for which the intervals are weakly compact. Now fix an arbitrary $x \in L_+$, and then select an order unit e such that $x \leq e$.

For each $0 < \alpha < 1$ let $e_\alpha = \alpha x + (1 - \alpha)e$. Clearly, each e_α is an order unit and $0 \leq e_\alpha \leq e$ holds for each $0 < \alpha < 1$. We consider the index set $(0,1)$ directed by the increasing order relation $\succeq$, i.e, $\alpha \succeq \beta$ in $(0,1)$ if and only if $\alpha \geq \beta$. Clearly, $e_\alpha \to x$. By Corollary 6.26 we know that $g(e_\alpha) = \mathcal{R}_f^{=}(e_\alpha)$ for each $0 < \alpha < 1$. Therefore, for each $0 < \alpha < 1$ there exists some $(x_1^\alpha, x_2^\alpha, \ldots, x_m^\alpha) \in L_+^m$ with $\sum_{i=1}^m x_i^\alpha = e_\alpha$ and

$$\alpha g(x) + (1-\alpha) g(e) = g(e_\alpha) = \sum_{i=1}^m f_i(x_i^\alpha)\,. \tag{$\dagger$}$$

Since $x_i^\alpha \in [0,e]$ for each $\alpha \in (0,1)$ and each $i = 1, 2, \ldots, m$ and the order interval $[0,e]$ is weakly compact, there exists a subnet of $\{x_1^\alpha, x_2^\alpha, \ldots, x_m^\alpha\}$ (which without loss of generality we can denote by $\{x_1^\alpha, x_2^\alpha, \ldots, x_m^\alpha\}$) such that $x_i^\alpha \xrightarrow{w} x_i \in [0,e]$ for each i. From $e_\alpha - x_i^\alpha \in [0,e]$ for each α and the weak closedness of $[0,e]$, we see that $0 \leq x - x_i \leq e$. So $0 \leq x_i \leq x$ and $\sum_{i=1}^m x_i = x$. Finally, letting $\alpha \to 1$ in ($\dagger$), it follows from the continuity of g and the weak continuity of each f_i that

$$g(x) = \sum_{i=1}^m f_i(x_i) \leq \sup\Big\{\sum_{i=1}^m f_i(y_i)\colon\ (y_1, y_2, \ldots, y_m) \in \mathcal{F}_x^m\Big\} \leq g(x)\,,$$

and the proof is finished. ■

The next result characterizes the Riesz decomposition property in terms of a lattice property of the topological dual.

Corollary 6.28. *Assume that an ordered vector space L has order units and is equipped with a Hausdorff locally convex topology under which the order intervals of L are weakly compact. Then the following statements are equivalent.*

(1) *For every pair $f, g \in L'$ the supremum $f \vee g$ exists in $L^\sim$.*

(2) *L has the Riesz decomposition property.*

Moreover, if this is the case, then $L^\sim$ is a Riesz space whose lattice operations are given by the Riesz–Kantorovich formulas.

Proof. If (2) is true, then the validity of (1) follows from Theorem 1.59. So, assume that (1) is true and fix $x, y \in L_+$. Clearly, $[0,x] + [0,y] \subseteq [0, x+y]$.

Now suppose by way of contradiction that there exists some $z \in [0, x+y]$ such that $z \notin [0,x] + [0,y]$. Since $[0,x] + [0,y]$ is a weakly compact convex set, it follows from the Strong Separating Hyperplane Theorem 8.17 that there exist some nonzero $f \in L'$ and some $\epsilon > 0$ such that

$$f(z) > \epsilon + f(u) + f(v)$$

for all $u \in [0,x]$ and all $v \in [0,y]$. Now a glance at Theorem 6.27 shows that

$$\begin{aligned}\sup_{0\le w\le x+y} f(w) &= f^+(x+y) = f^+(x)+f^+(y)\\ &= \sup_{0\le u\le x} f(u) + \sup_{0\le v\le y} f(v)\\ &\le f(z)-\epsilon\\ &< f(z),\end{aligned}$$

which is a contradiction. Hence, $[0,x]+[0,y]=[0,x+y]$ holds true, and so L has the Riesz decomposition property. ■

Recall that $L^\sim$ **separates the points** of L if $x \in L$ and $f(x)=0$ for all $f \in L^\sim$ implies $x=0$. In this case the weak topology $\sigma(L,L^\sim)$ is a Hausdorff locally convex topology on L. Therefore the following is a special case of Corollary 6.28.

Corollary 6.29. *Assume that L is an ordered vector space with order units and that $L^\sim$ separates the points of L. If the order intervals of L are also $\sigma(L,L^\sim)$-compact, then $L^\sim$ is a Riesz space if and only if L has the Riesz decomposition property.*

Let us now move to the case where L is an ordered vector space without an order unit. Recall that for any $x \in L_+$ we denote by L_x the (principal) **ideal** generated by x in L, i.e., the vector subspace $\bigcup_{n=1}^\infty n[-x,x]$. Under the order induced on L_x by L the ideal L_x is an ordered vector space having x as an order unit. As usual, the restriction of the positive cone L_+ to L_x is denoted by L_x^+, i.e., $L_x^+ = L_x \cap L_+$. Notice that for any $y \in L_x$ the order interval $[0,y]$ is contained in L_x^+, i.e, $[0,y] \subseteq L_x^+$.

The final result of this section presents one more connection between the supremum of linear functionals and their Riesz–Kantorovich functional.

Lemma 6.30. *If an ordered vector space L is equipped with a Hausdorff locally convex topology τ for which the order intervals of L are weakly compact, then the following two statements are equivalent.*

(1) *There is a pair of continuous linear functionals f and g such that $f\vee g$ exists in $L^\sim$ and $f\vee g \ne \mathcal{R}_{f,g}$.*

(2) *There exists a pair of continuous functionals f and g and a vector $x \in L_+$ such that $f\vee g$ exists in $L^\sim$ but the restriction $(f\vee g)|_{L_x}$ is not the supremum of $f|_{L_x}$ and $g|_{L_x}$ in the order dual $L_x^\sim$ of L_x.*

Proof. (1) $\Longrightarrow$ (2) Assume that $f,g \in L'$ is a pair of continuous functionals such that $f\vee g$ exists in $L^\sim$ and $f\vee g \ne \mathcal{R}_{f,g}$. Pick some $x \in L_+$ such that

$$(f\vee g)(x) > \sup_{0\le y\le x}\big[\,f(y)+g(x-y)\,\big]. \tag{$\star$}$$

Consider the ideal L_x (on which x is an order unit) and assume by way of contradiction that

$$(f \vee g)|_{L_x} = f|_{L_x} \vee g|_{L_x}\,, \tag{$\star\star$}$$

where the supremum on the right-hand side is taken in the order dual of L_x. Notice that τ restricted to L_x is a Hausdorff locally convex topology for which the order intervals of L_x are still weakly compact. So, according to Theorem 6.27, we have

$$\begin{aligned}(f|_{L_x} \vee g|_{L_x})(x) &= \sup_{0\le y\le x} \big[\, f|_{L_x}(y) + g|_{L_x}(x-y)\big] \\ &= \sup_{0\le y\le x} \big[\, f(y) + g(x-y)\,\big] \\ &< (f \vee g)(x) = \big[(f\vee g)|_{L_x}\big](x) \\ &= (f|_{L_x} \vee g|_{L_x})(x)\,,\end{aligned}$$

which is impossible, and the validity of (2) has been established.

(2) $\Longrightarrow$ (1) Let $f, g \in L'$ be a pair of continuous functions such that $f \vee g$ exists in $L^\sim$ and let $x \in L_+$ be a positive vector such that the restriction $(f \vee g)|_{L_x}$ is not the supremum of $f|_{L_x}$ and $g|_{L_x}$ in the order dual of L_x. We claim that $f \vee g \neq \mathcal{R}_{f,g}$.

To see this, assume by way of contradiction that for each $z \in L_+$ we have

$$(f \vee g)(z) = \mathcal{R}_{f,g}(z) = \sup_{0\le y\le z} \big[\, f(y) + g(z-y)\big]\,. \tag{$\dagger$}$$

Let $h \in L_x^\sim$ satisfy $h \ge f$ and $h \ge g$. From ($\dagger$) we see that for any $z \in L_x^+$ we have

$$h(z) \ge \sup_{0\le y\le z} \big[\, f|_{L_x}(y) + g|_{L_x}(z-y)\big] = (f \vee g)|_{L_x}(z)\,,$$

which implies $h \ge (f \vee g)|_{L_x}$. In particular, $(f \vee g)|_{L_x} = f|_{L_x} \vee g|_{L_x}$ holds in $L_x^\sim$, which contradicts our assumption, and the proof is finished. ■

Exercises

1. Prove Lemma 6.24.
2. Give an example of an ordered Hausdorff locally convex space L with a closed and generating cone and two vectors $0 < x, y \in L_+$ such that $[0, x]$ is weakly compact and $[0, y]$ is not weakly compact. [*Hint*: Let $L = \ell_\infty$ and let τ be the Hausdorff locally convex topology generated by the sup norm. If $x = (x_1, x_2, \ldots) \in \ell_\infty^+$, then show that the order interval $[0, x]$ is τ-compact if and only if $x_n \to 0$.]

3. Give an example of a Hausdorff locally convex ordered vector space L with weakly compact intervals whose cone is not closed. [*Hint*: For instance, let $L = \ell_1$ and consider L ordered by the subcone L_+ of ℓ_1^+ consisting of all eventually zero sequences. If τ is the locally convex Hausdorff topology induced by the ℓ_1-norm, then every interval of L is weakly compact (and in fact τ-compact) but the cone L_+ is not weakly closed (since it is clearly not τ-closed).]

4. Let L be a finite dimensional ordered vector space with a closed and generating cone. Show that for each $f \in L'$ the Riesz–Kantorovich functional $\mathcal{R}_f$ is additive if and only if L has the Riesz decomposition property.

5. Assume that L is an ordered Banach space with a closed cone. If L satisfies the approximate interpolation property, then show that $L^\sim$ is a vector lattice.

6. Assume that L is an ordered vector space such that for all $x, y \in L_+$ and each $0 \le \alpha \le 1$ we have

$$[0, \alpha x] + [0, (1-\alpha)y] = [0, \alpha x + (1-\alpha)y] .$$

Show that for each $f \in L^\sim$ the Riesz–Kantorovich functional $\mathcal{R}_f$ is an affine function, i.e., for all $x, y \in L_+$ and each $0 \le \alpha \le 1$ it satisfies $\mathcal{R}_f(\alpha x + (1-\alpha)y) = \alpha \mathcal{R}_f(x) + (1-\alpha)\mathcal{R}_f(y)$.

7. Let $L = C^\infty(\mathbb{R})$, the vector space of all real-valued C^∞-functions on $\mathbb{R}$ with compact support. With the pointwise ordering and the sup norm L is an ordered normed space.
(a) Is the Riesz–Kantorovich functional $\mathcal{R}_f$ additive for every $f \in L'$?
(b) Does L have the Riesz decomposition property?

8. Let L denote the vector space of all polynomials on $[0, 1]$. With the pointwise ordering and the sup norm L is an ordered normed space.
(a) Is the Riesz–Kantorovich functional $\mathcal{R}_f$ additive for every $f \in L'$?
(b) Does L have the Riesz decomposition property?

Chapter 7

Piecewise affine functions

Piecewise affine functions on $\mathbb{R}^m$ are real-valued continuous functions that agree with a finite number of affine functions. These functions arise in numerous applications, including estimation theory in statistics, combinatorial geometry, the theory of approximate fixed points, game theory, and the theory of linear complementarity.

The purpose of this chapter is to provide a treatment of these functions in terms of the theory of ordered vector space. In particular, we study the vector space of piecewise affine functions and show that it coincides with the Riesz subspace generated by the affine functions on $\mathbb{R}^m$.

We begin by studying the one-dimensional case in which continuous piecewise affine functions have a very special additive form (that is incidentally closely related to call and put options in finance). We then move to the multidimensional case where we show that these functions can always be represented using vector lattice operations on affine functions. So our interest here is in the space of continuous piecewise affine functions.

Our approach here is taken from the paper by C. D. Aliprantis, D. Harris, and R. Tourky [**17**] published in the journal *Macroeconomic Dynamics*. For some applications of piecewise affine functions to econometrics see [**18**].

7.1. One-dimensional piecewise affine functions

We present here a few properties and formulas dealing with continuous piecewise affine functions defined on $\mathbb{R}$ or on a closed interval of $\mathbb{R}$. Recall that any function $g\colon \mathbb{R} \to \mathbb{R}$ given by $g(t) = b + mt$ for each $t \in \mathbb{R}$, where b and m are constants, is known as an **affine function**.

Definition 7.1. *A function $f\colon \mathbb{R} \to \mathbb{R}$ is called* ***piecewise affine*** *(or* ***piecewise linear****) if there exist real numbers $-\infty < a_0 < a_1 < \cdots < a_k < \infty$ and pairs of real numbers (m_i, b_i), $i = 0, 1, \ldots, k, k+1$, such that*

$$f(t) = \begin{cases} m_i t + b_i & \text{if } a_{i-1} \le t \le a_i \text{ for some } 1 \le i \le k, \\ m_0 t + b_0 & \text{if } t \le a_0, \\ m_{k+1} t + b_{k+1} & \text{if } t \ge a_{k+1}\,. \end{cases}$$

The parameters $\{a_0, a_1, \ldots, a_k\}$ and the pairs (m_i, b_i), $i = 0, 1, \ldots, k, k+1$, are referred to as a ***representation*** *of f and the functions $f_i(t) = m_i t + b_i$ are called the* ***components*** *of the representation.*

Similarly, a function $f\colon [a, b] \to \mathbb{R}$, where $[a, b]$ is a closed interval of $\mathbb{R}$, is ***piecewise affine*** *if there exist a partition $a = a_0 < a_1 < \cdots < a_k = b$ of the interval $[a, b]$ and pairs of real numbers (m_i, b_i), $i = 1, \ldots, k$, such that $f(t) = m_i t + b_i$ for all $a_{i-1} \le t \le a_i$.*

Figure 1. The graph of a piecewise affine function

Notice that, according to these definitions, every piecewise affine function is automatically continuous. The following result should be obvious.

Lemma 7.2. *If $f\colon \mathbb{R} \to \mathbb{R}$ is piecewise affine, then its restriction to any closed interval of $\mathbb{R}$ is likewise piecewise affine. Moreover, if $[a, b]$ is any closed subinterval of $\mathbb{R}$, then the components of the piecewise affine function $f\colon [a, b] \to \mathbb{R}$ are among the components of $f\colon \mathbb{R} \to \mathbb{R}$.*

In addition, every piecewise affine function on a closed interval of $\mathbb{R}$ can be extended (in many ways) to a piecewise affine function to all of $\mathbb{R}$.

It should be clear that the collection of all piecewise affine functions is a vector space. We also have the following important approximation property of the piecewise affine functions defined on a closed interval.

Lemma 7.3. *The vector space of all piecewise affine functions defined on a closed interval $[a,b]$ is a Riesz subspace that is uniformly dense in $C[a,b]$.*

Proof. It is straightforward to verify that the vector space of all affine functions on $[a,b]$ is a Riesz subspace that contains the constant function one. Moreover, since the function $f(t)=t$ is piecewise affine and separates the points of the interval $[a,b]$, it follows at once that the Riesz subspace of all affine functions likewise separates the points of $[a,b]$. Its uniform denseness now follows immediately from the lattice version of the classical Stone–Weierstrass approximation theorem.[1] ■

The piecewise affine functions on a closed interval are characterized as follows. (Keep in mind that if a is a real number, then as usual we let $a^+ = \max\{a,0\}$.)

Lemma 7.4. *Let $f\colon [a,b] \to \mathbb{R}$ be an arbitrary piecewise affine function. If $\{a_0, a_1, \ldots, a_k\}$ and (m_i, b_i), $i = 1, \ldots, k$, is any representation of f, then for each $t \in \mathbb{R}$ we have*

$$f(t) = b_1 + m_1 t + \sum_{i=1}^{k-1} (m_{i+1} - m_i)(t - a_i)^+ \,.$$

In particular, a function $f\colon [a,b] \to \mathbb{R}$ is piecewise affine if and only if there exist a partition $a = a_0 < a_1 < \cdots < a_k = b$ of $[a,b]$ and constants $c, c_0, c_1, \ldots, c_k$ such that for each $t \in [a,b]$ we have

$$f(t) = c + \sum_{i=0}^{k} c_i (t - a_i)^+ \,.$$

Proof. Let $a \le t \le b$. If $a_0 \le t \le a_1$, then note that

$$b_1 + m_1 t + \sum_{i=1}^{k-1} (m_{i+1} - m_i)(t - a_i)^+ = b_1 + m_1 t = f(t)\,.$$

Consequently, we can assume that $a_{j-1} \le t \le a_j$ for some $1 < j \le k$. Notice that for each $1 < i \le k-1$ we have $m_i a_i + b_i = m_{i+1} a_i + b_{i+1}$ or

---

[1] Recall that the lattice version of the Stone–Weierstrass approximation theorem asserts the following: *If a Riesz subspace of some $C(\Omega)$-space, with Ω compact, separates the points of Ω and contains the constant function one, then it is uniformly dense in $C(\Omega)$.* For a proof see, for instance, [**16**, Theorem 11.3, p. 88].

$(m_{i+1} - m_i)a_i = -(b_{i+1} - b_i)$. This implies

$$\begin{aligned} b_1 \; &+ \; m_1 t + \sum_{i=1}^{k-1} (m_{i+1} - m_i)(t - a_i)^+ \\ &= \; b_1 + m_1 t + \sum_{i=1}^{j-1} (m_{i+1} - m_i)(t - a_i)^+ \\ &= \; b_1 + m_1 t + \sum_{i=1}^{j-1} (m_{i+1} - m_i)(t - a_i) \\ &= \; b_1 + m_1 t + \Big[\sum_{i=1}^{j-1} (m_{i+1} - m_i)\Big] t - \sum_{i=1}^{j-1} (m_{i+1} - m_i) a_i \\ &= \; b_1 + m_1 t + \Big[\sum_{i=1}^{j-1} (m_{i+1} - m_i)\Big] t + \sum_{i=1}^{j-1} (b_{i+1} - b_i) \\ &= \; b_1 + m_1 t + (m_j - m_1)t + (b_j - b_1) = m_j t + b_j = f(t)\,, \end{aligned}$$

and the proof is finished. ■

The idea is depicted in Figure 2.

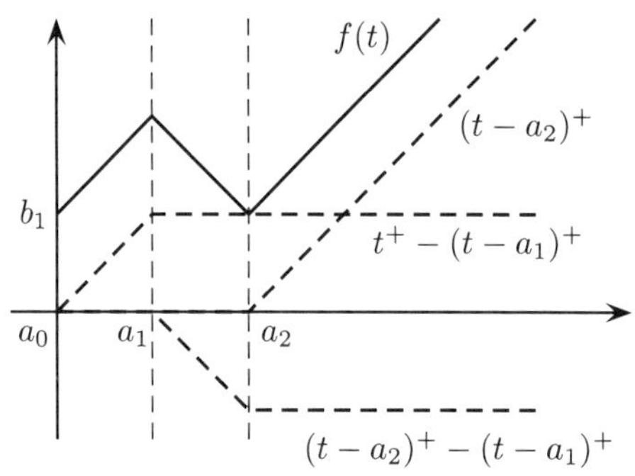

Figure 2. Notice that $f(t) = b_1 + t^+ - 2(t - a_1)^+ + 2(t - a_2)^+$.

The first consequence of Lemma 7.4 is the following result of D. J. Brown, C. B. Huijsmans, and B. de Pagter [**34**].

Corollary 7.5 (Brown–Huijsmans–de Pagter). *The vector subspace generated in $C[0,1]$ by the collection $\{\mathbf{1}, t\} \cup \{(\alpha - t)^+ \colon \ \alpha \in \mathbb{R}\}$ coincides with the Riesz subspace of all piecewise affine functions on $[0,1]$.*

Corollary 7.6. *Let $g\colon \mathbb{R} \to \mathbb{R}$ be a piecewise affine function. If $\{a_0, \ldots, a_k\}$ and (m_i, b_i), $i = 0, 1, \ldots, k, k+1$, is an arbitrary representation of g, then* *for each* $t \in \mathbb{R}$ *we have*

$$g(t) = b_0 + m_0 t + \sum_{i=0}^{k} (m_{i+1} - m_i)(t - a_i)^+ .$$

In particular, a function $f\colon \mathbb{R} \to \mathbb{R}$ *is piecewise affine if and only if there exist real constants* $m_0, b_0, a_0, a_1, \ldots, a_k$ *and* $c_0, c_1, \ldots, c_k$ *such that for each* $t \in \mathbb{R}$ *we have*

$$f(t) = b_0 + m_0 t + \sum_{i=0}^{k} c_i (t - a_i)^+ .$$

Proof. Consider the function $h\colon \mathbb{R} \to \mathbb{R}$ defined by

$$h(t) = b_1 + m_1 t + \sum_{i=1}^{k-1} (m_{i+1} - m_i)(t - a_i)^+ .$$

As in the proof of Lemma 7.4, it is easy to see that $h(t) = f(t)$ for all $a_0 \le t \le a_k$. Moreover, $h(t) = m_1 t + b_1$ for all $t \le a_0$ and $h(t) = m_k t + b_k$ for all $t \ge a_k$. Since

$$\begin{aligned}
m_1 t + b_1 + m_1(a_0 - t)^+ - m_0(a_0 - t)^+ &= b_0 + m_0 t \text{ for all } t \le a_0, \\
m_1 t + b_1 + m_1(a_0 - t)^+ - m_0(a_0 - t)^+ &= m_1 t + b_1 \text{ for all } t \ge a_0, \\
m_k t + b_k + (m_{k+1} - m_k)(t - a_k)^+ &= m_{k+1} + b_{k+1} \text{ for all } t \ge a_k, \\
m_k t + b_k + (m_{k+1} - m_k)(t - a_k)^+ &= m_k t + b_k \text{ for all } t \le a_k,
\end{aligned}$$

it follows that

$$\begin{aligned}
g(t) &= m_1(a_0 - t)^+ - m_0(a_0 - t)^+ h(t) + (m_{k+1} - m_k)(t - a_k)^+ \\
&= m_1(a_0 - t)^+ - m_0(a_0 - t)^+ + b_1 + m_1 t + \cdots \\
&\qquad \cdots + \sum_{i=1}^{k-1} (m_{i+1} - m_i)(t - a_i)^+ + (m_{k+1} - m_k)(t - a_k)^+ \\
&= m_1(a_0 - t)^+ - m_0(a_0 - t)^+ + b_1 + m_1 t + \sum_{i=1}^{k} (m_{i+1} - m_i)(t - a_i)^+ \\
&= b_0 + m_0 t + (m_1 - m_0)(t - a_0)^+ + \sum_{i=1}^{k} (m_{i+1} - m_i)(t - a_i)^+ \\
&= b_0 + m_0 t + \sum_{i=0}^{k} (m_{i+1} - m_i)(t - a_i)^+ ,
\end{aligned}$$

as desired. ∎

We close the section with two results, due to S. Ovchinnikov [**74**] , that will be useful for our study later.

Lemma 7.7 (Ovchinnikov). *Let $f\colon [a,b] \to \mathbb{R}$ be a piecewise affine function and let $\{a_0, a_1, \ldots, a_k\}$ and (m_i, b_i), $i = 1, \ldots, k$, be a representation of f. Also let $m = \frac{f(b)-f(a)}{b-a}$, the slope of the line segment joining the points $(a, f(a))$ and $(b, f(b))$.*

Then there exist some $1 \le i \le k$ with $m_i \ge m$ and some $a_{i-1} \le \xi \le a_i$ satisfying $f(\xi) = m(\xi - a) + f(a)$.

Proof. Assume by way of contradiction that the conclusion is false. That is, suppose that if $m_i \ge m$, then $f(t) \ne m(t-a)+f(a)$ for all $a_{i-1} \le t \le a_i$. In particular, we have $m_1 < m$. Given that for each $a \le t \le a_1$ we have $f(t) = m_1 t + b_1 = m_1(t-a)+f(a)$, the latter implies $f(t) < m(t-a)+f(a)$ for all $a < t \le a_1$. Notice that for each $a_1 \le t \le a_2$ we have

$$f(t) = m_2 t + b_2 = m_2(t - a_1) + f(a_1)\,.$$

So, if $m_2 < m$, then for each $a_1 \le t \le a_2$ we have $f(t) < m(t-a) + f(a)$. On the other hand, if $m_2 \ge m$, then for each $a_1 \le t \le a_2$ we must have $f(t) < m(t-a)+f(a)$; otherwise (by the intermediate value theorem) there should exist some $a_1 \le \xi \le a_2$ with $f(\xi) = m(\xi-a)+f(a)$, which contradicts our assumption. The same argument yields $f(t) < m(t-a) + f(a)$ for all $a_2 \le t \le a_3$. Continuing this way in an inductive manner, we see that $f(a_k) = f(b) < m(b-a) + f(a) = f(b)$, which is impossible. ∎

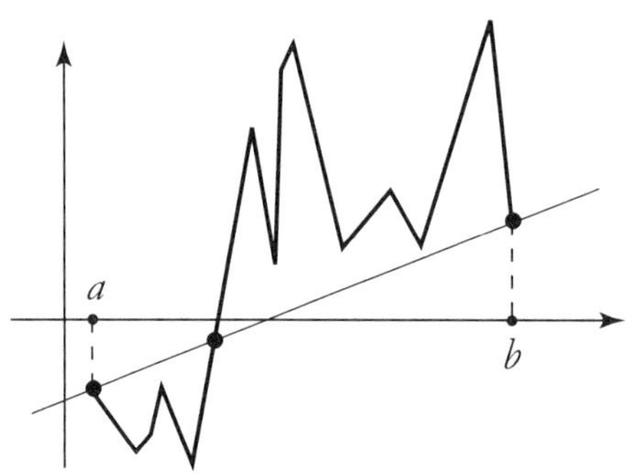

Figure 3. There exists a point ξ on the line segment joining $(a, f(a))$ and $(b, f(b))$ such that $(\xi, f(\xi))$ is on the line segment and the slope of f at ξ is greater than or equal to the slope of the line segment.

The geometric interpretation of Lemma 7.7 is depicted in Figure 3. As an immediate consequence we also get the following result.

Corollary 7.8 (Ovchinnikov). *Let $f\colon [a,b] \to \mathbb{R}$ be a piecewise affine function and let $\{a_0, a_1, \ldots, a_k\}$ and (m_i, b_i), $i = 1, \ldots, k$, be the parameters of a representation of f. Then there exists some $1 \le i \le k$ such that $f(a) \ge m_i a + b_i$ and $f(b) \le m_i b + b_i$.*

Proof. According to Lemma 7.7 there exist some $1 \le i \le k$ and some $a_{i-1} \le \xi \le a_i$ satisfying $m_i \ge m = \frac{f(b)-f(a)}{b-a}$ and $f(\xi) = m(\xi - a) + f(a)$. Note that for each $a_{i-1} \le t \le a_i$ we have $m_i t + b_i = m_i(t - \xi) + f(\xi)$ and that for all $a \le t \le b$ we have $m(t-a)+f(a) = m(t-\xi)+f(\xi)$. This implies $m_i t + b_i \le m(t-a) + f(a)$ for all $a \le t \le \xi$ and $m_i t + b_i \ge m(t-a) + f(a)$ for all $\xi \le t \le b$, and our conclusion follows. ■

Exercises

1. Show that every piecewise affine function as defined in Definition 7.1 is continuous.
2. Give a direct proof (i.e., without using the Stone–Weierstrass approximation theorem) of Lemma 7.3.
3. Prove Lemma 7.2.
4. Prove a version of Ovchinnikov's Lemma 7.7 for smooth one-dimensional functions. [*Hint*: Construct a sequence of piecewise affine functions using the derivatives.]

7.2. Multivariate piecewise affine functions

Recall that any function $f\colon \mathbb{R}^m \to \mathbb{R}$ of the form $f(x) = \alpha + a \cdot x$, where $\alpha \in \mathbb{R}$ is a constant and $a \in \mathbb{R}^m$ is a fixed vector, is called an **affine function**. As usual, an affine function f is **linear** if $\alpha = 0$, i.e., $f(x) = a \cdot x$. A function $f\colon S \to \mathbb{R}$, where S is a subset of $\mathbb{R}^m$, is said to be an **affine function** on S if it is the restriction of an affine function defined on $\mathbb{R}^m$. Let $\mathbf{Aff}$ denote the collection of all affine functions on $\mathbb{R}^m$ and note that $\mathbf{Aff}$ is a vector subspace of $C(\mathbb{R}^m)$.

Lemma 7.9. *Regarding affine functions, we have the following.*

(1) *The vector space* $\mathbf{Aff}$ *of all affine functions is the linear span in* $C(\mathbb{R}^m)$ *of the functions* $\{\mathbf{1}, e_1, e_2, \ldots, e_m\}$, *where* $\mathbf{1}(x) = 1$ *and* $e_i(x) = x_i$ *for all* $x = (x_1, x_2, \ldots, x_m) \in \mathbb{R}^m$. *That is, we have* $\mathbf{Aff} = \operatorname{Span}\{\mathbf{1}, e_1, e_2, \ldots, e_m\}$, *and so* $\mathbf{Aff}$ *is an* $(m+1)$*-dimensional vector space.*[2]

(2) *Two affine functions* $f, g \in \mathbf{Aff}$ *coincide if and only if* $f(x) = g(x)$ *for all* x *in a nonempty open subset of* $\mathbb{R}^m$. *In particular, if a subset* S *of* $\mathbb{R}^m$ *has an interior point, then any affine function on* S *is the restriction of a unique affine function defined on* $\mathbb{R}^m$.

[2] As a matter of fact, if we identify every vector $\mathbf{r} = (r_0, r_1, \ldots, r_m) \in \mathbb{R}^{m+1}$ with the affine function on $\mathbb{R}^m$ defined by $\mathbf{r}(x) = r_0 + r_1 x_1 + \cdots + r_m x_m$, then it is not difficult to see that we can identify $\mathbf{Aff}$ with the vector space $\mathbb{R}^{m+1}$.

Proof. The proof of part (1) is obvious. The proof of part (2) follows easily from the following simple property: *If a nonzero linear functional f satisfies $f(x) \geq \alpha$ for all x in a nonempty open set $\mathcal{O}$, then $f(x) > \alpha$ must be the case for all $x \in \mathcal{O}$.*

To see this, fix $x \in \mathcal{O}$ and assume that $f(x) = \alpha$. Since $\mathcal{O}$ is an open set, there exists some $\epsilon > 0$ such that $x + B(0, \epsilon) \subseteq \mathcal{O}$. So, for each $y \in B(0, \epsilon)$ we have $\alpha + f(y) = f(x + y) \geq \alpha$ or $f(y) \geq 0$. This implies $f(y) = 0$ for all $y \in B(0, \epsilon)$ and so $f = 0$, which is impossible. ■

We are now ready to introduce the general concept of piecewise affine function.

Definition 7.10. *A function $f: \mathbb{R}^m \to \mathbb{R}$ is called* ***piecewise affine*** *(or* ***piecewise linear****) if there exist distinct affine functions $f_1, f_2, \ldots, f_p$ and subsets $S_1, S_2, \ldots, S_p$ of $\mathbb{R}^m$ such that the following hold.*

(1) *Each S_i is closed with nonempty interior and $\overline{\operatorname{Int}(S_i)} = S_i$.*[3]
(2) *If $i \neq j$, then $\operatorname{Int}(S_i) \cap \operatorname{Int}(S_j) = \emptyset$.*
(3) $\bigcup_{i=1}^{p} S_i = \mathbb{R}^m$.
(4) *If $x \in S_i$, then $f(x) = f_i(x)$.*

We also introduce the following terminology and notation.

(a) *The sets S_i are called the* ***regions*** *of f and the functions f_i will be referred to as the* ***components*** *of f.*
(b) *The pairs $(S_1, f_1), \ldots, (S_p, f_p)$ are the* ***characteristic pairs*** *of f.*

We have a symbol for all piecewise affine functions.

Notation: The collection of all piecewise affine functions will be denoted by **PL**.

A remark is in order here. The same definition of a piecewise affine function can be given for **solid domains**, i.e., for closed convex subsets of $\mathbb{R}^m$ with nonempty interior. All results in this section hold true for piecewise affine functions with solid domains. We assume that our functions have domain $\mathbb{R}^m$ for the sole purpose of simplifying the exposition. The reader can verify directly that when $m = 1$ the definitions for piecewise affine functions given in Definitions 7.1 and 7.10 are equivalent; see also Corollary 7.18 below.

Here is an example of a piecewise affine function with a solid domain in $\mathbb{R}^2$.

[3] Observe that we do not assume that the sets S_i are connected.

Example 7.11. Consider the square

$$Q = [0,12] \times [0,12] = \{(x,y) \in \mathbb{R}^2\colon\ 0 \le x \le 12;\ 0 \le y \le 12\},$$

and define the function $f\colon Q \to \mathbb{R}$ by

$$f(x_1,x_2) = \begin{cases} x_1 - 5 & \text{if } x_2 \ge x_1 \text{ and } 2x_1 \ge 17 - x_2, \\ x_2 - 5 & \text{if } x_2 \le x_1 \text{ and } x_1 \ge 17 - 2x_2, \\ -x_1 - x_2 - 12 & \text{if } x_2 \ge x_1;\ 2x_1 \le 17 - x_2;\ 2x_1 \ge 17 - 2x_2, \\ & \text{or if } x_2 \le x_1;\ x_1 \le 17 - 2x_2;\ 2x_1 \ge 17 - 2x_2, \\ x_1 + x_2 - 5 & \text{if } 2x_1 \le 17 - 2x_2\,. \end{cases}$$

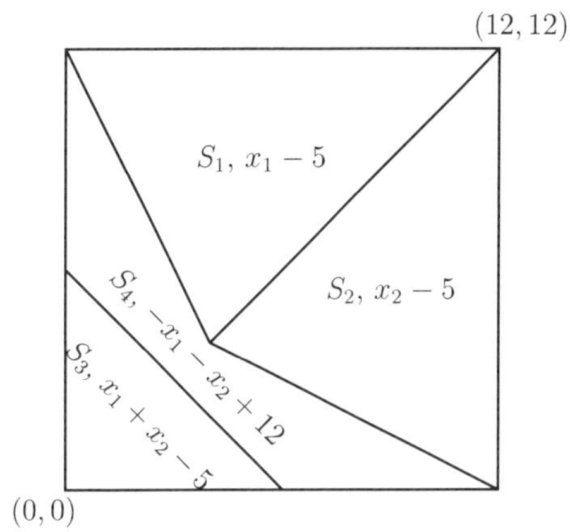

Figure 4. The regions of the function f.

The regions of this function are shown in Figure 4 and its graph is depicted in Figure 5. ■

The rest of the discussion in this section is devoted to the properties of piecewise affine functions.

Lemma 7.12. *Every piecewise affine function is continuous.*

Proof. Let $f\colon \mathbb{R}^m \to \mathbb{R}$ be piecewise affine and let $x_n \to x$. If $f(x_n) \not\to f(x)$, then (by passing to a subsequence) we can assume without loss of generality that there exists some $\epsilon > 0$ such that $|f(x_n) - f(x)| \ge \epsilon$ for each n. Now notice that there exist some i and a subsequence $\{y_n\}$ of $\{x_n\}$ satisfying $y_n \in S_i$ for each n. But then we have $\epsilon \le |f(y_n) - f(y)| = |f_i(y_n) - f_i(y)| \to 0$, which is impossible. This shows that f is continuous. ■

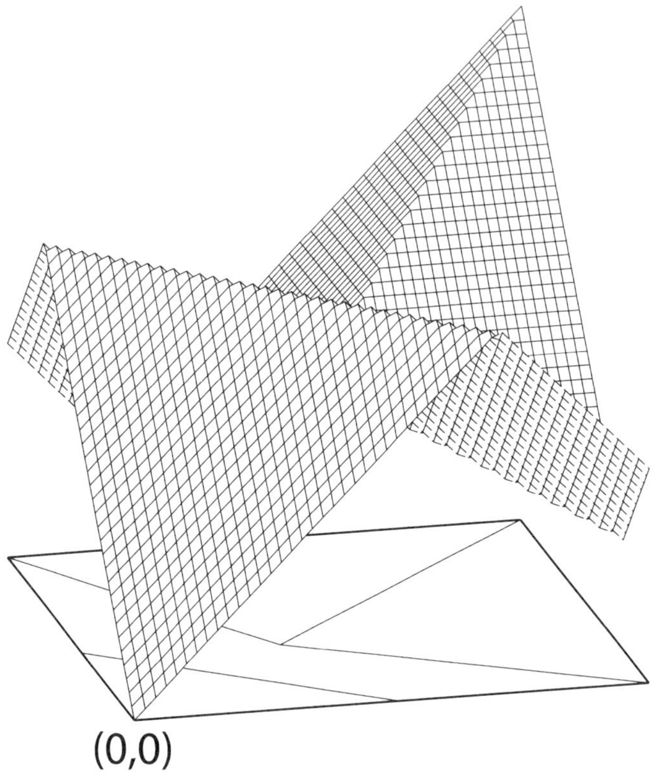

Figure 5. The graph of the function f of Example 7.11.

An extremely simple characterization of piecewise affine functions is included in the next result.

Theorem 7.13. *A continuous function* $f\colon \mathbb{R}^m \to \mathbb{R}$ *is piecewise affine if and only if there exist affine functions* $f_1, \ldots, f_k$ *such that for each* $x \in \mathbb{R}^m$ *there exists some* $1 \le i \le k$ *satisfying* $f(x) = f_i(x)$.

Moreover, the set of components of f *is a subcollection of the collection of affine functions* $\{f_1, \ldots, f_k\}$.

Proof. If f is piecewise affine, then the condition is trivially true. So, for the converse, assume that there exist affine functions $f_1, \ldots, f_k$ such that for each $x \in \mathbb{R}^m$ there exists some $1 \le i \le k$ such that $f(x) = f_i(x)$. We can assume that the affine functions $f_1, \ldots, f_k$ are distinct. We claim the following.

($\bullet$) *For each nonempty open subset* V *of* $\mathbb{R}^m$ *there exist a nonempty open subset* W *of* V *and some* $1 \le i \le k$ *such that* $f = f_i$ *on* W.

To see this, assume by way of contradiction that the claim is false. This implies that $f \neq f_1$ on V, i.e., $f_1(v) \neq f(v)$ for some $v \in V$. Since f and f_1 are continuous, there exists some nonempty open subset V_1 of V such that $f_1(x) \neq f(x)$ for all $x \in V_1$. Similarly, since (by our hypothesis) $f \neq f_2$ on V_1, there exists some nonempty open subset V_2 of V_1 such that $f_2(x) \neq f(x)$ for all $x \in V_2$. Continuing this way, we see that there exist nonempty open sets $V_k \subseteq V_{k-1} \subseteq \cdots \subseteq V_1 \subseteq V$ such that for each $1 \leq i \leq k$ we have $f_i(x) \neq f(x)$ for all $x \in V_i$. But then for each $x \in V_k$ we have $f(x) \neq f_i(x)$ for all $1 \leq i \leq k$, contrary to our assumption, and our claim has been established.

Now for each $1 \leq i \leq k$ let

$$\mathcal{O}_i = \bigcup \{U\colon \ U \text{ is an open subset of } \mathbb{R}^m \text{ and } f = f_i \text{ on } U\}.$$

That is, $\mathcal{O}_i$ is the largest open set on which $f = f_i$. By the preceding discussion $\mathcal{O}_i \neq \emptyset$ for at least one i. (To see this, take $V = \mathbb{R}^m$ and apply (•).) Deleting the $\mathcal{O}_i$ with $\mathcal{O}_i = \emptyset$, we can assume that $\mathcal{O}_i \neq \emptyset$ for each i. Put $S_i = \overline{\mathcal{O}_i}$, and note that $f = f_i$ on S_i. We shall verify that the closed sets $S_1, \ldots, S_k$ satisfy the conditions of Definition 7.10.

Start by observing that condition (4) is obvious. For (1) note that from $\mathcal{O}_i \subseteq S_i$, we get that $\operatorname{Int}(S_i) \neq \emptyset$ and that $\mathcal{O}_i \subseteq \operatorname{Int}(S_i)$. Moreover, $\mathcal{O}_i = \operatorname{Int}(S_i)$ must be the case, since otherwise the maximality property of $\mathcal{O}_i$ will be violated. The condition $\mathcal{O}_i \cap \mathcal{O}_j = \emptyset$ for $i \neq j$ should be obvious and the validity of (2) follows. If $\bigcup_{i=1}^k S_i \neq \mathbb{R}^m$, then by the above discussion there exists some nonempty open subset Q of $\mathbb{R}^m \setminus \bigcup_{i=1}^k S_i$ and some $1 \leq \ell \leq k$ such that $f = f_\ell$ on Q. But then the open set $\mathcal{O}_\ell \cup Q$ violates the maximality property of $\mathcal{O}_\ell$. Hence, $\bigcup_{i=1}^k S_i = \mathbb{R}^m$.

That the components of f are among the affine functions $f_1, \ldots, f_k$ should be obvious from the above discussion. ■

An immediate consequence of the preceding result is that **PL** is a Riesz subspace.

Corollary 7.14. *The collection of all piecewise affine functions on $\mathbb{R}^m$ is a Riesz subspace of $C(\mathbb{R}^m)$. In particular, we have*

$$\mathcal{R}(\mathbf{Aff}) = \mathbf{Aff}^{\vee\wedge} = \mathbf{Aff}^{\wedge\vee} \subseteq \mathbf{PL}.$$

Recall that an **affine transformation** from $\mathbb{R}^k$ to $\mathbb{R}^m$ is any function $T\colon \mathbb{R}^k \to \mathbb{R}^m$ of the form $T(t) = At + b$, where A is an $m \times k$ real matrix and $b \in \mathbb{R}^m$ is a fixed vector. Now if T is an affine transformation and $f\colon \mathbb{R}^m \to \mathbb{R}$ is an affine function, then the function $f \circ T\colon \mathbb{R}^k \to \mathbb{R}$ is also an affine function. To see this, assume that f is defined by $f(x) = \alpha + u \cdot x$

and note that for each $t \in \mathbb{R}^k$ we have

$$[f \circ T](t) = f(T(t)) = \alpha + u \cdot (At + b) = (\alpha + u \cdot b) + (A'u) \cdot t\,.$$

This conclusion in connection with Theorem 7.13 yields the following result.

Corollary 7.15. *Let $f\colon \mathbb{R}^m \to \mathbb{R}$ be a piecewise affine function and assume that $T\colon \mathbb{R}^k \to \mathbb{R}^m$ is an affine transformation. Then the function $f \circ T\colon \mathbb{R}^k \to \mathbb{R}$ is piecewise affine. Moreover, if f has the components $f_1, \ldots, f_p$, then the components of $f \circ T$ are among the affine functions $f_1 \circ T, \ldots, f_p \circ T$.*

In particular, for any two fixed vectors $a, b \in \mathbb{R}^m$ the function $\theta\colon \mathbb{R} \to \mathbb{R}$, defined by $\theta(t) = f(ta + (1-t)b)$, is (one-dimensional) piecewise affine.

A **hyperplane** of $\mathbb{R}^m$ is any subset of the form $H = \{x \in \mathbb{R}^m\colon\ a{\cdot}x = \alpha\}$, where $a \in \mathbb{R}^m$ is a nonzero vector and $\alpha \in \mathbb{R}$ is a constant. Clearly, every hyperplane is a closed set with Lebesgue measure zero. Notice that two affine functions $f, g\colon \mathbb{R}^m \to \mathbb{R}$ either do not agree at any point or the set where they agree is a hyperplane, i.e., the set $[f = g] = \{x \in \mathbb{R}^m\colon\ f(x) = g(x)\}$ is either empty or a hyperplane.

It turns out that the boundaries of the regions of a piecewise affine function are parts of hyperplanes.

Lemma 7.16. *Let $(S_1, f_1), \ldots, (S_p, f_p)$ be the characteristic pairs of a piecewise affine function $f\colon \mathbb{R}^m \to \mathbb{R}$. For each i let*

$$\mathcal{I}_i = \{j \in \{1, \ldots, p\}\colon\ j \neq i \text{ and } S_i \cap S_j \neq \emptyset\}\,.$$

Then the boundary of the region S_i has the following property:

$$\partial S_i = \bigcup_{j \in \mathcal{I}_i} S_i \cap S_j \subseteq \bigcup_{j \in \mathcal{I}_i} [f_i = f_j]\,.$$

In particular,

(a) *each boundary ∂S_i has Lebesgue measure zero and consists of "parts" of hyperplanes, and*

(b) *if $x \in \operatorname{Int}(S_i)$ for some i, then $x \notin S_j$ for all $j \neq i$.*

Proof. Let $x \in \partial S_i$. Since $B(x, \frac{1}{n}) \cap (\mathbb{R}^m \setminus S_i) \neq \emptyset$, there exists for each n some $x_n \in \bigcup_{r \neq i} S_r$ such that $x_n \in B(x, \frac{1}{n})$. It follows that for some $j \neq i$ we have $x_n \in S_j$ for infinitely many n. This implies $x \in \overline{S_j} = S_j$ and so $x \in S_i \cap S_j$.

Now assume that $x \in S_i \cap S_j$ for some $j \neq i$. If $x \notin \partial S_i$, then we have $x \in \operatorname{Int}(S_i)$ and so there exists some $\delta > 0$ such that $B(x, \delta) \subseteq \operatorname{Int}(S_i)$. Since $\operatorname{Int}(S_i) \cap \operatorname{Int}(S_j) = \emptyset$, we infer that $x \in \partial S_j$. From $\overline{\operatorname{Int}(S_j)} = S_j$, it follows that there exists some $y \in \operatorname{Int}(S_j)$ such that $y \in B(x, \delta)$. This implies $y \in \text{Int}(S_i) \cap \text{Int}(S_j)$, which is impossible. Consequently, $x \in \partial S_i$, and the proof is finished. ∎

As expected, the characteristic pairs of a piecewise affine function are uniquely determined.

Lemma 7.17. *The regions and the components of any piecewise affine function* $f\colon \mathbb{R}^m \to \mathbb{R}$ *are uniquely determined in the following sense: If another collection of pairs* $\{(S'_1, g_1), \ldots, (S'_q, g_q)\}$ *satisfies properties* (1)–(4) *of Definition* 7.10, *then* $q = p$ *and* $\{(S'_1, g_1), (S'_2, g_2), \ldots, (S'_q, g_q)\}$ *is a permutation of the collection of pairs* $\{(S_1, f_1), (S_2, f_2), \ldots, (S_p, f_p)\}$.

Proof. Fix some $1 \le i \le p$. Since $\text{Int}(S_i)$ is nonempty (and hence it has positive Lebesgue measure), it follows from Lemma 7.16 that there exists some $1 \le j \le q$ such that the open set $V = \text{Int}(S_i) \cap \text{Int}(S'_j)$ is nonempty. In particular, since $f_i(x) = g_j(x) = f(x)$ holds true for each $x \in V$, it follows from part (2) of Lemma 7.9 that $f_i = g_j$.

Now let $x \in \text{Int}(S_i)$. Fix $\delta > 0$ such that $B(x, \delta) \subseteq \text{Int}(S_i)$ and let $0 < \epsilon < \delta$. As above, $B(x, \epsilon) \cap \text{Int}(S'_r) \neq \emptyset$ must hold true for some index $1 \le r \le q$. But then (as above again) $g_j = f_i = g_r$ must be the case. Since the affine functions $g_1, \ldots, g_q$ are all distinct, we infer that $r = j$. Therefore, $B(x, \epsilon) \cap \text{Int}(S'_j) \neq \emptyset$ for all $0 < \epsilon < \delta$. This implies $x \in \overline{S'_j} = S'_j$, and so $\text{Int}(S_i) \subseteq S'_j$. Consequently, $S_i = \overline{\text{Int}(S_i)} \subseteq S'_j$.

By the symmetry of the situation, there exists some $1 \le t \le p$ such that $S'_j \subseteq S_t$. This implies $\text{Int}(S_i) \cap \text{Int}(S_t) = \text{Int}(S_i) \neq \emptyset$, from which it follows that $t = i$. Therefore, $S_i = S'_j$ and so $(S_i, f_i) = (S'_j, g_j)$. From the last result, the desired conclusion now easily follows. ∎

Another consequence of Theorem 7.13 is that for real functions defined on $\mathbb{R}$ the definitions for piecewise affine functions given in Definitions 7.1 and 7.10 are equivalent.

Corollary 7.18. *A function* $f\colon \mathbb{R} \to \mathbb{R}$ *is piecewise affine according to Definition* 7.1 *if and only if it is piecewise affine according to Definition* 7.10.

Proof. Let $f\colon \mathbb{R} \to \mathbb{R}$ be a function. If f is piecewise affine according to Definition 7.1, then f is clearly piecewise affine according to Definition 7.10.

For the converse, assume that f is piecewise affine according to Definition 7.10. Let $\{(S_1, f_1), (S_2, f_2), \ldots, (S_p, f_p)\}$ be the collection of characteristic pairs of f. Notice that every f_i is of the form $f_i(t) = m_i t + b_i$. So, every nonempty set of the form $[f_i = f_j]$ is simply a point of $\mathbb{R}$. This, in connection with Lemma 7.16, shows that the boundary of each S_i is a finite set. Now each $\text{Int}(S_i)$ is the union of an at most countable collection of pairwise disjoint open intervals. Since ∂S_i is a finite set, a moment's thought reveals that $\mathrm{Int}(S_i)$ is a union of a finite number of pairwise disjoint open intervals. From this it follows that S_i is the union of the closures of these intervals. Now it is easy to see that f is a piecewise affine function according to Definition 7.1. ■

Recall once more that any set of the form $H = \{x \in \mathbb{R}^m\colon\ a \cdot x = \alpha\}$, where $a \in \mathbb{R}^m$ is a nonzero fixed vector and $\alpha \in \mathbb{R}$ is a constant, is called a **hyperplane** of $\mathbb{R}^m$. We can assume without loss of generality that $\|a\| = 1$ and refer to a as a (unit) vector **normal** to H. Since we can also write $H = \{x \in \mathbb{R}^m\colon\ (-a) \cdot x = -\alpha\}$, we see that $-a$ is also another (unit) normal vector to H. In other words, H has essentially two unit normal vectors, each of which defines an **orientation** in the sense that it divides $\mathbb{R}^m$ into three parts: a "positive" part $\{x \in \mathbb{R}^m\colon\ a \cdot x > \alpha\}$, a "zero" part $\{x \in \mathbb{R}^m\colon\ a \cdot x = \alpha\}$, and a "negative" part $\{x \in \mathbb{R}^m\colon\ a \cdot x < \alpha\}$. Of course, if we let $H = \{x \in \mathbb{R}^m\colon\ (-a) \cdot x = -\alpha\}$, then the orientation changes: the positive part is now negative and the negative part is positive. Thus, writing H in the form $H = \{x \in \mathbb{R}^m\colon\ a \cdot x = \alpha\}$, the vector a defines automatically an orientation, and H is called an **oriented hyperplane**.[4]

Now let E be an arbitrary finite index set and let $\{H_e\}_{e \in E}$, where, as usual, $H_e = \{x \in \mathbb{R}^m\colon\ a_e \cdot x = \alpha_e\}$, be a family of (oriented) hyperplanes in $\mathbb{R}^m$. The family $\{H_e\}_{e \in E}$ is called an **oriented arrangement of hyperplanes** (or simply an **arrangement**). Every arrangement of hyperplanes $\{H_e\}_{e \in E}$ "almost" subdivides $\mathbb{R}^m$ into a finite number of nonempty convex regions. The subdivisions are obtained by means of the **sign mapping** $x \mapsto \sigma_x$, from $\mathbb{R}^m$ to $\{+, -, 0\}^E$, that is defined by

$$\sigma_x(e) = \begin{cases} + & \text{if} \quad a_e \cdot x > \alpha_e\,, \\ - & \text{if} \quad a_e \cdot x < \alpha_e\,, \\ 0 & \text{if} \quad a_e \cdot x = \alpha_e\,, \end{cases}$$

i.e., $\sigma_x = (\mathrm{Sign}\,(a_e \cdot x - \alpha_e))_{e \in E}$. Let $\mathcal{M}$ denote the range of the function σ, that is, $\mathcal{M} = \sigma(\mathbb{R}^m) \subseteq \{+, -, 0\}^E$.

A vector $T \in \mathcal{M}$ satisfying $T(e) \neq 0$ for all $e \in E$ is called a **tope** of $\mathcal{M}$. Note that σ_x is a tope if and only if $x \notin \bigcup_{e \in E} H_e$. Let $T_1, T_2 \ldots, T_J$ be an enumeration of the topes of $\mathcal{M}$. For each $1 \leq h \leq J$ let

$$K_h = \{x \in \mathbb{R}^m\colon\ \sigma_x = T_h\} = \sigma^{-1}(\{T_h\})\,.$$

Obviously, each K_h is a nonempty, open and convex set. Moreover, from the identity $\bigcup_{h=1}^J K_h = \mathbb{R}^m \setminus \bigcup_{e \in E} H_e$, we see that $\overline{\bigcup_{h=1}^J K_h} = \mathbb{R}^m$. The sets $K_1, K_2, \ldots, K_J$ are called the **cells** induced by the arrangement of the family

[4] For an extensive treatment of the theory of oriented arrangements of hyperplanes, which are well studied combinatorial constructions that are closely related to vector lattices and the simplex method in linear programming, see [**31**, Chapter 4].

of hyperplanes $\{H_e\}_{e\in E}$. It should not be difficult to see that the collection of cells $\{K_1, K_2, \dots, K_J\}$ is independent of the orientation of the family of hyperplanes $\{H_e\}_{e\in E}$, and so we can refer to $\{K_1, K_2, \dots, K_J\}$ as the **collection of cells generated** (or **induced**) by the family of hyperplanes $\{H_e\}_{e\in E}$. For an example of an arrangement of hyperplanes see Figure 6.

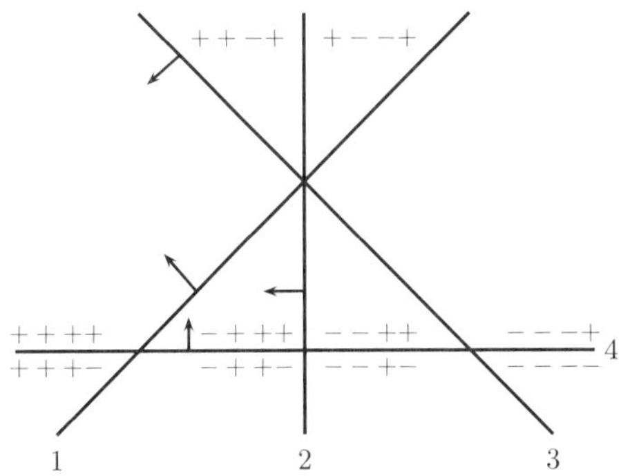

Figure 6. An arrangement of four oriented hyperplanes in $\mathbb{R}^2$.

Now let $\{f_1, \dots, f_p\}$, where $p \geq 2$, be a collection of distinct affine functions on $\mathbb{R}^m$. If for each $1 \leq i < j \leq p$ we let $H_{i,j} = [f_i = f_j]$, then the set

$$E = \{(i,j)\colon\ 1 \leq i < j \leq p \text{ and } H_{i,j} \neq \emptyset\}$$

is finite. Letting $H_e = [f_i = f_j] = \{x \in \mathbb{R}^m\colon\ a_e \cdot x = \alpha_e\}$ for each $e = (i,j) \in E$, we see that the family $\{H_e\}_{e\in E}$ is an arrangement of hyperplanes, called **an arrangement generated** by $\{f_1, \dots, f_p\}$. The collection of cells generated by $\{H_e\}_{e\in E}$ is called the **collection of cells generated** (or **induced**) by $\{f_1, \dots, f_p\}$.

With this terminology at hand, we are now ready to state several extra properties of piecewise affine functions.

Lemma 7.19. *Let $F = \{f_1, \dots, f_k\}$ be a finite collection of distinct affine functions of $\mathbb{R}^m$ and let $\{K_1, K_2, \dots, K_J\}$ be the collection of cells induced by F. Assume also that $f\colon \mathbb{R}^m \to \mathbb{R}$ is a continuous function such that for each $x \in \mathbb{R}^m$ there exists some $1 \leq i \leq k$ satisfying $f(x) = f_i(x)$.*[5] *Then for a vector $x \in K_h$ we have the following.*

(1) *If $f(x) = f_i(x)$, then $f(y) = f_i(y)$ for all $y \in K_h$.*
(2) *If $f(x) > f_i(x)$, then $f(y) > f_i(y)$ for all $y \in K_h$.*
(3) *If $f(x) < f_i(x)$, then $f(y) < f_i(y)$ for all $y \in K_h$.*

[5] Keep in mind that this implies (by Theorem 7.13) that f is piecewise linear.

Moreover, for each $1 \leq h \leq J$ there exists a unique $1 \leq i_h \leq k$ with $f = f_{i_h}$ on K_h.

Proof. We shall prove (1) first. To this end, suppose that some $x \in K_h$ satisfies $f(x) = f_i(x)$.

Let $X = \bigcup_{h=1}^{J} K_h$ and note that X is an open dense subset of $\mathbb{R}^m$. Notice that for each $z \in X$ and any pair of distinct functions $f_i, f_j \in F$ we have $f_i(z) \neq f_j(z)$. So for each $z \in X$ there exists a unique $1 \leq i_z \leq k$ such that $f(z) = f_{i_z}(z)$. Since f and the f_i are continuous functions and $f(z) = f_{i_z}(z) \neq f_j(z)$ for each $j \neq i_z$, there exists an open neighborhood $N_z \subseteq X$ of z such that for each $y \in N_z$ and all $j \neq i_z$ we have $f(y) \neq f_j(y)$ and $f_{i_z}(y) \neq f_j(y)$. This implies that for each $y \in N_z$ we have $f(y) = f_{i_z}(y)$, i.e., $i_y = i_z$.

Now fix $y \in K_h$. Let $L(x, y)$ be the line segment joining x and y and notice that since K_h is convex, $L(x, y) \subseteq K_h$. Since $L(x, y)$ is compact, there exists a finite set $Z = \{z_1, \ldots, z_r\} \subseteq L(x, y)$ such that $L(x, y) \subseteq \bigcup_{z \in Z} N_z$. We can assume that the neighborhoods $\{N_z \colon z \in Z\}$ form a chain, i.e., $N_{z_t} \cap N_{z_{t+1}} \neq \emptyset$ for each $t = 1, \ldots, r-1$; see [**4**, Problem 1.5.7, p. 50]. This easily implies that for each $z \in L(x, y)$ we have $f(z) = f_{i_x}(z) = f_{i_y}(z)$. In particular, we obtain $i_x = i_y$.

Therefore, we have shown that for each K_h there exists a unique index $1 \leq i_h \leq k$ such that $y \in K_h$ implies $f(y) = f_{i_h}(y)$. This proves (1) and the last part of the lemma.

To establish (2), assume that $f(x) > f_i(x)$ holds true for some $x \in K_h$ and that some other $y \in K_h$ satisfies $f(y) \leq f_i(y)$. If, on the other hand, $f(y) = f_i(y)$, then according to (1) we must have $f(x) = f_i(x)$, which is impossible. If $f(y) < f_i(y)$, then there exists some z in the line segment joining x and y (and hence $z \in K_h$) satisfying $f(z) = f_i(z)$. But then, according to (1) again, we get $f(x) = f_i(x)$, a contradiction. This establishes (2) and the validity of (3) can be proven in a similar fashion. ■

From Theorem 7.13 we know that if for a continuous function $f \colon \mathbb{R}^m \to \mathbb{R}$ and affine functions $f_1, \ldots, f_k$ for each $x \in \mathbb{R}^m$ there exists some $1 \leq i \leq k$ satisfying $f(x) = f_i(x)$, then f is piecewise affine. The next result constructs the characteristic pairs of such a piecewise affine function from a given collection of affine functions.

Theorem 7.20. *Assume that a continuous function $f \colon \mathbb{R}^m \to \mathbb{R}$ and a finite set of distinct affine functions $F = \{f_1, \ldots, f_k\}$ are such that for each $x \in \mathbb{R}^m$ there exists some $1 \leq i \leq k$ satisfying $f(x) = f_i(x)$. Let $\{K_1, K_2, \ldots, K_J\}$ be the cells generated by F. For each $1 \leq i \leq k$ let*

$$E_i = \big\{h \in \{1, \ldots, J\} \colon \ f = f_i \ \text{ on } K_h\big\},$$

and then define $S_i = \overline{\bigcup_{h\in E_i} K_h}$. *We have the following.*

(a) *If* $\{E_i\}_{i\in\mathcal{I}}$ *is the family of nonempty* E_i, *then the finite family* $\{(S_i, f_i)\}_{i\in\mathcal{I}}$ *is precisely the family of characteristic pairs of the piecewise linear function* f.

(b) *For each* $1 \le h \le J$ *there exists exactly one index* $i \in \mathcal{I}$ *such that* $K_h \subseteq \operatorname{Int}(S_i)$.

(c) *For each* $i \in \mathcal{I}$ *the nonempty set* $\operatorname{Int}(S_i)$ *is a union of a finite collection of pairwise disjoint nonempty open and connected subsets of* $\mathbb{R}^m$.

Proof. (a) We know from Theorem 7.13 that the function f is piecewise affine whose components are among the $f_1, \ldots, f_k$. The proof below will also present an alternate constructive proof of Theorem 7.13. Let $\{K_1, \ldots, K_J\}$ be the collection of cells generated by F and for each $1 \le i \le k$ define E_i and S_i as in the statement of this theorem.

According to Lemma 7.19, at least one of the E_i is nonempty; relabeling, we can assume that $E_1, \ldots, E_p$ are the nonempty E_i, i.e., $\mathcal{I} = \{1, \ldots, p\}$. Clearly, $f = f_i$ on S_i. Since the affine functions $f_1, \ldots, f_k$ are distinct, it follows from part (2) of Lemma 7.9 that $E_r \cap E_s = \emptyset$ for $r \ne s$ and from Lemma 7.19 we see that $\bigcup_{i=1}^p E_i = \{1, \ldots, J\}$. The latter yields

$$\bigcup_{i=1}^{p} S_i = \bigcup_{i=1}^{p} \overline{\bigcup_{h\in E_i} K_h} = \overline{\bigcup_{i=1}^{p} \bigcup_{h\in E_i} K_h} = \overline{\bigcup_{h=1}^{J} K_h} = \mathbb{R}^m .$$

Next notice that since for each $1 \le i \le p$ we have $\bigcup_{h\in E_i} K_h \subseteq \operatorname{Int}(S_i)$, it follows, on one hand, that $\operatorname{Int}(S_i) \ne \emptyset$ and, on the other hand, that $\overline{\operatorname{Int}(S_i)} = S_i$. Moreover, using part (2) of Lemma 7.9, it is not difficult to see that $\operatorname{Int}(S_r) \cap \operatorname{Int}(S_s) = \emptyset$ for $r \ne s$. Since $f = f_i$ holds true for each $1 \le i \le p$, it follows from Definition 7.10 that f is a piecewise affine function with characteristic pairs $\{(S_1, f_1), \ldots, (S_p, f_p)\}$.

(b) Now let $1 \le h \le J$. According to Lemma 7.19 there exists a unique $1 \le i_h \le k$ such that $f = f_{i_h}$ on K_h. This implies that $E_{i_h} \ne \emptyset$ and $K_h \subseteq \operatorname{Int}(S_{i_h})$.

(c) Observe that for each $i \in \mathcal{I}$ every component of $\operatorname{Int}(S_i)$, i.e., every maximal (with respect to $\supseteq$) nonempty and connected subset of $\operatorname{Int}(S_i)$, is open. Now notice that every $K_h \subseteq \operatorname{Int}(S_i)$ is open and connected (being a convex set) and so is included in some component of $\operatorname{Int}(S_i)$. Moreover, from the definition of S_i, we see that every component of $\operatorname{Int}(S_i)$ includes some K_h. Thus, the number of components of $\operatorname{Int}(S_i)$ is at most J, and the proof is finished. ∎

To continue our study, we need one more property of piecewise affine functions also due to S. Ovchinnikov [**74**].

Lemma 7.21 (Ovchinnikov). *If $f\colon \mathbb{R}^m \to \mathbb{R}$ is a piecewise affine function with components $f_1, \ldots, f_p$, then for any pair $a, b \in \mathbb{R}^m$ there exists a component f_i of f satisfying $f_i(a) \le f(a)$ and $f_i(b) \ge f(b)$.*

Proof. Fix $a, b \in \mathbb{R}^m$ and consider the continuous function $g\colon \mathbb{R} \to \mathbb{R}$ defined via the formula by $g(t) = f(tb + (1-t)a)$. By Corollary 7.15, g is a one-dimensional piecewise affine function whose components are among the affine functions $g_1, g_2, \ldots, g_p$, where $g_i(t) = f_i(tb + (1-t)a)$. Consider g restricted to $[0,1]$ and then use Lemma 7.2 in conjunction with Corollary 7.8 to see that there exists a component g_i satisfying $f(a) = g(0) \ge g_i(0) = f_i(a)$ and $f(b) = g(1) \le g_i(1) = f_i(b)$. ∎

The next result presents the basic structural properties of piecewise affine functions.

Theorem 7.22. *Assume that $f\colon \mathbb{R}^m \to \mathbb{R}$ is a piecewise affine function with characteristic pairs $\{(S_1, f_1), \ldots, (S_p, f_p)\}$ and let $\{K_1, K_2, \ldots, K_J\}$ be the set of cells induced by $\{f_1, \ldots, f_p\}$.*

(1) *If for each h we fix some $x_h \in K_h$ and let*

$$E_h = \big\{ i \in \{1, \ldots, p\} \colon \ f_i(x_h) \ge f(x_h) \big\},$$

then E_h is nonempty and

$$f = \bigvee_{h=1}^{J} \bigwedge_{i \in E_h} f_i \,.$$

In particular, $f \in \{f_1, f_2, \ldots, f_p\}^{\vee\wedge}$.

(2) *If J^* is the subset of $\{1, \ldots, J\}$ having the property that for each $1 \le h \le J$ there exists some $j \in J^*$ such that $E_j \subseteq E_h$, then we have*

$$f = \bigvee_{j \in J^*} \bigwedge_{i \in E_j} f_i \,.$$

Proof. (1) For each $1 \le h \le J$ fix some $x_h \in K_h$ and then use Theorem 7.20 to choose some $1 \le j \le p$ such that $K_h \subseteq \operatorname{Int}(S_j)$. Clearly, $f_j(x_h) = f(x_h)$. This implies that if for each $1 \le h \le J$ we let

$$E_h = \big\{ i \in \{1, \ldots, p\} \colon \ f_i(x_h) \ge f(x_h) \big\},$$

then, on one hand, $E_h \ne \emptyset$ and, on the other hand, a glance at Lemma 7.19 guarantees that for each $i \in E_h$ and each $y \in K_h$ we have $f_i(y) \ge f(y)$. Now for each $1 \le h \le J$ consider the function

$$F_h = \bigwedge_{i \in E_h} f_i\,, \qquad (\star)$$

and note that $F_h(y) \ge f(y)$ for each $y \in K_h$. Since for some $j \in E_h$ we have $f_j(x_h) = f(x_h)$, it follows from Lemma 7.19 that $f_j(y) = f(y)$ for all $y \in K_h$. Thus, $F_h(y) = f(y)$ for all $y \in K_h$.

Next, fix $y \in \mathbb{R}^m$. For each $1 \le h \le J$ there exists (according to Lemma 7.21) some f_j satisfying $f_j(y) \le f(y)$ and $f_j(x_h) \ge f(x_h)$. In particular, $j \in E_h$ and consequently $F_h(y) = \big[\bigwedge_{i \in E_h} f_i\big](y) \le f(y)$ for all $1 \le h \le J$. This implies $\big[\bigvee_{h=1}^{J} F_h\big](y) \le f(y)$ for all $y \in \mathbb{R}^m$.

On the other hand, since for each $x \in K_h$ we have $F_h(x) = f(x)$, it must be the case that

$$\bigvee_{h=1}^{J} F_h = \bigvee_{h=1}^{J} \bigwedge_{i \in E_h} f_i = f\,,$$

on $\bigcup_{h=1}^{J} K_h$. Since $\bigcup_{h=1}^{J} K_h$ is dense in $\mathbb{R}^m$, and $\bigvee_{h=1}^{J} \bigwedge_{i \in E_h} f_i$ and f are both continuous functions, we see that $\bigvee_{h=1}^{J} \bigwedge_{i \in E_h} f_i = f$ holds true on $\mathbb{R}^m$.

(2) To establish this identity, note first that if $E_j \subseteq E_h$, then

$$\bigwedge_{i \in E_h} f_i \le \bigwedge_{i \in E_j} f_i \le f\,.$$

This implies $\bigwedge_{i \in E_h} f_i \le \bigvee_{j \in J^*} \bigwedge_{i \in E_j} f_i \le f$ for each $1 \le h \le J$, and consequently we have $f = \bigvee_{h=1}^{J} \bigwedge_{i \in E_h} f_i \le \bigvee_{j \in J^*} \bigwedge_{i \in E_j} f_i \le f$, and the proof is finished. ■

Combining Corollary 7.14 and Theorem 7.22, we are now ready to state the fundamental result of this chapter.

Theorem 7.23. *The vector space* **PL** *of all piecewise affine functions is a vector sublattice of the Riesz space $C(\mathbb{R}^m)$ and coincides with* $\mathbf{Aff}^{\vee\wedge}$, *i.e.,*

$$\mathbf{PL} = \mathbf{Aff}^{\vee\wedge}\,.$$

In other words, **PL** *is precisely the Riesz subspace of the Riesz space $C(\mathbb{R}^m)$ that is generated by the $(m+1)$-dimensional vector subspace* **Aff** *of all affine functions.*

Noting that the set $\{f_1, f_2, \dots, f_p\}^{\vee\wedge}$ is finite, Theorem 7.23 yields also the following.

Corollary 7.24. *If $F = \{f_1, f_2, \dots, f_p\}$ is a finite set of affine functions on $\mathbb{R}^m$, then a function $f \in C(\mathbb{R}^m)$ is piecewise affine with components in F if and only if f belongs to the finite set $F^{\vee\wedge}$.*

Theorem 7.22 also provides a proceedure for constructing the sup-inf representation of a piecewise affine function with components $f_1, f_2, \ldots, f_p$ and unknown regions. The next example is a rudimentary procedure illustrating this.

Example 7.25 (From **PL** to $\mathbf{Aff}^{\vee\wedge}$). Consider a function $f \in \mathbf{PL}$ with components $f_1, f_2, \ldots, f_p$. According to Theorem 7.22 the function f can be reconstructed following the steps below:

Step I: Determine $E = \{(i,j)\colon\ 1 \le i < j \le p \text{ and } [f_i = f_j] \ne \emptyset\}$ and then for each $e = (i,j) \in E$ pick $\alpha_e \in \mathbb{R}$ and $a_e \in \mathbb{R}^m$ such that

$$H_e = \{x \in \mathbb{R}^m\colon\ a_e \cdot x = \alpha_e\} = [f_i = f_j]\,.$$

Step II: Using the hyperplane arrangement $\{H_e\}_{e\in E}$, determine the cells $K_1, \ldots, K_J$.

Step III: For each $h = 1, 2, \ldots, J$ choose some x_h from the cell K_h.

Step IV: For each $1 \le h \le J$ determine

$$E_h = \{i \in \{1, \ldots, p\}\colon\ f_i(x_h) \ge f(x_h)\}\,.$$

Step V: Select a "minimal" set $J^* \subseteq \{1, 2, \ldots, J\}$ so that it satisfies property (2) of Theorem 7.22. If this is the case, then

$$f = \bigvee_{j\in J^*} \bigwedge_{i\in E_j} f_i\,.$$

The above procedure gives a desired sup-inf representation of f. ■

The next example illustrates the preceding procedure. It also shows how in applying this procedure, we can restrict our attention to a closed convex domain with nonempty interior.

Example 7.26. Consider once again Example 7.11 but with the restricted domain shown in Figure 7. Take the four affine components of the function f:

$$\begin{aligned} f_1(x_1,x_2) &= x_1 - 5\,,\\ f_2(x_1,x_2) &= x_2 - 5\,,\\ f_3(x_1,x_2) &= x_1 + x_2 - 5\,,\\ f_4(x_1,x_2) &= -x_1 - x_2 + 12\,. \end{aligned}$$

These four affine functions induce eight cells. They are the eight regions of the arrangement in Step I of the procedure of Example 7.25 and they are depicted in Figure 7.

Notice that $E_1 = \{1,2,3\}$, $E_2 = \{1,2,3\}$, and

$$E_3 = E_4 = E_5 = E_6 = E_7 = E_8 = \{3,4\}\,.$$

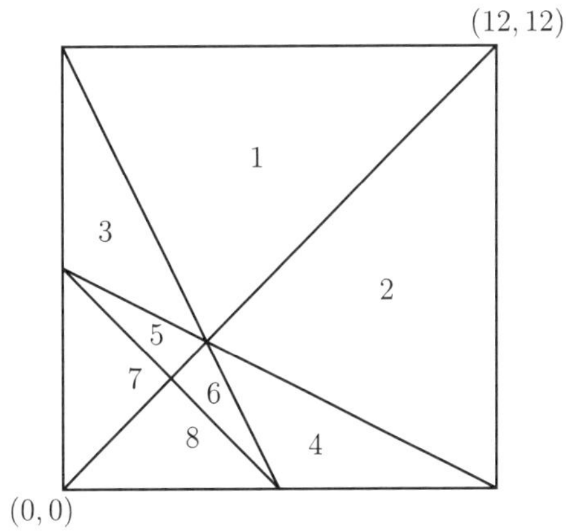

Figure 7. The eight regions of the hyperplane arrangement.

Therefore, if we take $J^* = \{1, 3\}$, then we can write

$$f = (f_3 \wedge f_4) \vee (f_1 \wedge f_2 \wedge f_3).$$

Since we have restricted the domain, we have $f = (f_3 \wedge f_4) \vee (f_1 \wedge f_2)$. ∎

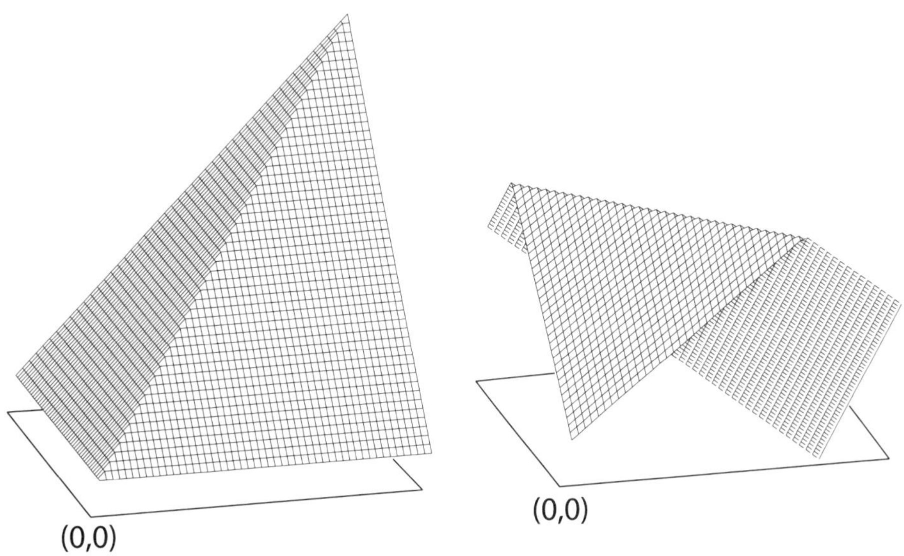

Figure 8. The graphs of $f_1 \wedge f_2$ and $f_3 \wedge f_4$.

A rudimentary and very inefficient procedure for computing the regions of the functions in $\mathbf{Aff}^{\vee\wedge}$ by means of Theorem 7.20 is presented next.

Example 7.27 (From $\mathbf{Aff}^{\vee\wedge}$ to **PL**). Take $f \in \{f_1, f_2, \ldots, f_p\}^{\vee\wedge}$, where as usual $f_1, f_2, \ldots, f_p$ are affine functions on $\mathbb{R}^m$. Following Theorem 7.20 the regions of the function f can be obtained following the steps below:

Step I: Determine $E = \{(i,j)\colon\ 1 \le i < j \le p \text{ and } [f_i = f_j] \ne \emptyset\}$ and then for each $e = (i,j) \in E$ pick $\alpha_e \in \mathbb{R}$ and $a_e \in \mathbb{R}^m$ such that

$$H_e = \{x \in \mathbb{R}^m\colon\ a_e \cdot x = \alpha_e\} = [f_i = f_j]\,.$$

Step II: Use the hyperplane arrangement $\{H_e\}_{e \in E}$ to determine the cells $K_1, \ldots, K_J$.

Step III: For each $h = 1, 2, \ldots, J$ choose some $x_h \in K_h$ and let

$$i_h = \min\{i \in \{1, \ldots, p\}\colon\ f_i(x_h) = f(x_h)\}\,.$$

Step IV: For each $h = 1, 2, \ldots, J$ determine the set

$$I_h = \{j \in \{1, \ldots, J\}\colon\ i_j = i_h\}\,.$$

Step V: For each $h = 1, 2, \ldots, J$ let $S_{i_h} = \overline{\bigcup_{i \in I_h} K_j}$.

The characteristic pairs of the piecewise affine function f are the distinct members of the family $\{(f_{i_h}, S_{i_h})\}_{h \in \{1,\ldots,J\}}$. ∎

Exercises

1. Complete the details of the proof of part (1) in Lemma 7.9.
2. Prove the following claim used in the proof of part (2) of Lemma 7.9: *If a nonzero linear functional f satisfies $f(x) \ge \alpha$ for all x in a nonempty open set $\mathcal{O}$, then $f(x) > \alpha$ must be the case for all $x \in \mathcal{O}$.*
3. Verify that the function defined in Example 7.11 is indeed a piecewise affine function and establish that its regions are as shown in Figure 4.
4. If $f\colon \mathbb{R}^m \to \mathbb{R}$ is a piecewise affine function, then for any two fixed vectors $a, b \in \mathbb{R}^m$ show that the function $\theta\colon \mathbb{R} \to \mathbb{R}$, defined via the formula $\theta(t) = f(ta + (1-t)b)$, is (one-dimensional) piecewise linear.
5. Prove that two affine functions $f, g\colon \mathbb{R}^m \to \mathbb{R}$ either do not agree at any point or the set where they agree is a hyperplane, i.e., show that the set $[f = g] = \{x \in \mathbb{R}^m\colon\ f(x) = g(x)\}$ is either empty or a hyperplane.
6. Complete the details of the proof of Corollary 7.18.
7. Let $\{H_e\}_{e \in E}$ be an oriented arrangement of hyperplanes. Show that for some $x \in \mathbb{R}^m$ the vector σ_x is a tope if and only if $x \notin \bigcup_{e \in E} H_e$.
8. Show that the collection of cells induced by a family of hyperplanes is independent of the orientation of the hyperplanes.
9. Prove part (3) of Lemma 7.19.
10. In Example 7.26 verify that indeed the cells are as shown in Figure 7.

Chapter 8

Appendix: linear topologies

The objective of this chapter is to list (mainly without proofs) some basic properties of linear topologies that are used (with or without explicit mention) throughout this monograph. As a rule, we do not present proofs of results that are readily accessible in the literature. However, we include proofs of several results that (in our opinion) are not quite readily available in the literature. For complete and extensive treatments of the theory of topological vector spaces, we refer the reader to [**12, 33, 37, 51, 59, 85**]. For a comprehensive exposition of locally solid Riesz spaces and their applications in economics we refer to [**14**].

The material in this chapter has been spread out over eight sections. The first section introduces linear topologies on vector spaces and presents their basic structure while the second one deals with duality theory of locally convex spaces. The third section studies $\mathfrak{S}$-topologies and the fourth exhibits the basic separation theorems of convex sets in vector spaces.

Normed and Banach spaces are discussed in the fifth section and the sixth section specializes to finite dimensional vector spaces. The seventh section introduces and presents proofs of the open and closed graph theorems in the setting of completely metrizable topological vector spaces. The eighth, and final, section of this chapter, presents the surprising properties of the bounded weak* topology on the norm dual of a Banach space.

As mentioned in the previous chapters, all vector spaces in this monograph will be *real* vector spaces. Also, as usual, $\mathbb{R}$ will denote the set of real numbers, $\mathbb{N}$ the set of natural numbers, $\mathbb{Q}$ the set of rational numbers, and $\mathbb{Z}$ the set of integers. The letter L will be used explicitly in this chapter to designate our generic real vector space.

8.1. Linear topologies on vector spaces

A **linear topology** on a vector space L is any topology τ on L that makes the addition and the scalar multiplication continuous, that is, the topology τ makes the two functions

$$\begin{array}{llll} (1) \ (x,y) \mapsto x+y & \text{from} & L\times L \ \text{ to } \ L \text{ and} \\ (2) \ (\lambda,x) \mapsto \lambda x & \text{from} & \mathbb{R}\times L \ \text{ to } \ L \end{array}$$

continuous. A **topological vector space** (L,τ) is a vector space L equipped with a linear topology τ.

Two topological vector spaces (L,τ) and (L_1,τ_1) are called **isomorphic** if there exists a one-to-one and onto operator $T\colon L\to L_1$ that is at the same time a topological homeomorphism, in other words, if there exists a one-to-one mapping from L onto L_1 preserving both structures, the algebraic and the topological.

Every linear topology τ on any vector space L has a base $\mathcal{N}$ for the τ-neighborhoods at zero satisfying the following properties.

(a) Each $V\in\mathcal{N}$ is a **circled** (or a **balanced**) **set**, that is, $\lambda u\in V$ holds for all $u\in V$ and every $\lambda\in\mathbb{R}$ with $|\lambda|\le 1$.

(b) Each $V\in\mathcal{N}$ is an **absorbing set**, that is, for every $u\in L$, there exists some $\lambda_0>0$ such that $\lambda u\in V$ for all λ with $|\lambda|\le\lambda_0$.

(c) For each $V\in\mathcal{N}$ there exists some $W\in\mathcal{N}$ with $W+W\subseteq V$.

Conversely, if a nonempty collection $\mathcal{B}$ of subsets of a vector space L is a **filter base**[1] satisfying properties (a), (b), and (c), then there exists a unique linear topology τ on L whose neighborhoods at zero have $\mathcal{B}$ as a base. (For details see [**12**, Theorem 5.6, p. 169].)

If L is a topological vector space and $\mathcal{N}$ is a base for the neighborhoods at zero, then for each $x\in L$ the collection of sets $x+\mathcal{N}=\{x+V\colon\ V\in\mathcal{N}\}$ is a base for the neighborhoods at x. This is a simple consequence of the fact that the mapping $y\mapsto x+y$ is a linear homeomorphism. Thus, the topological structure of a topological vector space is completely determined by any base of its neighborhood system at zero. An easy inductive argument in conjunction with property (c) also guarantees the following.

Lemma 8.1. *For any neighborhood V of zero in a topological vector space there exists a sequence $\{V_n\}$ of neighborhoods of zero satisfying $V_1=V$ and $V_{n+1}+V_{n+1}\subseteq V_n$ for each n.*

---

[1] This means that every $V\in\mathcal{B}$ is nonempty and for every pair $V,W\in\mathcal{B}$ there exists some $U\in\mathcal{B}$ with $U\subseteq V\cap W$.

A net $\{u_\alpha\}$ in a topological vector space (L, τ) is called τ-**Cauchy** if for every τ-neighborhood V of zero there is an index α_1 such that $u_\alpha - u_\beta \in V$ for all $\alpha, \beta \succeq \alpha_1$.

A subset A of (L, τ) is called τ-**complete** if every τ-Cauchy net of A is τ-convergent to some element in A. In particular, (L, τ) is called **topologically complete** (or briefly τ-**complete**) if every τ-Cauchy net of L is τ-convergent in L.

A topological vector space (L, τ) is called **metrizable** if there exists a metric generating the topology τ. The metrizable topological vector spaces are characterized as follows. (For a proof see [**12**, Section 5.3, p. 172].)

Theorem 8.2. *A Hausdorff topological vector space (L, τ) is metrizable if and only if τ has a countable neighborhood base at zero.*

The locally convex topologies are defined as follows.

Definition 8.3. *A* ***locally convex topology*** *on a vector space is a linear topology that has a base at zero consisting of convex sets.*[2]

A ***locally convex space*** *is a vector space equipped with a locally convex topology.*

It is not difficult to see that every locally convex space has a base at zero consisting of convex, circled, and absorbing sets.

A **seminorm** ρ on a vector space L is a function $\rho\colon L \to \mathbb{R}$ satisfying the following three properties.

(1) $\rho(u) \geq 0$ for all $u \in L$.
(2) $\rho(\lambda u) = |\lambda|\rho(u)$ for all $\lambda \in \mathbb{R}$ and $u \in L$.
(3) $\rho(u+v) \leq \rho(u) + \rho(v)$ for all $u, v \in L$ (the **triangle inequality**).

If ρ is a seminorm on a vector space L, then for any $\epsilon > 0$ the set $\{u \in L\colon\ \rho(u) \leq \epsilon\}$, the ϵ-**closed ball** of ρ at zero, is a convex subset of L.

Every family $\{\rho_\alpha\}_{\alpha \in A}$ of seminorms on a vector space L defines a locally convex topology on L by taking as a base for the neighborhoods at zero the finite intersections of the closed balls of the seminorms ρ_α. In other words, a subset V of L is a neighborhood of zero for this topology if and only if there exist some $\epsilon > 0$ and a finite number of indices $\alpha_1, \ldots, \alpha_n$ such that

$$\big\{u \in L\colon\ \rho_{\alpha_m}(u) \leq \epsilon \text{ for each } m = 1, \ldots, n\big\} \subseteq V\,.$$

This linear topology is called the **locally convex topology generated by the family of seminorms** $\{\rho_\alpha\}_{\alpha \in A}$.

[2] As usual, a subset A of L is called **convex** if $\lambda u + (1-\lambda)v \in A$ holds for all $u, v \in A$ and all $0 \leq \lambda \leq 1$.

It turns out that the locally convex topologies are precisely the linear topologies that are generated by families of seminorms.

Theorem 8.4. *A linear topology τ on a vector space L is locally convex if and only if there is a family $\{\rho_\alpha\}_{\alpha\in A}$ of seminorms on L that generates τ.*

Proof. Assume that τ is a locally convex topology on a vector space L. Then τ has a base $\mathcal{N}$ at zero consisting of convex, circled, and absorbing sets. For each $V \in \mathcal{N}$, let $\|\cdot\|_V$ denote the Minkowski functional of V.[3] It is a routine matter to verify that the family of seminorms $\{\|\cdot\|_V\colon\ V \in \mathcal{N}\}$ generates the locally convex topology τ. ■

A subset A in a topological vector space (L, τ) is called **topologically bounded** (or simply τ-**bounded**) if for every τ-neighborhood V of zero there exists some $\lambda > 0$ such that $\lambda A \subseteq V$. Note that a subset A in a locally convex space (L, τ) whose topology τ is generated by a family of seminorms $\{\rho_\alpha\}_{\alpha\in A}$ is τ-bounded if and only if $\rho_\alpha(A)$ is a bounded subset of $\mathbb{R}$ for each α.

The **topological dual** L' of a topological vector space (L, τ) is the collection of all τ-continuous linear functionals on L. With the usual algebraic operations of addition and scalar multiplication, L' is a vector space; actually L' is a vector subspace of the **algebraic dual** L^*, the vector space of all linear functionals on L. It should be noted that a linear functional $f \in L^*$ is τ-continuous if and only if it is bounded on a τ-neighborhood of zero.

The **weak topology** $\sigma(L, L')$ of a locally convex space (L, τ) is the locally convex topology generated on the vector space L by the family of seminorms $\{p_f\colon\ f \in L'\}$, where $p_f(u) = |f(u)|$ for $u \in L$. The weak topology $\sigma(L, L')$ is also called the **topology of pointwise convergence** on L since a net $\{u_\alpha\}$ in L satisfies $u_\alpha \xrightarrow{\sigma(L,L')} u$ if and only if $f(u_\alpha) \to f(u)$ holds in $\mathbb{R}$ for each $f \in L'$. The weak topology $\sigma(L, L')$ is the smallest locally convex topology on L for which every $f \in L'$ is continuous. That is, we have the following result.

Theorem 8.5. *The topological dual of $\big(L, \sigma(L, L')\big)$ is precisely L'.*

Similarly, the weak topology $\sigma(L', L)$ is the locally convex topology on L' generated by the family of seminorms $\{p_u\colon\ u \in L\}$, where $p_u(f) = |f(u)|$ for $f \in L'$. The topology $\sigma(L', L)$ is also called the **topology of pointwise convergence** on L' since a net $\{f_\alpha\}$ in L' satisfies $f_\alpha \xrightarrow{\sigma(L',L)} f$ in L' if and only if $f_\alpha(u) \to f(u)$ holds in $\mathbb{R}$ for each $u \in L$.

---

[3]The **Minkowski functional** (or the **gauge**) of any subset A of a vector space L is the function $\|\cdot\|_A\colon L \to [0, \infty]$ defined by $\|x\|_A = \inf\{\lambda > 0\colon\ x \in \lambda A\}$ for each $x \in L$. If A is absorbing, then $\|\cdot\|_A$ is real-valued. If A is nonempty, convex, circled, and absorbing, then $\|\cdot\|_A$ is a seminorm of L.

8.2. Duality theory

The theory of locally convex spaces is usually studied in conjunction with duality theory. This connection is commonly achieved in the framework of dual systems.

Definition 8.6. *A* ***dual pair*** *(or a* ***dual system****) is a pair of vector spaces $\langle L, L' \rangle$ "glued" together by means of a bilinear mapping*[4] $\langle\cdot,\cdot\rangle\colon L \times L' \to \mathbb{R}$ *that satisfies the two properties*

(1) *if* $\langle x, x' \rangle = 0$ *for all* $x' \in L'$, *then* $x = 0$,

(2) *if* $\langle x, x' \rangle = 0$ *for all* $x \in L$, *then* $x' = 0$.

The bilinear function $\langle\cdot,\cdot\rangle$ *is called the* ***bilinearity*** *of the pair* $\langle L, L' \rangle$.

The typical dual pair is a pair of the form $\langle L, L' \rangle$, where L is a Hausdorff locally convex space, L' is its topological dual, and the bilinearity of the pair is defined for each $(x, x') \in L \times L'$ by

$$\langle x, x' \rangle = x'(x)\,.$$

For the rest of the discussion in this section, $\langle L, L' \rangle$ will denote a fixed dual pair. By property (1) of the definition of the dual pair, it follows that every $x' \in L'$ defines a (unique) linear functional on L via the formula $x \mapsto \langle x, x' \rangle$. This means that the vector space L' can be viewed as a vector subspace of $\mathbb{R}^L$. Similarly, the vector space L can be viewed as a vector subspace of $\mathbb{R}^{L'}$. By means of these identifications of L' and L, we can define the weak topologies on L and L'. Let us recall next how the weak topologies are defined.

For any nonempty set X the real vector space $\mathbb{R}^X$ equipped with its product topology becomes a locally convex vector space (actually a Hausdorff locally convex-solid Riesz space). The product topology on $\mathbb{R}^X$ is called the **topology of pointwise convergence** and is generated by the family of Riesz seminorms $\{p_x\colon\ x \in X\}$, where for each $f \in \mathbb{R}^X$ we have

$$p_x(f) = |f(x)|\,.$$

For any nonempty set A of real-valued functions defined on X (i.e., $A \subseteq \mathbb{R}^X$), the **weak topology** $\sigma(A, X)$ on A (also called the **topology of pointwise convergence on** A) is the Hausdorff topology induced on A by the product topology of $\mathbb{R}^X$. Note that indeed a net $\{f_\alpha\} \subseteq A$ satisfies $f_\alpha \xrightarrow{\sigma(A,X)} f$ in A if and only if $f_\alpha(x) \to f(x)$ holds in $\mathbb{R}$ for each $x \in X$. If A is a vector subspace of $\mathbb{R}^X$, then $\sigma(A, X)$ is a Hausdorff linear topology on A.

[4] A **bilinear mapping** is any function $f\colon X \times Y \to \mathbb{R}$, where X and Y are vector spaces, such that $f(\cdot, y)$ is linear for each fixed $y \in Y$ and $f(x, \cdot)$ is linear for each fixed $x \in X$.

If L is a vector space, then its algebraic dual L^* (the vector space of all linear functionals on L) is clearly a vector subspace of $\mathbb{R}^L$. Obviously, L^* is a closed vector subspace of $\mathbb{R}^L$. Moreover, for every vector subspace L' of L^* we have the following vector subspace inclusions: $L' \subseteq L^* \subseteq \mathbb{R}^L$. This implies that the weak topology $\sigma(L', L)$ is nothing more than the restriction of the product topology of $\mathbb{R}^L$ to L'. Similarly, considering L as a vector subspace of $\mathbb{R}^{L'}$, we see that the weak topology $\sigma(L, L')$ on L is the restriction of the product topology of $\mathbb{R}^{L'}$ to L.

For a subset A of L the **polar** (also called the **absolute polar** or the **two-sided polar**) of A is defined by

$$A^{\circ} = \left\{x' \in L' \colon \ |\langle x, x'\rangle| \leq 1 \text{ for all } x \in A\right\}.$$

Clearly, A° is a $\sigma(L', L)$-closed set. Similarly, if A is a nonempty subset of L', then its **polar** is defined by

$$A^{\circ} = \left\{x \in L \colon \ |\langle x, x'\rangle| \leq 1 \text{ for all } x' \in A\right\}.$$

An important base for the $\sigma(L', L)$-neighborhoods at zero can be described in terms of polars as follows.

Lemma 8.7. *If $\langle L, L'\rangle$ is a dual system, then the collection of polars*

$$\{F^{\circ} \colon \ F \text{ is a finite subset of } L\}$$

is a base for the $\sigma(L', L)$-neighborhoods at zero. A similar result is true for the $\sigma(L, L')$-neighborhoods of zero.

A locally convex topology τ on L is said to be **consistent** (or **compatible**) with the dual pair $\langle L, L'\rangle$ if $(L, \tau)' = L'$, i.e., if the topological dual of (L, τ) coincides with the vector space L'.

Now fix a compatible locally convex topology τ of L. If V is an arbitrary τ-neighborhood of zero, then V° is pointwise closed in $\mathbb{R}^L$. Indeed, if a net $\{\phi_\alpha\} \subseteq V^{\circ}$ satisfies $\phi_\alpha \to \phi$ in $\mathbb{R}^L$, then $\phi \in L^*$ and $|\phi(u)| \leq 1$ holds for every $u \in V$ and hence $\phi \in L'$, since ϕ is bounded on V. On the other hand, as is readily seen, V° is topologically bounded in $\mathbb{R}^L$ and thus V° is a compact subset of $\mathbb{R}^L$ [and hence of L' for $\sigma(L', L)$] by Tychonoff's classical compactness theorem. We have, thus, proven the famous Alaoglu–Bourbaki theorem.

Theorem 8.8 (Alaoglu–Bourbaki). *Let $\langle L, L'\rangle$ be a dual pair and assume that τ is a consistent locally convex topology on L. Then for any τ-neighborhood V of zero, its polar V° is a $\sigma(L', L)$-compact subset of L'.*

The next result describes three basic properties of consistent topologies.

Theorem 8.9 (Mackey). *If $\langle L, L'\rangle$ is a dual pair, then all consistent topologies on L have*

(1) *the same closed convex sets,*

(2) *the same topologically bounded sets, and*

(3) *the same upper semicontinuous quasiconcave real functions.*

The **bipolar** of a subset A of L is the set $(A^{\circ})^{\circ}$, i.e., the polar of A°, and is denoted $A^{\circ\circ}$. The following basic result (known as the Bipolar Theorem) can be found in every book on topological vector spaces and is a very useful tool.

Theorem 8.10 (The Bipolar Theorem). *Let $\langle L, L'\rangle$ be a dual pair and let A be a nonempty subset of L. Then the bipolar $A^{\circ\circ}$ is the balanced, convex, $\sigma(L, L')$-closed hull of A.*

In particular, if A is a balanced, convex, and $\sigma(L, L')$-closed subset of L, then $A = A^{\circ\circ}$.

There is also the notion of the one-sided bipolar of a set that was introduced in Definition 3.30. If $\langle L, L'\rangle$ is a dual pair and A is a nonempty subset of L, then the **one-sided polar** of A, denoted $A^{\bullet}$, is defined by

$$A^{\bullet} = \{x' \in L' \colon \ \langle x, x'\rangle \leq 1 \text{ for all } x \in A\}.$$

Likewise, for a nonempty subset B of L' we let

$$B^{\bullet} = \{x \in L \colon \ \langle x, x'\rangle \leq 1 \text{ for all } x' \in B\}.$$

Lemma 8.11. *The one-sided polar $A^{\bullet}$ is nonempty, convex, $\sigma(L', L)$-closed, and contains zero.*

The **one-sided bipolar** of a nonempty subset A of L is the set $(A^{\bullet})^{\bullet}$, i.e., the one-sided polar of $A^{\bullet}$, and is denoted $A^{\bullet\bullet}$. The corresponding one-sided bipolar theorem (that was stated and proved as Theorem 3.31) can be restated as follows.

Theorem 8.12 (The One-Sided Bipolar Theorem). *If $\langle L, L'\rangle$ is a dual pair and A is a nonempty subset A of L, then the one-sided bipolar $A^{\bullet\bullet}$ is the convex $\sigma(L, L')$-closed hull of the set $A \cup \{0\}$. (In particular, if A is convex, $\sigma(L, L')$-closed, and contains zero, then $A = A^{\bullet\bullet}$.)*

8.3. $\mathfrak{S}$-topologies

We remind the reader here "the polar route" to characterizing consistent locally convex topologies for a dual pair $\langle L, L'\rangle$. So, for our discussion $\langle L, L'\rangle$ will denote a fixed dual pair. We start with any nonempty $\sigma(L', L)$-bounded subset A of L' and note that the formula

$$q_A(x) = \sup\{|\langle x, x'\rangle| \colon \ x' \in A\}$$

defines a seminorm on L. Furthermore, $\{x \in L\colon\ q_A(x) \le 1\} = A^\circ$, and we have the identity

$$q_A(x) = \inf\bigl\{\alpha > 0\colon\ x \in \alpha A^\circ\bigr\} = p_{A^\circ}(x)\,,$$

where p_{A° is the Minkowski functional (or the gauge) of the polar set A°.

To see that $q_A = p_{A^\circ}$ is indeed the case, fix $x \in L$. If x belongs to αA°, then write $x = \alpha y$ with $\alpha > 0$ and $y \in A^\circ$. Note that $|\langle x, x'\rangle| = \alpha|\langle y, x'\rangle| \le \alpha$ for all $x' \in A$. Hence $q_A(x) \le \alpha$, from which we see that $q_A(x) \le p_{A^\circ}(x)$. To prove the reverse inequality, note that $\frac{x}{q_A(x)+\epsilon}$ belongs to A° for every $\epsilon > 0$. Thus $p_{A^\circ}\bigl(\frac{x}{q_A(x)+\epsilon}\bigr) = \frac{p_{A^\circ}(x)}{q_A(x)+\epsilon} \le 1$, so $p_{A^\circ}(x) \le q_A(x) + \epsilon$ for all $\epsilon > 0$. Hence $p_{A^\circ}(x) \le q_A(x)$, and we are done.

In other words, q_A is a seminorm that coincides with the gauge A°. By the Bipolar Theorem 8.10, the set $A^{\circ\circ}$ is the convex circled $\sigma(L', L)$-closed hull of A. Since $A^\circ = (A^{\circ\circ})^\circ$, we see that $q_A = q_{A^{\circ\circ}}$.

Now let $\mathfrak{S}$ be a family of nonempty $\sigma(L', L)$-bounded subsets of L'.[5] The corresponding **$\mathfrak{S}$-topology** on L is the locally convex topology generated by the family of seminorms $\{q_A\colon\ A \in \mathfrak{S}\}$. Equivalently, it is the topology generated by the neighborhood subbase $\{\epsilon A^\circ\colon\ A \in \mathfrak{S} \text{ and } \epsilon > 0\}$ at zero. Thus we may enlarge $\mathfrak{S}$ to $\widehat{\mathfrak{S}} = \{\epsilon A : A \in \mathfrak{S} \text{ and } \epsilon > 0\}$ and still generate the same topology on L. In other words, the neighborhood base at zero for the $\mathfrak{S}$-topology consists of all sets of the form

$$A_1^\circ \cap \cdots \cap A_n^\circ,$$

where $A_1, \dots, A_n \in \widehat{\mathfrak{S}}$. Also note that since $q_A = q_{A^{\circ\circ}}$, we may restrict our attention to families of convex circled sets. The $\mathfrak{S}$-topology is Hausdorff if and only if the linear span of the set $\bigcup_{A \in \mathfrak{S}} A$ is $\sigma(L', L)$-dense in L'.

Since $x_\alpha \xrightarrow{\mathfrak{S}} x$ in L if and only if $q_A(x_\alpha - x) \to 0$ for every $A \in \mathfrak{S}$ and since $q_A(x_\alpha - x) \to 0$ for every $A \in \mathfrak{S}$ if and only if $\{x_\alpha\}$ converges uniformly to x on each member of $\mathfrak{S}$, the $\mathfrak{S}$-topology is also called the **topology of uniform convergence** on the members of $\mathfrak{S}$.

Remarkably, every consistent locally convex topology on L (or on L') is an $\mathfrak{S}$-topology. This important result is known as the Mackey–Arens theorem and characterizes the locally convex topologies that are consistent with a given dual pair.

Theorem 8.13 (Mackey–Arens). *A locally convex topology τ on L is consistent with the dual pair $\langle L, L'\rangle$ if and only if τ is the $\mathfrak{S}$-topology for a family $\mathfrak{S}$ of convex, circled, and $\sigma(L', L)$-compact subsets of L' with $\bigcup_{A \in \mathfrak{S}} A = L'$.*

Observe that the weak topology $\sigma(L, L')$ is the $\mathfrak{S}$-topology for the collection $\mathfrak{S} = \bigl\{\{x'\}\colon\ x' \in L'\bigr\}$. The weak topology $\sigma(L, L')$ is the smallest

[5]The symbol $\mathfrak{S}$ is well established. It is an upper case "S" in the old German fraktur alphabet.

locally convex topology on L consistent with $\langle L, L' \rangle$. The largest consistent locally convex topology on L is by Theorem 8.13 the $\mathfrak{S}$-topology for the family $\mathfrak{S}$ consisting of all convex, circled, and $\sigma(L', L)$-compact subsets of L'. This important topology is called the **Mackey topology** and is denoted $\tau(L, L')$. The Mackey topology $\tau(L', L)$ is defined analogously.

The Mackey–Arens Theorem 8.13 can be restated in terms of the weak topology and the Mackey topology as follows.

Theorem 8.14. *A locally convex topology τ on L is consistent with the dual pair $\langle L, L' \rangle$ if and only if $\sigma(L, L') \subseteq \tau \subseteq \tau(L, L')$.*

Similarly, a locally convex topology τ' on L' is consistent with the dual pair $\langle L, L' \rangle$ if and only if $\sigma(L', L) \subseteq \tau' \subseteq \tau(L', L)$.

The **strong topology** $\beta(L', L)$ on L' is the locally convex topology generated by the family of seminorms $\{\rho_A\colon\ A \in \mathcal{B}\}$, where

$$\rho_A(x') = \sup\big\{|x'(x)|\colon\ x \in A\big\}$$

and $\mathcal{B}$ is the collection of all $\sigma(L, L')$-bounded subsets of L. The strong topology $\beta(L', L)$ on L' is the $\mathfrak{S}$-topology of uniform convergence on every $\sigma(L, L')$-bounded subset of L. The topology $\beta(L, L')$ is defined similarly.

8.4. The separation of convex sets

Amongst the important theorems of functional analysis that are useful in many branches of science are the ones dealing with the separation of convex sets. Recall that two nonempty sets A and B of a vector space are said to be **separated** by a nonzero linear functional f if $f(b) \le f(a)$ holds for all $a \in A$ and all $b \in B$. For the geometric meaning of separation see the accompanying figure. Our purpose here is to state the five basic separation theorems of convex sets that are used extensively in applications.

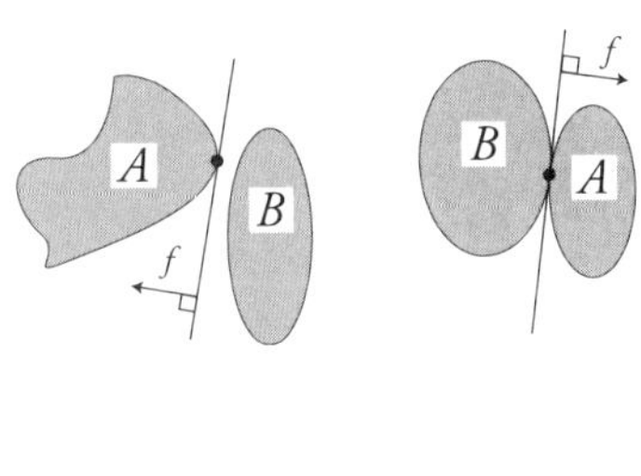

The first one is "algebraic" in nature and requires no topology. Recall that a vector $a \in A$, where A is a subset of a vector space L, is called an **internal point** of A if for each $x \in L$ there exists some $\lambda_0 > 0$ such that $a + \lambda x \in A$ for each $-\lambda_0 \le \lambda \le \lambda_0$ (or, equivalently, if for each x there exists some $\lambda_0 > 0$ such that $a + \lambda x \in A$ for all $0 \le \lambda \le \lambda_0$).

For the proof of the next basic result see [**12**, Theorem 5.61, p. 200].

Theorem 8.15 (Algebraic Separating Hyperplane Theorem)**.** *If two nonempty convex sets of a vector space are disjoint and one of them has an internal point, then they can be separated by a nonzero linear functional.*

For a proof of the next separation result see [**12**, Theorem 5.67, p. 202].

Theorem 8.16 (Interior Separating Hyperplane Theorem). *If in a topological vector space the interior of a convex set is nonempty and is disjoint from another nonempty convex set, then the closures of the convex sets can be separated by a nonzero continuous linear functional.*

Recall that two nonempty sets A and B of a vector space are said to be **strongly separated** by a nonzero linear functional f if there exist a constant $c \in \mathbb{R}$ and some $\epsilon > 0$ such that $f(b) \leq c < c + \epsilon \leq f(a)$ holds for all $a \in A$ and all $b \in B$. The geometric meaning of strong separation is shown in the accompanying figure.

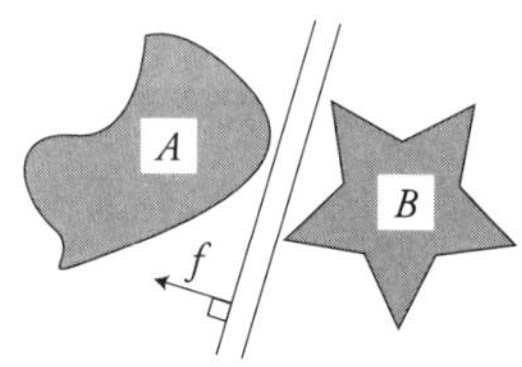

For a proof of the next "strong" separation result see [**12**, Theorem 5.79, p. 207].

Theorem 8.17 (Strong Separating Hyperplane Theorem). *If two nonempty convex sets of a (not necessarily Hausdorff) locally convex space are disjoint and one of them is closed and the other is compact, then they can be strongly separated by a nonzero continuous linear functional.*

A special but important case of the preceding separation theorem is the following.

Corollary 8.18 (The Separation of Points from Closed Convex Sets). *Every point that lies outside of a closed convex subset of a (not necessarily Hausdorff) locally convex space can be strongly separated by a nonzero continuous linear functional from the set.*

Two nonempty disjoint closed convex sets in a Hausdorff locally convex space need not be separated by a continuous linear functional—for an example see [**12**, Theorem 5.69, p. 203]. However, in a finite dimensional vector space any pair of disjoint nonempty convex sets can be separated. For a proof of the next result see [**12**, Theorem 7.35, p. 279].

Theorem 8.19 (Finite Dimensional Separating Hyperplane Theorem). *Any pair of nonempty disjoint convex sets in a finite dimensional vector space can be separated by a nonzero linear functional.*

8.5. Normed and Banach spaces

A function $\|\cdot\|: L \to \mathbb{R}$ defined on a vector space is said to be a **norm** if it satisfies the following properties.

(1) $\|x\| \geq 0$ for each $x \in L$ and $\|x\| = 0$ if and only if $x = 0$.

(2) $\|\alpha x\| = |\alpha| \cdot \|x\|$ for all $x \in L$ and all $\alpha \in \mathbb{R}$.

(3) $\|x + y\| \leq \|x\| + \|y\|$ for all $x, y \in L$ (the **triangle inequality**).

A **normed space** is a vector space equipped with a norm. Every normed space has a natural metric defined by means of its norm via the formula $d(x, y) = \|x - y\|$. This natural metric is called the **metric induced** by the norm. The topology generated by d is called the **norm topology**. The norm topology is Hausdorff and locally convex. Thus, normed spaces are examples of Hausdorff locally convex spaces. A **Banach space** is a normed space that is a complete metric space under the induced metric generated by the norm.

Now let L be a normed space. The **closed unit ball** of a normed space L is the set $U = \{x \in L\colon \ \|x\| \leq 1\}$. The **norm dual** L' of L is the topological dual of L endowed with its norm topology. It is a Banach space under the norm $\|\cdot\|$ defined by

$$\|x'\| = \sup_{x \in U} |x'(x)|\,.$$

The **double dual** L'' of a normed space L is the norm dual of L'. The closed unit balls of L' and L'' are usually denoted by U' and U''. That is,

$$U' = \{x' \in L'\colon \ \|x'\| \leq 1\} \quad \text{and} \quad U'' = \{x'' \in L''\colon \ \|x''\| \leq 1\}\,.$$

We have the following important extension theorem of continuous linear functionals known as the Hahn–Banach extension theorem.

Theorem 8.20 (Hahn–Banach). *If X is a vector subspace of a normed space L, then every continuous linear functional on X has a continuous extension to all of L that preserves its norm.*

With every normed space L there are two natural dual pairs associated with L. These are the pairs $\langle L, L'\rangle$ and $\langle L', L''\rangle$. Regarding the pair $\langle L, L'\rangle$, we call the topology $\sigma(L, L')$ the **weak topology** of L and we shall denote it by w, i.e., $w = \sigma(L, L')$. Likewise, we shall call the topology $\sigma(L', L)$ on L' the **weak* topology** of L' and denote it by w^*, i.e., $w^* = \sigma(L', L)$.

If L is a normed space, then there is a natural norm linear isometry $x \mapsto \hat{x}$ from L to L'' defined by

$$\hat{x}(x') = x'(x)$$

for each $x' \in L'$. This means that for each $x \in L$ we have

$$\|x\| = \|\hat{x}\| = \sup_{x' \in U'} |x'(x)|\,.$$

Under this linear isometry we can identify L with a vector subspace of L''. A Banach space L is said to be **reflexive** if $x \mapsto \hat{x}$ is surjective. The following result characterizes reflexive Banach spaces.

Theorem 8.21. *A Banach space is reflexive if and only if its closed unit ball is weakly compact.*

We also have the following result.

Theorem 8.22. *If L is a normed space, then the Mackey topologies $\tau(L, L')$ and $\tau(L', L)$ coincide with the norm topologies of L and L', respectively.*

Moreover, we have $U' = U^\circ$, the polar of the closed unit ball U of L with respect to the dual pair $\langle L, L' \rangle$ (and so by the Alaoglu–Bourbaki Theorem 8.8 the closed unit ball U' of L' is w^*-compact).

We list below some of the basic classical results regarding Banach spaces that are useful for this work.

Theorem 8.23. *A normed space L is separable if and only if the closed unit ball U' of its norm dual L' is w^*-metrizable.*

Theorem 8.24. *The dual L' of a Banach space L is separable if and only if the closed unit ball of L is weakly metrizable.*

Among the classical examples of Banach spaces are the Banach spaces of type $C(\Omega)$, where Ω is a Hausdorff compact topological space and the norm is the sup norm $\|\cdot\|_\infty$, i.e., $\|x\|_\infty = \sup_{\omega\in\Omega} |x(\omega)|$. The following result (whose proof uses Theorem 8.23) is very useful.

Theorem 8.25. *A Hausdorff compact topological space Ω is metrizable if and only if the Banach space $C(\Omega)$ is separable.*

Theorem 8.26 (Banach–Steinhaus). *Let L be a Banach space. Then a subset A of L' is norm bounded if and only if A is w^*-bounded.*[6]

Theorem 8.27 (Mazur). *In a Banach space, the closed convex hull of any compact subset is likewise compact.*

Theorem 8.28 (Krein–Šmulian). *In a Banach space, the convex circled hull and the convex hull of a relatively weakly compact set are both relatively weakly compact sets.*[7]

[6] This theorem is also known as the **Principle of Uniform Boundedness** and is the Banach space version of part (2) of Theorem 8.9.

[7] We need to recall a few things here. Let A be a nonempty subset of a vector space X. Its **convex hull** $\operatorname{co} A$ is the smallest (with respect to inclusion) subset of X that includes A. It is precisely the set consisting of all **convex combinations** of A, i.e., of all vectors of the form $x = \sum_{i=1}^n \alpha_i x_i$, where $x_i \in A$ and $\lambda_i \in [0,1]$ for each i, and $\sum_{i=1}^n \lambda_i = 1$. If now X is also a topological vector space, then the **closed convex hull of a set** of A is the smallest closed convex set that includes A. This is precisely the set $\overline{\operatorname{co} A}$ and is denoted $\overline{\operatorname{co}}\, A$, i.e., $\overline{\operatorname{co}}\, A = \overline{\operatorname{co} A}$.

The **circled hull** and **closed circled hull** of A are defined similarly.

Theorem 8.29 (Eberlein–Šmulian). *A subset of a Banach space is weakly compact if and only if every sequence of the set has a subsequence that converges weakly to a point in the set.*

Theorem 8.30 (James). *A nonempty weakly closed bounded subset of a Banach space is weakly compact if and only if every continuous linear functional attains a maximum on the set.*

Our final result contains two useful characterizations of norm and weakly compact subsets of a Banach space.

Theorem 8.31 (Grothendieck). *A subset A of a Banach space L is*

(1) *norm totally bounded if and only if it is included in the closed convex hull of a sequence that is norm convergent to zero,*

(2) *relatively weakly compact if and only if for each $\epsilon > 0$ there exists a weakly compact set W such that $A \subseteq W + \epsilon U$, where U denotes the closed unit ball of L.*

Proof. The proof of part (1) goes as follows. If there exists a sequence $\{x_n\}$ of L with $\|x_n\| \to 0$ and $A \subseteq \overline{\mathrm{co}}\,\{x_n\}$, then by Theorem 8.27 the set A is norm totally bounded.

For the converse, assume that A is a norm totally bounded set. Let U denote the closed unit ball of L. The desired sequence $\{x_n\}$ will be constructed by an inductive argument as follows. Put $A_0 = A$ and $k_0 = 0$. Pick a finite subset $\Phi_1 = \{x_1, \ldots, x_{k_1}\}$ of $2A_0$ with $2A_0 \subseteq \Phi_1 + 2^{-1}U$. Then, put $A_1 = (2A_0 - \Phi_1) \cap 2^{-1}U$, and note that the set A_1 is norm totally bounded. Now for the induction argument, assume that $\Phi_n = \{x_{k_{n-1}+1}, \ldots, x_{k_n}\}$ is a finite subset of the totally bounded set $2A_{n-1}$ with $2A_{n-1} \subseteq \Phi_n + 2^{-n}U$. Put $A_n = (2A_{n-1} - \Phi_n) \cap 2^{-n}U$, and note that A_n is a norm totally bounded set. Choose a finite subset $\Phi_{n+1} = \{x_{k_n+1}, \ldots, x_{k_{n+1}}\}$ of $2A_n$ with the property $2A_n \subseteq \Phi_{n+1} + 2^{-n-1}U$.

Now we claim that the sequence $\{x_n\}$ satisfies $A \subseteq \overline{\mathrm{co}}\,\{x_n\}$. To see this, note first that for each $x \in A_n$ we have $\|x\| \leq 2^{-n}$, and so $\|x_n\| \to 0$. On the other hand, if $x \in A$, then an easy argument shows that there exist integers $m_1 < m_2 < \cdots$ with $k_{i-1} < m_i \leq k_i$ such that the element $\frac{x_{m_1}}{2} + \cdots + \frac{x_{m_n}}{2^n} \in \overline{\mathrm{co}}\,\{x_n\}$ satisfies

$$\Big\| x - \Big(\frac{x_{m_1}}{2} + \cdots + \frac{x_{m_n}}{2^n}\Big)\Big\| \leq \frac{1}{4^n}\,.$$

This implies that $x \in \overline{\mathrm{co}}\,\{x_n\}$. That is, $A \subseteq \overline{\mathrm{co}}\,\{x_n\}$ holds, and the proof is finished. ■

8.6. Finite dimensional topological vector spaces

There are a few results regarding finite dimensional vector spaces that we shall mention in this section. So, for our discussion here, L will denote an ℓ-dimensional real vector space. It is easy to see that L is algebraically isomorphic to $\mathbb{R}^\ell$. On $\mathbb{R}^\ell$ the familiar **Euclidean norm** for each $x = (x_1, \dots .x_\ell)$ is defined by

$$\|x\|_2 = \Big[\sum_{i=1}^{\ell} x_i^2\Big]^{\frac{1}{2}}.$$

The topology it generates is called the **Euclidean topology** of $\mathbb{R}^\ell$. This topology is also called the **Euclidean topology** of L.

The first basic property asserts that any two norms on a finite dimensional vector space are equivalent.

Theorem 8.32. *The Euclidean topology on a finite dimensional vector space is the only Hausdorff linear topology that the space can carry.*

In particular, every finite dimensional vector space is automatically a Banach lattice—and so a complete metrizable topological vector space.

Proof. We can suppose that $L = \mathbb{R}^\ell$. Now let τ_1 be a Hausdorff linear topology on $\mathbb{R}^\ell$ and let τ be the Hausdorff linear topology generated by the Euclidean norm.

We must show that the identity operator $I\colon (\mathbb{R}^\ell, \tau) \to (\mathbb{R}^\ell, \tau_1)$ is a homeomorphism. Notice first that $x_\alpha \xrightarrow{\tau} 0$, where $x_\alpha = (x_1^\alpha, \dots, x_\ell^\alpha) \in \mathbb{R}^\ell$, is equivalent to $x_i^\alpha \to 0$ in $\mathbb{R}$ for each $i = 1, \dots, \ell$. Since addition and scalar multiplication are continuous for τ_1, it follows that $x_\alpha \xrightarrow{\tau} 0$ implies

$$x_\alpha = \sum_{i=1}^{\ell} x_i^\alpha e_i \xrightarrow{\tau_1} \sum_{i=1}^{\ell} 0e_i = 0\,,$$

where e_i denotes the ith coordinate unit vector of $\mathbb{R}^\ell$. This shows that the identity operator $I\colon (\mathbb{R}^\ell, \tau) \to (\mathbb{R}^\ell, \tau_1)$ is continuous, and so $\tau_1 \subseteq \tau$.

To see that I is a homeomorphism, let $U = \{x \in \mathbb{R}^\ell\colon\ \|x\|_2 \le 1\}$ and $S = \{x \in \mathbb{R}^\ell\colon\ \|x\|_2 = 1\}$. Since S is a τ-compact set, it follows from $\tau_1 \subseteq \tau$ that S is also a τ_1-compact set. Since τ_1 is Hausdorff, we obtain that S is a closed set. In view of $0 \notin S$, there exists a circled τ_1-neighborhood V of zero such that $V \cap S = \emptyset$. We claim that $V \subseteq U$. To see this, notice that if some $x \in V$ satisfies $x \notin U$ (i.e., $\|x\|_2 > 1$), then we have $\frac{x}{\|x\|_2} \in S \cap V$, which is impossible. Hence, $V \subseteq U$ is true and so U is a τ_1-neighborhood of zero. This implies $\tau \subseteq \tau_1$, and consequently $\tau_1 = \tau$. ■

Corollary 8.33. *Every finite dimensional subspace of a Hausdorff topological vector space is closed.*

Proof. Let X be a finite dimensional vector subspace of a Hausdorff topological vector space (L, τ). Then, according to Theorem 8.32, the topology τ induces the Euclidean topology on X. So, if a net $\{x_\alpha\} \subseteq X$ satisfies $x_\alpha \xrightarrow{\tau} x$ in L, then $\{x_\alpha\}$ is a τ-Cauchy net of (X, τ). Since (X, τ) is complete, there exists some $y \in X$ such that $x_\alpha \xrightarrow{\tau} y$. It follows that $x = y \in X$, and so X is a closed subspace of L. ■

Corollary 8.34. *If $\{e_1, \ldots, e_\ell\}$ is a basis in an ℓ-dimensional vector space L, then for each $\epsilon > 0$ the set*

$$\Big\{\sum_{i=1}^{\ell} \lambda_i e_i \colon \sum_{i=1}^{\ell} |\lambda_i| < \epsilon\Big\}$$

is an open neighborhood of zero for the Euclidean topology on L.

Proof. Note that the real function $\|\cdot\|\colon L \to \mathbb{R}$ defined via the formula $\|\sum_{i=1}^{\ell} \lambda_i e_i\| = \sum_{i=1}^{\ell} |\lambda_i|$ is a norm on L, which (according to Theorem 8.32) must generate the Euclidean topology of L. ■

Finally, we state the following characterization of finite dimensional vector spaces. Its proof can be found in [**12**, Theorem 5.26, p. 180].

Theorem 8.35 (F. Riesz). *A Hausdorff topological vector space is locally compact if and only if it is finite dimensional.*

8.7. The open mapping and the closed graph theorems

We shall state (with proofs) in the setting of complete metrizable topological vector spaces the two basic theorems of functional analysis known as the *Open Mapping Theorem* and the *Closed Graph Theorem*. For generalizations to general topological vector spaces and more on these theorems we refer the reader to [**52**].

Recall that a linear function $T\colon X \to Y$ between two vector spaces is called an **operator**, i.e., T is an operator if $T(\alpha x + \beta y) = \alpha T(x) + \beta T(y)$ holds for all $\alpha, \beta \in \mathbb{R}$ and all $x, y \in X$. Also recall that a mapping $f\colon X \to Y$ between topological spaces is said to be **open** if $f(V)$ is an open subset of Y for each open subset V of X.

Theorem 8.36 (The Open Mapping Theorem). *A surjective continuous operator between two completely metrizable topological vector spaces is an open mapping.*

Proof. Let $T\colon (L_1, \tau_1) \to (L_2, \tau_2)$ be a surjective continuous operator between two complete metrizable topological vector spaces and let U be a circled τ_1-closed neighborhood of zero. It suffices to show that the set $T(U)$ is a τ_2-neighborhood of zero.

We shall establish first the following claim.

- *For any τ_1-neighborhood W of zero in L_1 there exists a τ_2-neighborhood V of zero in L_2 satisfying $V \subseteq \overline{T(W)}$.*

To see this, let W and W_0 be two circled τ_1-neighborhoods of zero such that $W_0 + W_0 \subseteq W$. From $L_1 = \bigcup_{n=1}^{\infty} nW_0$ and the fact that T is surjective, it follows that $L_2 = T(L_1) = \bigcup_{n=1}^{\infty} nT(W_0)$ and so by the Baire category theorem for some n the set $\overline{nT(W_0)} = n\overline{T(W_0)}$ must have an interior point. This implies that there exists some $y \in \overline{T(W_0)}$ and some circled τ_2-neighborhood V of zero with $y + V \subseteq \overline{T(W_0)}$. Since $\overline{T(W_0)}$ is symmetric, we see that $v - y \in \overline{T(W_0)}$ for each $v \in V$. Thus, if $v \in V$, then it follows from $v = (v - y) + y \in \overline{T(W_0)} + \overline{T(W_0)} \subseteq \overline{T(W)}$ that $v \in \overline{T(W)}$, and so $V \subseteq \overline{T(W)}$.

Now pick a countable base $\{W_n\}$ at zero for τ_1 consisting of circled and τ_1-closed sets satisfying $W_{n+1} + W_{n+1} \subseteq W_n$ for all $n = 1, 2, \ldots$ and $W_1 + W_1 \subseteq U$. The above established claim and an easy inductive argument guarantee the existence of a countable base $\{V_n\}$ at zero for τ_2 consisting of circled and τ_2-closed sets satisfying $V_{n+1} + V_{n+1} \subseteq V_n$ and $V_n \subseteq \overline{T(W_n)}$ for all $n = 1, 2, \ldots$.

We shall finish the proof by showing that $V_1 \subseteq T(U)$. To this end, let $y \in V_1$. From $V_1 \subseteq \overline{T(W_1)}$ and the fact that $y + V_2$ is a τ_2-neighborhood of y, it follows that there exists some $w_1 \in W_1$ with $y - T(w_1) \in V_2$, and so $y - T(w_1) \in \overline{T(W_2)}$. Now, by an inductive argument, we can construct a sequence $\{w_n\}$ of L_1 such that for each $n = 1, 2, \ldots$ we have $w_n \in W_n$ and

$$y - \sum_{i=1}^{n} T(w_i) = y - T\Big(\sum_{i=1}^{n} w_i\Big) \in V_{n+1}\,. \tag{$\star$}$$

Next, let $x_n = \sum_{i=1}^{n} w_i$ and note that from

$$x_{n+p} - x_n = \sum_{i=n+1}^{n+p} w_i \in W_{n+1} + W_{n+2} + \cdots + W_{n+p} \subseteq W_n\,,$$

we see that $\{x_n\}$ is a τ_1-Cauchy sequence. Since (L_1, τ_1) is complete, there exists some $x \in L_1$ such that $x_n \xrightarrow{\tau_1} x$. Rewriting ($\star$) as $y - T(x_n) \in V_{n+1}$ for each n, we see that $y - T(x_n) \xrightarrow{\tau_2} 0$ in L_2. On the other hand, the continuity of T yields $T(x_n) \xrightarrow{\tau_2} T(x)$, and from this we get $y = T(x)$.

Finally, from $x_n = \sum_{i=1}^{n} w_i \in W_1 + W_2 + \cdots + W_n \subseteq W_1 + W_1 \subseteq U$ and the τ_1-closedness of U, we easily infer that $x \in U$, and so $y = T(x) \in T(U)$. In other words $V_1 \subseteq T(U)$ holds true, and the proof is finished. ■

Corollary 8.37. *A surjective continuous one-to-one operator between two completely metrizable topological vector spaces is a homeomoprhism.*

Recall that the **graph** of a function $f\colon A \to B$ is simply the subset of the Cartesian product $A \times B$ defined by

$$G(f) = \{(a, f(a))\colon\ a \in A\}.$$

Notice that if $T\colon X \to Y$ is an operator between two vector spaces, then the graph $G(T)$ of T is a vector subspace of $X \times Y$.

The next result that characterizes the continuity of linear operators is known as the closed graph theorem.

Theorem 8.38 (The Closed Graph Theorem). *An operator between two completely metrizable topological vector spaces is continuous if and only if it has a closed graph.*

Proof. Assume that $T\colon (L_1, \tau_1) \to (L_2, \tau_2)$ is an operator between two arbitrary complete metrizable topological vector spaces such that its graph $G(T) = \{(x, T(x))\colon\ x \in L_1\}$ is a closed subspace of $L_1 \times L_2$. It follows that $G(T)$ (with the induced product topology from $L_1 \times L_2$) is also a complete metrizable topological vector space. Since the mapping $S\colon G(T) \to L_1$, defined by $S(x, T(x)) = x$, is a surjective continuous one-to-one operator, it follows from Corollary 8.37 that S is a homeomorphism. In particular, the operator $x \mapsto (x, T(x)) = S^{-1}(x)$, from L_1 to $G(T)$, is continuous. Since the projection $P_2\colon L_1 \times L_2 \to L_2$, defined by $P_2(x_1, x_2) = x_2$, is continuous, we see that the operator $T = P_2 S^{-1}$ is likewise continuous. ■

8.8. The bounded weak* topology

In this section, L will denote a Banach space and, as before, the closed unit balls of L and L' will be denoted by U and U', respectively. That is,

$$U = \{x \in L\colon\ \|x\| \le 1\} \quad \text{and} \quad U' = \{x' \in L'\colon\ \|x'\| \le 1\}.$$

We start with the introduction of the bounded weak* topology on L'.

Definition 8.39. *The **bounded weak* topology** bw^* on L' is the finest topology on L' that induces the weak* topology (i.e., the $\sigma(L', L)$-topology) on every closed ball of L'.*

Our objective here is to prove the following useful result.

Theorem 8.40. *The bounded weak* topology is a locally convex topology on L' that is consistent with the dual pair $\langle L, L'\rangle$, i.e., $(L', bw^*)' = L$.*

The proof of this theorem will be accomplished with a series of lemmas. We begin by observing that from Definition 8.39 the following characterization of the bounded weak* topology should be obvious.

Lemma 8.41. *The bounded weak* topology is the topology on L' given by*

$$bw^* = \{A \subseteq L' \colon \; A \cap \rho U' \;\; \text{is } w^*\text{-relatively open in } \rho U' \; \forall \; \rho \geq 0\}.$$

The reader should verify that the above collection of subsets of L' is indeed a topology that is the finest topology on L' that induces the weak* topology on every closed ball of L'. This characterization of the bounded weak* topology, coupled with the identity $(L \setminus C) \cap \rho U' = \rho U' \setminus (C \cap \rho U')$, allows us to characterize the bw^*-closed sets as follows.

Lemma 8.42. *A subset C of L' is bw^*-closed (i.e., $L' \setminus C \in bw^*$) if and only if the set $C \cap \rho U'$ is a w^*-relatively closed subset of $\rho U'$ for each $\rho \geq 0$.*

Since for each $\rho > 0$ the mapping $h \colon (\rho U', w^*) \to (U', w^*)$, defined by $h(x') = \frac{1}{\rho}x'$, is a homeomorphism, we easily obtain from Lemma 8.41 the following.

Lemma 8.43. *A function $\phi \colon (L', bw^*) \to (X, \tau)$, where (X, τ) is an arbitrary topological vector space, is continuous if and only if the restriction of ϕ to U', i.e., the function $\phi \colon (U', w^*) \to (X, \tau)$, is continuous.*

At this stage we have no clue as to whether or not the topology bw^* is linear, and even in case bw^* is linear, there is no indication suggesting that it is locally convex let alone consistent with the dual pair $\langle L, L' \rangle$!

To continue our discussion, let $\mathfrak{S}$ denote the collection of all nonempty compact subsets of L. As usual, we shall also denote by $\mathfrak{S}$ the locally convex topology on L' of uniform convergence on the members of $\mathfrak{S}$. That is, a net $\{x'_\alpha\}$ of L' satisfies $x'_\alpha \xrightarrow{\mathfrak{S}} x'$ if and only if the net $\{x'_\alpha\}$ converges uniformly to x' on every compact subset of L. This locally convex topology $\mathfrak{S}$ is generated by the family of seminorms $\{q_A\}_{A \in \mathfrak{S}}$, where

$$q_A(x') = \sup_{x \in A} |x'(x)|$$

for each $x' \in L'$ and each $A \in \mathfrak{S}$.

A glance at part (1) of Theorem 8.31 shows that we can generate the topology $\mathfrak{S}$ by restricting ourselves to a very nice class of totally bounded subsets of L, namely, to the class of all sequences convergent to zero. That is, if we replace $\mathfrak{S}$ by the collection $\mathfrak{S}_1$ of all sequences convergent to zero, then the two topologies $\mathfrak{S}$ and $\mathfrak{S}_1$ coincide. Thus, if $\mathfrak{S}$ consists of all sequences $\sigma = (\sigma_1, \sigma_2, \ldots)$ of L that satisfy $\|\sigma_n\| \to 0$ and we define

$$q_\sigma(x') = \sup_{n \in \mathbb{N}} |x'(\sigma_n)|,$$

then the family of seminorms $\{q_\sigma\}_{\sigma\in\mathfrak{S}_1}$ generates the topology $\mathfrak{S}$. So, we can assume (for all practical purposes) that $\mathfrak{S}$ is simply the collection of all sequences that converge in norm to zero.

It turns out that our $\mathfrak{S}$-topology of uniform convergence on the compact subsets of A induces the weak* topology on the closed unit balls of L'.

Lemma 8.44. *The $\mathfrak{S}$-topology of uniform convergence on the compact subsets of L induces the w^*-topology on every closed ball of L'. In particular, we have $\mathfrak{S}\subseteq bw^*$.*

Proof. Fix some $\rho>0$ and consider a net $\{x'_\alpha\}$ of $\rho U'$. We must establish that $x'_\alpha \xrightarrow{w^*} x'$ if and only if $x'_\alpha \xrightarrow{\mathfrak{S}} x'$.

To this end, assume first that $x'_\alpha \xrightarrow{w^*} x'$ holds true in $\rho U'$ and suppose $\sigma=(\sigma_1,\sigma_2,\ldots)$ is a sequence of L satisfying $\|\sigma_n\|\to 0$. Fix $\epsilon>0$. From $|x'_\alpha(\sigma_n)|\le\|x'_\alpha\|\cdot\|\sigma_n\|\le\rho\|\sigma_n\|\to 0$, we see that there exists some n_0 such that $|x'_\alpha(\sigma_n)|<\epsilon$ holds for all $n>n_0$ and all α. Since for each fixed $x\in L$ we have $x'_\alpha(x)\to 0$, there exists some index α_0 such that $|x'_\alpha(\sigma_n)|<\epsilon$ holds for all $n=1,\ldots,n_0$ and all $\alpha\succeq\alpha_0$. But then we have $q_\sigma(x'_\alpha)\le\epsilon$ for all $\alpha\succeq\alpha_0$. This shows that $x'_\alpha \xrightarrow{\mathfrak{S}} x'$ holds true in $\rho U'$.

For the converse, assume that $x'_\alpha \xrightarrow{\mathfrak{S}} x'$ holds in $\rho U'$ and fix any nonzero vector $x\in L$. If we let $\sigma=(x,0,0,0,\ldots)$, then σ is a null sequence of L and so from $|[x'_\alpha-x'](x)|=q_\sigma(x'_\alpha-x')\to 0$ we get $x'_\alpha(x)\to x'(x)$. This shows that $x'_\alpha \xrightarrow{w^*} x'$ holds in $\rho U'$, and the proof is finished. ■

Our next lemma shows that the bounded weak* topology is indeed the $\mathfrak{S}$-topology of uniform convergence on the compact subsets of L.

Lemma 8.45. *The topology of uniform convergence on the compact sets coincides with the bounded weak* topology, i.e., $\mathfrak{S}=bw^*$.*

In particular, the bounded weak topology is a Hausdorff locally convex topology on L'.*

Proof. In view of Lemma 8.44, it suffices to show that the identity operator $I\colon(L',\mathfrak{S})\to(L',bw^*)$ is continuous. To this end, let $\{x'_\alpha\}\subseteq L'$ be a net satisfying $x'_\alpha \xrightarrow{\mathfrak{S}} x'$ in L' and assume by way of contradiction that the net $\{x'_\alpha\}$ does not converge to x' with respect to the topology bw^*. This means that there exists a bw^*-open neighborhood V of x' and a subnet of $\{x'_\alpha\}$ (which we can denote without loss of generality by $\{x'_\alpha\}$ again) such that $x'_\alpha\notin V$ for each α.

Fix some $\rho>1$ with $x'\in V\cap\rho U'$. We claim that there exists a sequence $\{A_0,A_1,A_2,\ldots\}$ of finite nonempty subsets of L such that

(1) if $n\ge 1$, then $\|x\|\le\frac{1}{\rho+n-1}$ for each $x\in A_n$ and

(2) if for each $n \geq 0$ we let $F_n = \bigcup_{i=0}^{n} A_i$, then its polar set satisfies

$$(x' + F_n^{\circ}) \cap [(\rho(\rho+n)+1)U'] \subseteq V\,.$$

The existence of the sequence $\{A_0, A_1, A_2, \ldots\}$ will be established by induction. For $n = 0$, note that since $V \in bw^*$, $x' \in V \cap (\rho^2+1)U'$ and because of the fact that bw^* induces w^* on $(\rho^2+1)U'$, there exists a nonempty finite subset A_0 of L such that

$$(x' + A_0^{\circ}) \cap \big[(\rho^2+1)U'\big] \subseteq V \cap \big[(\rho^2+1)U'\big] \subseteq V\,.$$

(See Lemma 8.7.) For the inductive step, assume that the nonempty finite sets $A_0, A_1, \ldots, A_n$ have been selected so that properties (1) and (2) are true. Let $\mathcal{G}$ denote the collection of all nonempty finite subsets G of L such that $\|x\| \leq \frac{1}{\rho+n}$ holds for all $x \in G$. Now assume by way of contradiction that for every $G \in \mathcal{G}$ we have

$$W_G =: [x' + (F_n \cup G)^{\circ}] \cap [(\rho(\rho+n+1)+1)U'] \cap (L' \setminus V) \neq \emptyset\,. \qquad (\star)$$

Since $L' \setminus V$ is bw^*-closed and the set $(\rho(\rho+n+1)+1)U'$ is (according to Theorem 8.22) w^*-compact, it follows that each W_G is a w^*-closed set. Moreover, from the polar identity $\bigcap_{i=1}^{k}(F_n \cup G_i)^{\circ} = (F_n \cup G_1 \cup \cdots \cup G_k)^{\circ}$, it is easy to see that the family of w^*-closed sets $\mathcal{F} = \{W_G\colon\ G \in \mathcal{G}\}$ has the finite intersection property. Consequently, $\mathcal{F}$ has a nonempty intersection. In particular, there exists some $z' \in (x'+F_n^{\circ}) \cap [(\rho(\rho+n+1)+1)U'] \cap (L' \setminus V)$ that satisfies $|z'(x)| \leq |x'(x)| + 1$ for all $x \in L$ with $\|x\| \leq \frac{1}{\rho+n}$. The latter implies $\|z'\| \leq \rho(\rho+n)+1$ and so $z' \in (\rho(\rho+n)+1)U'$. But then $z' \in (x' + F_n^{\circ}) \cap [(\rho(\rho+n)+1)]U' \subseteq V$, contradicting $z' \in L' \setminus V$. Hence, there exists some $A_{n+1} = G \in \mathcal{G}$ so that $W_G = \emptyset$. This set A_{n+1} satisfies properties (1) and (2) and completes the inductive argument for the existence of the sequence $\{A_0, A_1, A_2, \ldots\}$.

Next, for each $n \geq 1$ let $A_n = \{a_1^n, \ldots, a_{k_n}^n\}$ and consider the sequence

$$\sigma = (\sigma_1, \sigma_2, \sigma_3, \ldots) = (a_1^1, \ldots, a_{k_1}^1, a_1^2, \ldots, a_{k_2}^2, a_1^3, \ldots, a_{k_3}^3, a_1^4, \ldots)\,.$$

Clearly, $\|\sigma_n\| \to 0$ and so $\sigma \in \mathfrak{S}$.

Finally, fix an index α and then pick some n so that $\|x'_\alpha\| \leq \rho(\rho+n)+1$. Since $x'_\alpha \notin V$, it follows from (2) that $x'_\alpha - x' \notin F_n^{\circ}$. Therefore, for some σ_k we must have $|(x'_\alpha - x')(\sigma_k)| > 1$, and consequently $q_\sigma(x'_\alpha - x') > 1$ for each α. However, the latter contradicts the fact that $x'_\alpha \xrightarrow{\mathfrak{S}} x'$ in L', and the proof is finished. ■

The final property missing from the proof of Theorem 8.40 is the consistency of bw^*. This is taken care of in the next lemma.

Lemma 8.46. *A linear functional $f\colon L' \to \mathbb{R}$ on a Banach space is weak* continuous if and only if f is bounded weak*-continuous. That is, we have $(L', bw^*)' = (L', w^*)' = L$.*

Proof. Let $f\colon L' \to \mathbb{R}$ be a linear functional on a Banach space. If f is weak* continuous, then it easily follows from Lemma 8.43 that f is bounded weak* continuous. For the converse, assume that f is bw^*-continuous.

Since bw^* coincides with the $\mathfrak{S}$-topology, there exists a norm null sequence $\sigma = (\sigma_1, \sigma_2, \ldots)$ such that

$$|f(x')| \leq q_\sigma(x) = \sup_{n\in\mathbb{N}} |x'(\sigma_n)| \tag{$\dagger$}$$

holds for each $x' \in L'$. Now consider the operator $T\colon L' \to c_0$ defined by

$$Tx' = (x'(\sigma_1), x'(\sigma_2), \ldots),$$

and let R_T denote its range. On this range R_T we define the linear functional $\phi\colon R_T \to \mathbb{R}$ via the formula

$$\phi(Tx') = f(x').$$

Notice that if two linear functionals $x', y' \in L'$ satisfy $x'(\sigma_n) = y'(\sigma_n)$ for each n, then it follows from ($\dagger$) that $f(x') = f(y')$. This implies that the linear functional ϕ is well defined, and moreover using ($\dagger$) once more, we see that

$$|\phi(Tx')| = |f(x')| \leq \sup_{n\in\mathbb{N}} |x'(\sigma_n)| = \|Tx'\|$$

holds true for each $x' \in L'$. This implies that the linear functional ϕ is norm continuous on c_0 and so (by the Hahn–Banach Extension Theorem 8.20) it has a continuous linear extension to all of c_0, which we denote by ϕ again. Given that $c_0' = \ell_1$, there exists a sequence $(\alpha_1, \alpha_2, \ldots) \in \ell_1$ such that

$$\phi(z) = \sum_{n=1}^{\infty} \alpha_n z_n$$

holds true for all $z \in c_0$. Letting $z = Tx'$ yields

$$f(x') = \phi(Tx') = \sum_{n=1}^{\infty} \alpha_n x'(\sigma_n) = x'\Big(\sum_{n=1}^{\infty} \alpha_n \sigma_n\Big),$$

for all $x' \in L'$. So, if we let $x_0 = \sum_{n=1}^{\infty} \alpha_n \sigma_n \in L$, then $f(x') = x'(x_0)$ holds for each $x' \in L'$. This proves that f is weak*-continuous, as desired. ■

Given that all consistent topologies on L' have the same closed convex sets (see Theorem 8.9), we get the following special case of this result. (This special case was employed in the proof of Andô's Theorem 2.47.)

Corollary 8.47 (Krein–Šmulian). *A convex subset of L' is weak* closed if and only if it is closed for the bounded weak* topology.*

In other words, a convex subset A of L' is weak-closed if and only if for every norm bounded net $\{x'_\alpha\}$ of A that satisfies $x'_\alpha \xrightarrow{w^*} x'$ we have $x' \in A$.*

Finally, we close the section by rephrasing Lemma 8.46 as follows.

Theorem 8.48. *A linear functional $f\colon L' \to \mathbb{R}$ is weak*-continuous if and only if f is weak*-continuous when restricted to the closed unit ball U' of L'.*

This formulation of Lemma 8.46 has a long history. It was proven first by S. Banach [**27**] for the separable case and subsequently it was extended to locally convex spaces. The interested reader can find more on this theorem in [**33**, Chapter IV, § 3], [**51**, Section 11, p. 247], and [**85**, Section 6, p. 147].

Bibliography

1. Y. A. Abramovich, Injective envelopes of normed lattices, *Dokl. Akad. Nauk USSR* **197** (1971), 743–745.
2. Y. A. Abramovich, When each continuous operator is regular, *Functional Analysis, Optimization, and Mathematical Economics*, A memorial volume for L. V. Kantorovich, Oxford Univ. Press, 1990, pp. 133–140.
3. Y. A. Abramovich and C. D. Aliprantis, *An Invitation to Operator Theory*, Vol. 50, Graduate Studies in Mathematics, Amer. Math. Soc., Providence, RI, 2002.
4. Y. A. Abramovich and C. D. Aliprantis, *Problems in Operator Theory*, Vol. 51, Graduate Studies in Mathematics, Amer. Math. Soc., Providence, RI, 2002.
5. Y. A. Abramovich, C. D. Aliprantis, and O. Burkinshaw, Positive operators on Krein spaces, *Acta Appl. Math.* **27** (1992), 1–22.
6. Y. A. Abramovich, C. D. Aliprantis, and I. A. Polyrakis, Lattice-subspaces and positive projections, *Proc. Royal Irish Academy* **94**A (1994), 237-253.
7. Y. A. Abramovich and V. A. Geyler, On one question of D. H. Fremlin, *Colloq. Math.* **46** (1982), 15–17.
8. Y. A. Abramovich and A. W. Wickstead, Regular operators from and into small Riesz spaces, *Indag. Math. N.S.* **2** (1991), 257–274.
9. Y. A. Abramovich and A. W. Wickstead, The regularity of order bounded operators into $C(K)$, II, *Quart. J. Math. Oxford Ser.* (2), **44** (1993), 257–270.
10. Y. A. Abramovich and A. W. Wickstead, When each continuous operator is regular, II, *Indag. Math.* **8** (1997), 281–294.
11. P. Alexandroff and P. Urysohn, Mémoire sur les espaces topologique compacts, *Nederl. Akad. Wetensch. Proc. Ser. A* **14** (1929), 1–96.
12. C. D. Aliprantis and K. C. Border, *Infinite Dimensional Analysis: A Hitchhiker's Guide*, 3rd Edition, Springer–Verlag, Heidelberg and New York, 2006.
13. C. D. Aliprantis, D. J. Brown, I. A. Polyrakis, and J. Werner, Yudin cones and inductive limit topologies, *Atti. Sem. Mat. Fis. Univ. Modena* **46** (1998), 389–412.
14. C. D. Aliprantis and O. Burkinshaw, *Locally Solid Riesz Spaces with Applications to Economics*, 2nd Edition, Mathematical Surveys and Monographs, Volume # 105, American Mathematical Society, Providence, RI, 2003.

15. C. D. Aliprantis and O. Burkinshaw, *Positive Operators*, Springer–Verlag, The Netherlands, 2006. (This is a reprint from the original published by Academic Press in the Pure and Applied Math. Series, # 76, New York and London, 1985.)

16. C. D. Aliprantis and O. Burkinshaw, *Principles of Real Analysis*, 3rd Edition, Academic Press, New York and London, 1998.

17. C. D. Aliprantis, D. Harris, and R. Tourky, Piecewise linear functions, *Macroeconomic Dynamics* **10** (2006), 77–99.

18. C. D. Aliprantis, D. Harris, and R. Tourky, Riesz estimators, *J. Econometrics* **136** (2007), 431–456.

19. C. D. Aliprantis and R. Tourky, The super order dual of an ordered vector space and the Riesz–Kantorovich formula, *Trans. Amer. Math. Soc.* **354** (2002), 2055–2077.

20. C. D. Aliprantis, R. Tourky, and N. C. Yannelis, Cone conditions in general equilibrium theory, *J. Econom. Theory* **92** (2000), 55–76.

21. C. D. Aliprantis, R. Tourky, and N. C. Yannelis, The Riesz–Kantorovich formula and general equilibrium theory, *J. Math. Econom.* **34** (2000), 55–76.

22. C. D. Aliprantis, R. Tourky, and N. C. Yannelis, A theory of value: equilibrium analysis beyond vector lattices, *J. Economic Theory* **100** (2001), 22–72.

23. T. Andô, On fundamental properties of a Banach space with a cone, *Pacific J. Math.* **12** (1962), 1163–1169.

24. I. A. Bakhtin, On the existence of eigenvectors of linear positive non-compact operators, *Math. USSR-Sb. (N.S.)* **64** (1964), 102–114. (Russian.)

25. I. A. Bakhtin, On the existence of common eigenvector for a commutative family of linear positive operators, *Math. USSR-Sb. (N.S.)* **67** (1965), 267–278. (Russian.)

26. I. A. Bakhtin, M. A. Krasnoselsky and V. Ya. Stezenko, Continuity of linear positive operators, *Sibirsk. Math. Ž.* **3** (1962), 156–160.

27. S. Banach, *Théorie des Opérations Linéaires*, Monografje Matematyczne, Warsaw, 1932. (Reprinted by Chelsea Publishing Co., New York, 1955.)

28. H. Bauer, Sur le prolongement des formes linéaires positives dans un espace vectoriel ordonné, *C. R. Acad. Sci. Paris* **244** (1957), 289–292.

29. R. Becker, *Convex Cones in Analysis*, Hermann, Collection Travaux en Cours, # 67, Paris, France, 2006.

30. G. Birkhoff, *Lattice Theory*, 3rd Edition, Amer. Math. Soc. Colloq. Publ., # **25**, Providence, RI, 1967.

31. A. Björner, M. Las Vergas, B. Sturmfels, N. White, and G. Ziegler, *Oriented Matroids*, 2nd Edition, Encyclopedia of Mathematics and its Applications, # 46, Cambridge University Press, Cambridge, UK, 1999.

32. K. C. Border, A polynomial cone counterexample, California Institute of Technology manuscript, December 12, 2003.

33. N. Bourbaki, *Topological Vector Spaces*, Chapters 1–5, Springer–Verlag, New York and Heidelberg, 1987.

34. D. J. Brown, C. B. Huijsmans, and B. de Pagter, Approximating derivative securities in f-algebras, in: C. D. Aliprantis, K. C. Border, and W. A. J. Luxemburg, eds., *Positive Operators, Riesz Spaces, and Economics*, Springer–Verlag, Heidelberg and New York, 1991, pp. 171–177.

35. D. E. Catlin, *Estimation, Control, and the Discrete Kalman Filter*, Applied Mathematical Sciences, # 71, Springer–Verlag, New York, 1989.

36. R. W. Cottle, J.-S. Pang, and R. E. Stone, *The Linear Complementarity Problem*, Academic Press, Computer Science and Scientific Computing, New York and London, 1992.

37. N. Dunford and J. T. Schwartz, *Linear Operators I*, Wiley (Interscience), New York, 1958.

38. F. Facchinei and J.-S. Pang, *Finite-Dimensional Variational Inequalities and Complementarity Problems*, Springer series in Operations Research, New York and Heidelberg, 2003.

39. K. Fan, I. L. Glicksberg, and A. J. Hoffman, Systems of inequalities involving convex functions, *Proc. Amer. Math. Soc.* **8** (1957), 617–622.

40. J. Farkas, Über die Theorie der einfachen Ungleichung, *J. Reigne Angew. Math.* **124** (1902), 1–24.

41. M. Florenzano and C. Le Van, *Finite Dimensional Convexity and Optimization*, Studies in Economic Theory, # 13, Springer–Verlag, New York and Heidelberg, 2001.

42. J. L. Franklin, *Methods of Mathematical Economics: Linear and Nonlinear Programming, Fixed Point Theorems*, Classics in Applied Mathematics, # 37, SIAM, Philadelphia, PA, 2002. (Reprint of the 1980 edition published by Springer–Verlag in its 'Undergraduate Texts in Mathematics' series.)

43. G. Frobenius, Über Matrizen aus nicht-negativen Elementen, *Sitz. Berichte Kgl. Preuß. Akad. Wiss. Berlin*, 456–477, 1912.

44. L. Fuchs, On partially ordered vector spaces with the Riesz interpolation property, *Publ. Math. Debrecen* **12** (1965), 335–343.

45. L. Fuchs, Riesz groups, *Ann. Scuola Norm. Sup. Pisa* **19** (1965), 1–34.

46. L. Fuchs, *Riesz Vector Spaces and Riesz Algebras*, Queen's papers in pure and applied mathematics, No. 1, Queen's University, Kingston, Ontario, Canada, 1966.

47. A. Göpfert, H. Riahi, C. Tammer and C. Zălinescu, *Variational Methods in Partially Ordered Spaces*, Springer–Verlag, New York and Heidelberg, 2003.

48. B. Grünbaum, *Convex Polytopes*, 2nd Edition, Graduate Text in Mathematics, Vol. 221, Springer, New York and Heidelberg, 2003.

49. D. Guo and V. Lakshmikantham, *Nonlinear Problems in Abstract Cones*, Notes and Reports in Mathematics in Science and Engineering, Vol. 5, Academic Press, New York and London, 1988.

50. H. Hahn, Über lineare gleichungssysteme in linearen räumen, *J. Reine Angew. Math.* **157** (1927), 214–229.

51. J. Horváth, *Topological Vector Spaces and Distributions*, I, Addison-Wesley, Reading, Massachusetts, 1966.

52. T. Husain, *The Open Mapping and Closed Graph Theorems in Topological Vector Spaces*, Oxford Mathematical Monographs, Oxford University Press, London, 1965.

53. G. Isac, *Topological Methods in Complementarity Theory*, Kluwer Academic Publishers, Hingham, MA, 2000.

54. G. Isac, V. A. Bulavsky, and V. V. Kalashnikov, *Complementarity, Equilibrium, Efficiency and Economics*, Kluwer Academic Publishers, Hingham, MA, 2002.

55. G. Jameson, *Ordered Linear Spaces*, Lecture Notes in Mathematics, # **141**, Springer–Verlag, Berlin and New York, 1970.

56. L. V. Kantorovich, On partially ordered linear spaces and their applications in the theory of linear operators, *Dokl. Akad. Nauk SSSR* **4** (1935), 13–16. (Russian.)

57. L. V. Kantorovich, On the moment problem for a finite interval, *Dokl. Akad. Nauk SSR* **14** (1937), 531–537. (Russian.)

58. L. V. Kantorovich, Linear operators in semi-ordered spaces, *Math. Sbornik* **49** (1940), 209–284. MR **2**, 317; Zbl **23**, 328

59. J. L. Kelley, I. Namioka, et al., *Linear Topological Spaces*, Graduate Texts in Mathematics, # 36, Springer–Verlag, New York and Berlin, 1976. (Reprint of its 1963 edition published in the 'University Series in Higher Mathematics' by Nan Nostrand.)

60. A. K. Kitover, The spectral properties of weighted homomorphisms in algebras of continuous functions and their applications, *Zap. Naučn. Sem. Leningrad Otdel. Mat. Inst. Steklov. (LOMI)* **107** (1982), 89–103.

61. V. L. Klee, Jr., Extremal structure of convex sets, *Arch. Math.* **8** (1957), 234–240.

62. M. G. Krein, Propriétés fondamentales des ensembles coniques normaux dans l'espace de Banach, *Dokl. Akad. Nauk SSSR* **28** (1940), 13–17.

63. M. G. Krein and M. A. Rutman, Linear operators leaving invariant a cone in a Banach space, *Uspekhi Mat. Nauk* **3** (1948), 3–95. (Russian.) Also, *Amer. Math. Soc. Transl.*, No. 26, 1950.

64. D. G. Luenberger, *Optimization by Vector Space Methods*, John Wiley and Sons, New York and London, 1969.

65. D. G. Luenberger, *Linear and Nonlinear Programming*, 2nd Edition, Kluwer Academic Publishers, Boston, MA, 2003.

66. W. A. J. Luxemburg and A. C. Zaanen, *Riesz Spaces*, I, North-Holland, Amsterdam, 1971.

67. A. A. Markov, Some theorems on Abelian collections, *Dokl. Akad. Nauk USSR* **10** (1936), 311–313.

68. A. Meyer-Nieberg, *Banach Lattices*, Springer–Verlag, New York and Heidelberg, 1991.

69. S. Miyajima, Structure of Banach quasi-sublattices, *Hokkaido Math. J.* **12** (1983), 83–91.

70. L. Nachbin, On the continuity of positive linear transformations, *Proc. Internat. Congress of Math.*, pp. 464–465, 1950.

71. I. Namioka, Partially ordered linear topological spaces, *Mem. Amer. Math. Soc.*, Vol. 24, 1957.

72. H. Nikaidô, *Convex Structures and Economic Theory*, Mathematics in Science and Engineering, # 61, Academic Press, New York and London, 1968.

73. T. Oikhberg and V. G. Troitsky, A theorem of Krein revisited, *Rocky Mountain J. Math.* **35** (2005), 195–210.

74. S. Ovchinnikov, Max-min representation of piecewise linear functions, *Beiträge Algebra Geom.* **43** (2002), 297–302.

75. A. L. Peressini, *Ordered Topological Vector Spaces*, Harper and Row, New York and London, 1967.

76. A. L. Peressini, F. E. Sullivan, and J. J. Uhl, Jr., *The Mathematics of Nonlinear Programming*, Undergraduate Texts in Mathematics, Springer–Verlag, Berlin and New York, 1988.

77. O. Perron, Zur Theorie der Matrizen, *Math. Ann.* **64** (1907), 248–263.

78. I. A. Polyrakis, Lattice-subspaces of $C[0,1]$ and positive bases, *J. Math. Anal. Appl.* **184** (1994), 1–18.

79. I. A. Polyrakis, Finite-dimensional lattice-subspaces of $C(\Omega)$ and curves of $\mathbb{R}^n$, *Trans. Amer. Math. Soc.* **348** (1996), 2793–2810.

80. F. Riesz, Sur la décomposition des opérations linéaires, *Atti. Congr. Internaz. Mat. Bologna* **3** (1930), 143–148.

81. F. Riesz, Sur quelques notions fondamentals dans la theorie générale des opérations linéaires, *Ann. of Math.* **41** (1940), 174–206. (This work was published first in 1937 in Hungarian.)

82. R. T. Rockafellar, *Convex Analysis*, Princeton University Press, Princeton, NJ, 1972.

83. A. C. M. van Rooij, When do the regular operators between two Riesz spaces form a vector lattice?, *Indag. Math.* **47** (1985), 95–98.

84. H. H. Schaefer, Halbgeordnete lokalkonvexe Vektorräume, II, *Math. Ann.* **135** (1958), 115–141.

85. H. H. Schaefer, *Topological Vector Spaces*, Springer–Verlag, Berlin and New York, 1974.

86. H. H. Schaefer, *Banach Lattices and Positive Operators*, Springer–Verlag, Berlin and New York, 1974.

87. E. Stiemke, Über positive Lösungen homogener linearer Gleichugen, *Math. Ann.* **76** (1915).

88. B. Z. Vulikh, *Introduction to the Theory of Partially Ordered Spaces*, Wolter–Hoordhoff, Groningen, Netherlands, 1967. (English translation from the Russian.)

89. B. Z. Vulikh, *Special Topics in Geometry of Cones in Normed Spaces*, Kalinin State University, 1978.

90. H. Weyl, The elementary theory of convex polyhedra, in: H. W. Kuhn and A. W. Tucker, eds., *Contributions to the Theory of Games I*, Annals of Mathematics Studies, Vol. 24 (Princeton University Press, Princeton, NJ, 1950), pp. 3–18. [Translation by H. W. Kuhn of the original paper by H. Weyl: Elementare Theorie der konvexen Polyeder, *Coment. Math. Helv.* **7** (1934-35), 290–306.]

91. Y. C. Wong and K. F. Ng, *Partially Ordered Topological Vector Spaces*, Clarendon Press, Oxford, 1973.

92. A. I. Yudin, A solution of two problems in the theory of partially ordered spaces, *Doklady Akad. Nauk SSSR* **23** (1939), 418–422.

93. A. C. Zaanen, *Riesz Spaces II*, North-Holland, Amsterdam, 1983.

94. G. M. Ziegler, *Lectures on Polytopes*, Graduate Texts in Mathematics, # 152, Springer–Verlag, New York and Heidelberg, 1995.

Index

Titles in This Series

46 **Joseph L. Taylor,** Several complex variables with connections to algebraic geometry and Lie groups, 2002

45 **Inder K. Rana,** An introduction to measure and integration, second edition, 2002

44 **Jim Agler and John E. McCarthy,** Pick interpolation and Hilbert function spaces, 2002

43 **N. V. Krylov,** Introduction to the theory of random processes, 2002

42 **Jin Hong and Seok-Jin Kang,** Introduction to quantum groups and crystal bases, 2002

41 **Georgi V. Smirnov,** Introduction to the theory of differential inclusions, 2002

40 **Robert E. Greene and Steven G. Krantz,** Function theory of one complex variable, third edition, 2006

39 **Larry C. Grove,** Classical groups and geometric algebra, 2002

38 **Elton P. Hsu,** Stochastic analysis on manifolds, 2002

37 **Hershel M. Farkas and Irwin Kra,** Theta constants, Riemann surfaces and the modular group, 2001

36 **Martin Schechter,** Principles of functional analysis, second edition, 2002

35 **James F. Davis and Paul Kirk,** Lecture notes in algebraic topology, 2001

34 **Sigurdur Helgason,** Differential geometry, Lie groups, and symmetric spaces, 2001

33 **Dmitri Burago, Yuri Burago, and Sergei Ivanov,** A course in metric geometry, 2001

32 **Robert G. Bartle,** A modern theory of integration, 2001

31 **Ralf Korn and Elke Korn,** Option pricing and portfolio optimization: Modern methods of financial mathematics, 2001

30 **J. C. McConnell and J. C. Robson,** Noncommutative Noetherian rings, 2001

29 **Javier Duoandikoetxea,** Fourier analysis, 2001

28 **Liviu I. Nicolaescu,** Notes on Seiberg-Witten theory, 2000

27 **Thierry Aubin,** A course in differential geometry, 2001

26 **Rolf Berndt,** An introduction to symplectic geometry, 2001

25 **Thomas Friedrich,** Dirac operators in Riemannian geometry, 2000

24 **Helmut Koch,** Number theory: Algebraic numbers and functions, 2000

23 **Alberto Candel and Lawrence Conlon,** Foliations I, 2000

22 **Günter R. Krause and Thomas H. Lenagan,** Growth of algebras and Gelfand-Kirillov dimension, 2000

21 **John B. Conway,** A course in operator theory, 2000

20 **Robert E. Gompf and András I. Stipsicz,** 4-manifolds and Kirby calculus, 1999

19 **Lawrence C. Evans,** Partial differential equations, 1998

18 **Winfried Just and Martin Weese,** Discovering modern set theory. II: Set-theoretic tools for every mathematician, 1997

17 **Henryk Iwaniec,** Topics in classical automorphic forms, 1997

16 **Richard V. Kadison and John R. Ringrose,** Fundamentals of the theory of operator algebras. Volume II: Advanced theory, 1997

15 **Richard V. Kadison and John R. Ringrose,** Fundamentals of the theory of operator algebras. Volume I: Elementary theory, 1997

14 **Elliott H. Lieb and Michael Loss,** Analysis, 1997

13 **Paul C. Shields,** The ergodic theory of discrete sample paths, 1996

12 **N. V. Krylov,** Lectures on elliptic and parabolic equations in Hölder spaces, 1996

11 **Jacques Dixmier,** Enveloping algebras, 1996 Printing

For a complete list of titles in this series, visit the AMS Bookstore at **www.ams.org/bookstore/**.